T0237911

Lecture Notes in Physics

The Lecture Notes in Physics

The series Lecture Notes in Physics (LNP), founded in 1969, reports new developments in physics research and teaching – quickly and informally, but with a high quality and the explicit aim to summarize and communicate current knowledge in an accessible way. Books published in this series are conceived as bridging material between advanced graduate textbooks and the forefront of research to serve the following purposes:

• to be a compact and modern up-to-date source of reference on a well-defined topic;

• to serve as an accessible introduction to the field to postgraduate students and nonspecialist researchers from related areas;

• to be a source of advanced teaching material for specialized seminars, courses and schools.

Both monographs and multi-author volumes will be considered for publication. Edited volumes should, however, consist of a very limited number of contributions only. Proceedings will not be considered for LNP.

Volumes published in LNP are disseminated both in print and in electronic formats, the electronic archive is available at springerlink.com. The series content is indexed, abstracted and referenced by many abstracting and information services, bibliographic networks, subscription agencies, library networks, and consortia.

Proposals should be sent to a member of the Editorial Board, or directly to the managing editor at Springer:

Dr. Christian Caron
Springer Heidelberg
Physics Editorial Department I
Tiergartenstrasse 17
69121 Heidelberg/Germany
christian.caron@springer.com

J. Dolinšek   M. Vilfan   S. Žumer   (Eds.)

Novel NMR
and EPR Techniques

Editors

Professor Dr. Janez Dolinšek[+]
Professor Dr. Marija Vilfan
Professor Dr. Slobodan Žumer[+]

[+] Physics Department
Faculty of Mathematics and Physics
University Ljubljana
Jadranska 19
1000 Ljubljana
Slovenia

Jozef Stefan Institute
Jamova 39
1000 Ljubljana
Slovenia

E-mails:

jani.dolinsek@ijs.si
mika.vilfan@ijs.si
slobodan.zumer@fmf.uni-lj.si

J. Dolinšek et al., Novel NMR and EPR Techniques
Lect. Notes Phys. 684 (Springer, Berlin Heidelberg 2006), DOI 10.1007/b11540830

ISSN 0075-8450

ISBN 978-3-642-06907-9   e-ISBN 978-3-540-32627-4

Springer is a part of Springer Science+Business Media
springer.com
© Springer-Verlag Berlin Heidelberg 2006
Softcover reprint of the hardcover 1st edition 2006

Cover design: design & production GmbH, Heidelberg

This book is dedicated to
Professor Robert Blinc,
with respect and kind regards

Professor Robert Blinc

Preface

This book is a collection of scientific articles on current developments of NMR and ESR techniques and their applications in physics and chemistry. It is dedicated to Professor Robert Blinc, on the occasion of his seventieth birthday, in appreciation of his remarkable scientific accomplishments in the NMR of condensed matter. He is a physicist commanding deep respect and a ection from those who had the opportunity to work with him.

Robert Blinc was born on October 31, 1933, in Ljubljana, Slovenia. He graduated in 1958 and completed his Ph.D in 1959 in physics at the University of Ljubljana. His doctoral research on proton tunneling in ferroelectrics with short hydrogen bonds was supervised by Professor Du˘ san Hadˇzi. After a postdoctoral year spent in the group of Professor John Waugh at M.I.T., Cambridge, Mass., Robert Blinc was appointed as a professor of physics at the University of Ljubljana at a time when there was scarce research in the field of condensed matter in Slovenia. With his far-sighted mind, Robert Blinc, together with Ivan Zupan˘ ciˇc, started the NMR laboratory at the Jozef Stefan Institute in Ljubljana. He immediately realized the enormous potential of NMR methods in the research of structure, dynamics, and phase transitions in solids. In the subsequent years he made significant contributions in applying magnetic resonance to the research of ice, ferroelectric materials, liquid crystals, incommensurate systems, spin glasses, relaxors, fullerenes, and fullerene nanomagnets. His work led to the detailed understanding of the microscopic nature and properties of those materials. To mention only a few: Robert Blinc and coworkers elucidated the isotopic e ect in ferroelectric crystals, predicted the Goldstone mode in ferroelectric liquid crystals, studied the impact of collective orientational fluctuations on spin relaxation, detected solitons and phasons in incommensurate systems using NMR, and determined the Edwards-Anderson order parameter in glasses and relaxors. He also pioneered the application of NMR to the nondestructive oil-content measurements in seeds and the development of the NMR measurements of the self-di usion coe cient in broad-line materials. In the early stage of the double resonance technique, he succeeded in obtaining the first nitrogen NMR spectra in nucleic acids and peptides. An important achievement of Professor Robert Blinc, which attracted considerable attention in the broad scientific community, is the book   Soft Modes in Ferroelectrics and Antiferroelectrics (North Holland), written by him and Boˇ   stjan Žekš in 1974. The book was translated into Russian (1975) and Chinese (1982) and belongs to the 600 most-cited scientific books in the world. Another of his books, written together with Igor Muˇ sevič and Boˇstjan Žekš, The Physics of Ferroelectric and Antiferroelectric Liquid Crystals   , was published by World Scientific in 2000.

Apart from being professor of physics at the University of Ljubljana, the head of the Condensed Matter Physics Department at Joˇ   zef Stefan Institute and a member (and vice-president in the years 1980–1999) of Academy of Science and Arts of Slovenia, Robert Blinc maintained a wide range of contacts with scientists worldwide. There is an amazingly long list of his international scientific activities. To mention only a few of them, he was a visiting professor at the University of Washington in Seattle; ETH Zurich; Federal University of Minas Gerais in Belo Horizonte, Brazil; University of Vienna in Austria; University of Utah in Salt Lake City; Kent State University in Ohio, Argon National Laboratory, and several others. In the years 1988–1994 he was the president of the Groupement AMPERE (Atomes et Molecules Par Etudes Radio Electrique), and president of the European Steering Committee on Ferroelectrics (1990–1999). He is a member of seven foreign Academies of Sciences and has received several national and international scientific prizes and medals.

As the head of the Condensed Matter Physics Department at the Joˇ   zef Stefan Institute for more than 40 years, Robert Blinc promoted, in addition to NMR, other experimental techniques: ESR, dielectric measurements, optical spectroscopy. He also atomic force microscopy. He also took active part in solving theoretical problems related to the systems under study. He was the supervisor of 67 diploma works and 35 Ph. D. theses in the field of condensed matter physics in Ljubljana. He can therefore be recognized as the founder and tireless promoter of the condensed matter physics research in Slovenia.

Most of Robert Blinc's research is tightly related to nuclear magnetic resonance. Therefore we invited a number of prominent researchers in this field to write chapters on the recent condensed matter physics research based on new NMR and ESR techniques. The book covers:

- Adiabatic and nonadiabatic magnetization caused by rotation of solids with dipole-dipole coupled spins.
- Magnetic resonance techniques for studying spin-to-spin pair correlation in multi-spin systems.
- Studies of selectively deuterated semisolid materials and anisotropic liquids by deuterium NMR.
- Initial steps toward quantum computing with electron and nuclear spins in crystalline solids.

- Laser radiation-induced increase of the spin polarization in various magnetic resonance experiments.
- Multiple-photon processes in cw and pulse electron paramagnetice resonance spectroscopy.
- NMR and EPR for the determination of ion localization and charge transfer in metallo-endofullerenes.
- NMR shifts in metal nanoparticles of silver, platinum, and rhodium.
- NMR relaxation studies of di erent superconducting systems.
- Investigations of static and dynamic properties of low dimensional magnetic systems by NMR.
- NMR-NQR relaxation studies of spin fluctuations in two-dimensional quantum Heisenberg antiferromagnets.
- The dynamics of the deuteron glass in KDP type crystals studied by various one-dimensional and two-dimensional NMR techniques.
- Nuclear Magnetic Resonance cryoporometry based on depression of the melting temperature of liquids confined in pores.

We congratulate Professor Robert Blinc on his great scientific achievements and also express our deep gratitude for his continuous e orts in stimulating and supporting the NMR and condensed matter physics community.

Ljubljana

June 2005

Janez Dolin˘sek

Marija Vilfan

Slobodan Žumer

Contents

List of Contributors

M.P. Augustine
Department of Chemistry
One Shields Avenue
University of California
Davis, CA 95616
augustin@chem.ucdavis.edu

F. Borsa
Dipartimento di Fisica "A.Volta"
e Unita' INFM, Universita'
di Pavia, 27100 Pavia, Italy
and
Department of Physics
and Astronomy and Ames
Laboratory
Iowa State University
Ames, IA 50011
borsa@ameslab.gov

P. Carretta
Department of Physics
"A.Volta" and Unit` a INFM
University of Pavia
Via Bassi n 6
I-27100, Pavia (Italy)

K.-P. Dinse
Physical Chemistry III
Darmstadt University of Technology
Petersenstrasse 20
D-64287 Darmstadt, Germany
dinse@chemie.tu-darmstadt.de

M. Fedin
Laboratory for Physical Chemistry
ETH Zurich, CH-8093 Zurich
Switzerland

Y. Furukawa
Division of Physics
Graduate School of Science
Hokkaido University
Sapporo 060-0810, Japan

I. Gromov
Laboratory for Physical Chemistry
ETH Zurich, CH-8093 Zurich
Switzerland

J. Gutschank
Universit¨at Dortmund
Fachbereich Physik
44221 Dortmund, Germany

E.L. Hahn
Department of Physics
University of California
Berkeley, CA 94720
hahn@socrates.berkeley.edu

G. Jeschke
Max Planck Institute
for Polymer Research
Postfach 3148

55021 Mainz, Germany
jeschke@mpip-mainz.mpg.de

T. Kato
Institute for Molecular
Science
Myodaiji Okazaki 444-8585, Japan
present address: Department of
Chemistry
Josai University, 1-1 Keyakidai
Sakado 350-0295, Japan
rik@josai.ac.jp

M. K¨alin
Laboratory for Physical Chemistry
ETH Zurich, CH-8093 Zurich
Switzerland

R. Kind
Institute of Quantum Electronics
ETH-Hoenggerberg, CH-8093 Zurich
Switzerland
kindrd@phys.ethz.ch

J.J. van der Klink
Institut de Physique
des Nanostructures
EPFL, CH-1015
Lausanne, Switzerland
jacques.vanderklink@epfl.ch

A. Lascialfari
Dipartimento di Fisica "A.Volta"
e Unita' INFM, Universita'
di Pavia, 27100 Pavia, Italy
lascialfari@fisicavolta.
unipv.it

M. Mehring
Physikalisches Institut
University Stuttgart
D-70550 Stuttgart, Germany
m.mehring@physik.uni-
stuttgart.de

J. Mende
Physikalisches Institut
University Stuttgart
D-70550 Stuttgart, Germany

J. Mitchell
Department of Physics
University of Surrey
Surrey, UK, GU2 7XH
j.mitchell@surrey.ac.uk

K. M¨uller
Institut f¨ur Physikalische Chemie
Universit¨at Stuttgart
Pfa enwaldring 55
D-70569 Stuttgart, Germany
k.mueller@ipc.uni-stuttgart.de

N. Papinutto
Department of Physics
"A.Volta" and Unit` a INFM
University of Pavia
Via Bassi n 6
I-27100, Pavia (Italy)

A. Rigamonti
Department of Physics
"A.Volta" and Unit` a INFM
University of Pavia
Via Bassi n 6
I-27100, Pavia (Italy)
rigamonti@fisicavolta.unipv.it

W. Scherer
Physikalisches Institut
University Stuttgart
D-70550 Stuttgart, Germany

A. Schweiger
Laboratory for Physical Chemistry
ETH Zurich, CH-8093 Zurich
Switzerland
schweiger@esr.phys.chem.ethz.ch

C.P. Slichter
Department of Physics
and Frederick Seitz Materials
Research Laboratory
University of Illinois
Urbana, IL 61801
cslichte@uiuc.edu

D.F. Smith
Department of Physics
and Frederick Seitz Materials
Research Laboratory
University of Illinois
Urbana, IL 61801

H.W. Spiess
Max Planck Institute
for Polymer Research
Postfach 3148
55021 Mainz, Germany
spiess@mpip-mainz.mpg.de

J.H. Strange
School of Physical Sciences
University of Kent
Canterbury, Kent
UK, CT2 7NR,
j.h.strange@kent.ac.uk

D. Suter
Universit¨at Dortmund
Fachbereich Physik
44221 Dortmund, Germany
dieter.suter@physik.uni-
dortmund.de

B.K. Tenn
Department of Chemistry
One Shields Avenue
University of California
Davis, CA 95616

J. Villanueva-Garibay
Institut f¨ur Physikalische Chemie
Universit¨at Stuttgart
Pfa enwaldring 55
D-70569 Stuttgart, Germany

Nuclear Spin Analogues of Gyromagnetism: Case of the Zero-Field Barnett Eect

E.L. Hahn [1], B.K. Tenn [2] and M.P. Augustine [2]

[1] Department of Physics, University of California, Berkeley, CA 94720
hahn@socrates.berkeley.edu
[2] Department of Chemistry, One Shields Avenue, University of California, Davis
CA 95616
augustin@chem.ucdavis.edu

Abstract. A short review of the history and elementary principles of gyromagnetic eects is presented. The Barnett eect is considered as a mechanism for inducing nuclear spin magnetization in solids by sample spinning in zero and low field. Simulations of rotation induced adiabatic and non-adiabatic magnetization derived from initial dipolar order in homonuclear dipole-dipole coupled spins are carried out. Aspects of the converse Einstein-de Haas eect are included.

1 Personal Tribute

We thank the editors for the invitation to write in honor of Robert Blinc and to celebrate his 70 [th] birthday. Over his many years of international research collaborations and leadership of NMR research groups at the Josef Stefan Institute, he and his colleagues have generated a body of comprehensive experimental data leading to new concepts and clarifications concerning unusual solid state structures. These have involved topics such as ferroelectrics, disordered systems, and liquid crystals, often connected with phase transitions and modal behavior.

While continually devoting himself to his group as a pioneering research physicist working out new interpretations of experiments, Professor Blinc also served as a leader of Slovenian science, maintaining many personal contacts as a virtual "Science Ambassador" in Europe. During the Cold War and through its waning years, because of Robert's international connections in the East, he was able to arrange contacts with people from both the East and the West to attend conferences in Slovenia and on the Dalmation Coast. He has made it possible for many NMR research people from the Eastern block to interact with those of us from the West. His international influence has been unique in making the world a better place for scientific cooperation. Personally I can testify how much we have enjoyed Robert's hospitality, friendship, and

E.L. Hahn et al.: Nuclear Spin Analogues of Gyromagnetism: Case of the Zero-field Barnett
Eect , Lect. Notes Phys. 684 , 1–19 (2006)
www.springerlink.com   c Springer-Verlag Berlin Heidelberg 2006

stimulation provided by visitations to interact with his NMR research group at the Jozef Stefan Institute for which we are grateful. We wish him and his spouse many happy and fruitful years and that he should remain active and not really retire.

2 Introduction

The motivation of this work is the possibility of the ultra-sensitive detection of pico and femto-Tesla fields from the nuclear spin polarization induced in spinning solids at zero field caused by the Barnett e ect. Both the Superconducting Quantum Interference Device (SQUID) [1] and non-linear optical Faraday rotation methods [2] of measuring magnetic fields with an ultimate sensitivity of about one femto-Tesla Hz $^{-1/2}$ or 10^{-11} G Hz$^{-1/2}$ promise to detect these small fields generated by diamagnetic solids.

A brief review of well known elementary principles of gyromagnetic experiments [3, 4] sets the stage for discussion of coupling mechanisms in the Barnett e ect that may account for momentum transfer from a mechanically spinning macroscopic body to microscopic nuclear spins within the body. Only a minute fraction of the total mechanical angular momentum of the spinning sample is transferred to the oriented macroscopic magnetic spin angular momentum, thus conserving the total angular momentum of the system. In the absence of diamagnetic e ects, the ratio of the change of the macroscopic magnetism of a rotating body to this corresponding change in angular momentum of the body is well known as the e ective gyromagnetic ratio (q/2mc) g where q is the fundamental charge, m is the electron mass, c is the speed of light, and g is the empirical g factor.

The Barnett e ect was first observed in 1914 [4] by detecting the magnetism due to the polarization of electron spins caused by rotation of a cylinder of soft unmagnetized iron. Although the Barnett e ect is looked upon today as an archaic experiment, it was an important experiment of the old physics era. Today many people are not aware of the Barnett e ect because it is referred to so little in the literature. It is interesting that even though the concept of electron spin did not exist at that time, the original Barnett experiment provided the first evidence that the electron had an anomalous magnetic moment with a g factor of 2. Today it is well known that the electron spin g factor can also di er significantly from the value of 2 (ignoring the small radiative correction) because of spin orbit coupling. Barnett concluded in 1914 that he measured the gyromagnetic ratio of classical rotating charges q to be q/mc , an anomalous value twice the classical value he expected.

In 1915 Einstein and de Haas [5] carried out the converse of the Barnett experiment. The reversal of an initially known magnetization M_0, or the growth of M_0 from zero, of an iron cylinder produces a small mechanical rotation of the cylinder. In contrast to Barnett's experiment, there is an apparent transfer of spin angular momentum from M $_0$ into mechanical rotation. Curiously

Einstein and de Haas reported   g = 1 from their measurements, apparently rejecting data that deviated from the expected classical value of   g = 1 to account for the orbital magnetism. In 1820 Ampere established that the current due to a charge q and mass m rotating in a circle of radius   r, multiplied by circle area r ^{2}, expresses the classical orbital magnetic moment. This picture supported the idea of hidden classical Amperian currents in permanent magnets, a view that held sway into the early years of the 20 th century. But this view of classical magnetism, and ultimately the Einstein and de Haas   g = 1 experimental interpretation, was first challenged by a theorem formulated by Miss van Leeuwen [6] and Bohr, namely, that any confined configuration of free charges obeying classical laws of motion and precessing in any magnetic field must yield zero magnetic susceptibility. Finally the advent of momentum quantization and the concept of magnetic spin made possible a break   away from the invalid classical picture of magnetism. The classical Amperian magnetic moment was replaced by the non-classical entity of magnetism, the Bohr magneton,

$$\mu = \frac{e}{2mc} = \qquad . \tag{1}$$

3 Parameter Rules for Interpretation of Gyromagnetic Experiments

Let the ratio nμ/n   = M/   =   = e/ 2mc be defined from (1), where   n is the number of polarized spins, or circulating charges in the old picture, lined up to define a macroscopic magnetic moment   M = nμ. The corresponding angular momentum is given by   = n . By itself this ratio is a trivial identity, given that the Bohr magneton of every particle with   L = 1 is μ = e / 2mc. The terms n and   contained in   always cancel, implying in first order that the macroscopic body must display the same   as a single spin would, and provide a measure of e/ 2mc multiplied by any anomalous   g factor. However this argument deserves a better physical justification, relating phenomenologically and still somewhat obscurely to the response of a gyroscope to torque. Sample rotation at a given frequency $_r$ may be viewed as equivalent to a Larmor precession caused by a magnetic field   H . As shown in Fig. 1, the imposed torque due to   H tends to line up the spins. Changes in   M and evolve coaxially. They precess about   H independent of the angle between   H and M or   . A real magnetic field   H causes spin precession of M about the direction of H while M develops and finally reaches equilibrium because of spin-lattice relaxation. However, if the sample is not left to rotate freely as in the Einstein-de Haas experiment, there can be no direct evidence of any mechanical exchange of momentum   no matter how minute. Except for certain special circumstances of macroscopic radiation damping, theories of spin relaxation keep track of energy degrees of freedom but not of elusive internal mechanisms of spin lattice momentum transfer.

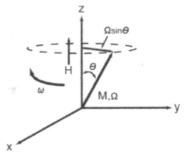

Fig. 1.   Relationship of the magnetization   M , rotational angular momentum   , and static magnetic field   H used to discuss magnetomechanical rotation experiments

Before the development of gyromagnetic experiments, in 1861 Maxwell perceived Amperian currents as hidden gyroscopic sources of permanent magnetism. He tried to detect the precession of a permanent magnet in response to an outside torque, but the e ect is too small to detect. His attempt relates to Fig. 1. In place of a mechanical torque, the magnetic field   H subjects the magnetization M to the torque   T = M sin H =   sin $_r$. Including the empirical factor g the ratio

$$\frac{M}{H} = \frac{_r}{} = \frac{e}{2mc} g \tag{2}$$

defines values of M and   as final values representing changes from zero. As a gyromagnetic rule,   M/   should be written as the ratio of changes   M/ at any time in the evolution of the spin alignment. The Einstein-de Haas experiment measures the ratio of any imposed   M change to the resulting sample rotation angular momentum which is observed. The Barnett experiment measures the ratio of $_r$ to a calibrated H field that produces the same   M caused by sample rotation at the frequency $_r$. Generation of a calibrated   H in the pico to femto-Tesla range from a stable current source would be an extremely di cult requirement. No real field is present when the Barnett e ect takes place. Instead, the field   H in (2) acts like a "ghost" field   H$_{\mathrm{ghost}}$, having the same e ect as a real field. Its definition relates to Larmor's theorem, where H$_{\mathrm{ghost}}$ = $_r$/   is defined as an equivalent field. Equation (2) must follow from energy conservation arguments. A sample rotating at the rate $_r$ and with moment of inertia I is endowed with rotational energy   U = I $_r^2$/ 2. Any small momentum transfer   = I $_r$ that might take place to the spins would require a corresponding increase in spin energy   MH . The total energy transferred between spin and rotation is then   U = I $_r$ $_r$ = $_r$ = MH , a relation that immediately rearranges to express the gyroscopic rule given by (2).

4 Nuclear Spin Analogues

Prior to 1940, gyromagnetic experiments served as a measure of   g values of electron spin systems in ferromagnetic and paramagnetic substances [4]. The Barnett magnetization, although at least a thousand times or more greater than nuclear spin magnetization in samples of comparable size, is very small, di cult to measure, and easily obscured by instrumental instabilities and stray fields. Even large magnetic fields in those days could not be measured to better than a fraction of a percent by rotating pick up coils. These methods are now obsolete, superseded by the application of magnetic resonance detection methods [7, 8, 9]. As a physical mechanism, the nuclear Barnett e ect was invoked later by Purcell [10] to account for the observation of weakly polarized starlight. In that account, Purcell discusses the mechanism of di erential light scattering from fast "suprathermal" rotating grains in interstellar space. Because of their rotation it is postulated that these grains become magnetized due to the nuclear Barnett e ect. The common directivity and polarization of light scattering by the grains occurs over vast distances because the polarized grains in turn precess about the direction of weak interstellar magnetic fields.

Rather than discuss parameters of this very special unearthly case [11] of Barnett polarization, consider a more representative and yet marginal case on Earth. Here a $1\,cm^3$ sample of $N = 10^{22}$ nuclear Bohr magnetons where $\mu_B = (9.27/1840) \times 10^{-21}\,erg\,G^{-1}$ is rotated at the rate $_r/2$ $= 4\,kHz$ in zero applied magnetic field at $T = 300\,K$. Assume that the sample acquires an equilibrium magnetization $M_0 = N\mu_B(\,_r/kT)$ because of spin lattice relaxation in the ghost field $H_{ghost} = \,_r/$. The resulting polarization field in the sample is about 4 M_0 10^{-10} G or about 1–10 femto-Tesla. Clearly rotation at higher speeds and at lower temperature could provide M_0 values 10 to 100 times larger, providing an extra margin for weak field detection [1, 2].

5 Homonuclear Dipole-Dipole Coupling

The crude estimate of the field due to a Barnett induced magnetization mentioned above assumes that the spins polarize in a ghost field $H_{ghost} = \,_r/$ during a spin-lattice relaxation process as though it were a real field. However, the complexity of many momentum transfer relaxation mechanisms between spin and lattice thermal reservoirs is too di cult to handle. Some understanding can be gained from a specific example of momentum conservation by simulating the e ect of sample spinning on dipolar interactions among nuclear spins. A rigid lattice firmament of spins is assigned only a spin temperature with lattice coordinates $,r$ and   independent of time in the absence of sample spinning. Since there is no lattice thermal reservoir in this picture, the source of magnetization is obtained from previously prepared dipolar order in zero field that is converted to magnetization by sample rotation. A simple starting point considers two identical spins I_1 and I_2

separated by the distance r and coupled via their dipolar fields. A real DC magnetic field H_0, or sample spinning at the rate ω_r may be applied separately or simultaneously, axially or non-axially. The dipolar coupling is defined as $H_D = \omega_D \sum_{q=-2}^{2}(-1)^q T_q^{(2)}(I_1,I_2)R_{-q}^{(2)}(\theta,\phi)$ where $\omega_D = \gamma_1\gamma_2/r^3$ and the irreducible components $T_q^{(2)}(I_1,I_2)$ and $R_q^{(2)}(\theta,\phi)$ are given by

$$T_0^{(2)}(I_1,I_2) = \tfrac{1}{\sqrt{6}}(3I_{z,1}I_{z,2} - I_1 \cdot I_2) \qquad R_0^{(2)}(\theta,\phi) = \tfrac{\sqrt{6}}{2}(1 - 3\cos^2\theta),$$

$$T_{\pm1}^{(2)}(I_1,I_2) = \mp\tfrac{1}{2}(I_{\pm,1}I_{z,2} + I_{z,1}I_{\pm,2}) \qquad R_{\pm1}^{(2)}(\theta,\phi) = \pm 3\sin\theta\,\cos\theta\,e^{\pm i\phi}, \qquad (3)$$

$$T_{\pm2}^{(2)}(I_1,I_2) = I_{\pm,1}I_{\pm,2} \qquad R_{\pm2}^{(2)}(\theta,\phi) = \tfrac{3}{2}\sin^2\theta\,e^{\pm 2i\phi}.$$

When written in this way it should be clear that H_D not only couples the spins I_1 and I_2 to each other with the $T_q^{(2)}(I_1,I_2)$ operators, but also to the lattice with the $R_q^{(2)}(\theta,\phi)$ coefficients. The lattice degrees of freedom with large heat capacity are usually assumed to be at constant temperature in diagonal states of the density matrix, while the spin temperature may change due to relaxation. For this reason the understanding of NMR experiments in solids considers only the effect of the lattice on the spins, while effects of the spins on the lattice are usually neglected. However gyromagnetic experiments confirm that the lattice hooked to the rotor does exchange angular momentum with the spins, showing that the spins have a momentum effect on the lattice. In connection with nuclear spin diffusion in zero field, Sodickson and Waugh [12] introduced the interesting and related question about momentum exchange and conservation between nuclear spins and the lattice in zero field. Here the components of the angular momentum operators $L = r \times p$ are given by

$$L_x = i\hbar\left(\sin\phi\,\frac{\partial}{\partial\theta} + \cot\theta\,\cos\phi\,\frac{\partial}{\partial\phi}\right),$$

$$L_y = i\hbar\left(-\cos\phi\,\frac{\partial}{\partial\theta} + \cot\theta\,\sin\phi\,\frac{\partial}{\partial\phi}\right), \qquad (4)$$

$$L_z = -i\hbar\frac{\partial}{\partial\phi},$$

which in combination with Ehrenfest's theorem can be used to show that the time derivative of the expectation value of the total spin angular momentum $J = I_1 + I_2$ is given by

$$\frac{d}{dt}J = I_1 \times H_{D,2} + I_2 \times H_{D,1} = i[H_D,J] = -\frac{d}{dt}L, \qquad (5)$$

where $H_{D,1}$ and $H_{D,2}$ are the dipolar fields from spin 1 and 2. This reformulation of Bloch's equation shows that the total angular momentum, spin plus lattice, is conserved. This angular momentum conservation relationship is at the heart of recovering Zeeman order from dipolar order by sample rotation and can be used to determine the effect of the spins on the lattice. As an example consider an ensemble of identical dipole-dipole coupled two spin systems at thermal equilibrium in zero applied magnetic field. The thermal equilibrium density operator corresponding to this situation is given by $\rho_{eq} = \exp(-H_D/kT) \approx 1 - H_D/kT$ where k is the Boltzmann constant and the high temperature approximation has been applied. Provided that there are no real magnetic fields present, the expectation value $\langle J \rangle = Tr\{\rho_{eq} J\}$ is zero at all times. In order to verify (5) or equivalently show that $\langle L \rangle = 0$ in this example, a rigorous quantum mechanical treatment of the expectation value of the lattice angular momentum $\langle L \rangle$ is needed. This treatment requires that the rotational motion of the lattice be quantized in direct analogy to the rotational level structure in molecular H_2 gas [7] or in tunneling methyl groups [13]. Adopting this approach requires some knowledge of the partition function corresponding to rotation in addition to the calculation of the matrix elements of the $R_q^{(2)}(\Omega,\Theta)$ coefficients from the $|L,m\rangle = Y_m^{(L)}(\Omega,\Theta)$ spherical harmonic basis functions. In the case of H_2 gas in a molecular beam where the moment of inertia I is small, molecular rotation is fast, and the temperature is low, it is safe to truncate the basis set to a finite number of rotational energy levels. However, in the case of a real macroscopic sample at room temperature, I is large and the sample rotation is slow. This means that an untractably large number of very closely spaced energy levels will be populated at thermal equilibrium. The inability to define the density matrix of the rotor in this case can be circumvented by realizing that the expectation value for the lattice angular momentum $\langle L \rangle$ must reduce to the classical result $I\omega_r$ for a macroscopic object. In this way in zero field (5) reduces to

$$\frac{d}{dt}(I\omega_r) = -\frac{d}{dt}\langle J \rangle,\tag{6}$$

therefore in the absence of sample rotation, $\omega_r = 0$ and $\langle L \rangle = I\omega_r = 0$.

Equation (6) contains a very important result, namely, any change in sample rotation rate ω_r will lead to a corresponding change in sample magnetization. If a stationary sample is initially in zero magnetic field, (6) indicates that a jump in rotational frequency from zero to a final value ω_r will lead to a corresponding change in sample magnetization from zero to a final value. In certain special cases like the ensemble of identical two spin systems mentioned above, (6) can be used in combination with the solution to the Liouville-von Neumann equation during sample spinning along the $+z$ direction to determine the dynamics of both the formation of magnetization $\langle J \rangle$ and the change in the spin rate ω_r due to this magnetization. Here the static dipolar coupling H_D mentioned above becomes time dependent in $\rho(t)$ as $H_D(t) = \omega_D \sum (-1)^q T_q^{(2)}(I_1,I_2)R_{-q}^{(2)}(\Omega,\Theta)e^{iq\omega_r t}$ since the internuclear vector r now precesses about the rotation direction taken along $+z$. This particular time dependent form for the laboratory frame Hamiltonian $H_{lab} = H_D(t)$ is not convenient for practical calculations and does not offer much insight into how magnetization develops due to sample spinning. The symmetry between the $T_q^{(2)}(I_1,I_2)$ and $R_q^{(2)}(\Omega,\Theta)$

products in $H_D(t)$ permit the $\exp(i\,q_r t)$ time dependence to be grouped together with the spin operators as $T_q^{(2)}(I_1,I_2)\exp(i\,q_r t)$ instead of with the
$R_q^{(2)}(\,,\,)$ spatial terms that describe the orientation of r as a function of
time. In this way it is clear that transformation to a rotating frame in spin
space by rotation operator $U(_r t)=e^{iq I_z \,_r t}$ about the laboratory $+ z$ axis as
$U(_r t)\,T_q^{(2)}(I_1,I_2)\,U^\dagger(_r t)=T_q^{(2)}(I_1,I_2)\exp(i\,q_r t)$ yields the time independent rotating frame Hamiltonian

$$H_{rot} = {}_r(I_{z,1}+I_{z,2})+ {}_D \quad (-1)^q T_q^{(2)}(I_1,I_2)R_{-q}^{(2)}(\,,\,)\,, \qquad (7)$$

where the ghost field $H_{ghost} = {}_r/\quad$ appears. As expectation values of the
total spin magnetization J are independent of unitary transformations in
the trace, the fictitious term in the rotating frame generates the same magnetization that a real field $H_0 = H_{ghost}$ would develop in the laboratory frame.
Furthermore, since (5) and (6) relate traces of spin and lattice momentum
operators, these relations are also frame independent.

A combination of the solution to the Liouville-von Neumann equation in
the rotating frame using H_{rot} in (7) with the angular momentum conservation
rule in (6) allows the dynamics of the magnetization and the rotor frequency
to be determined starting from a stationary sample in zero magnetic field.
Figure 2 shows the effect of instantaneously switching on an $_r/2 = 10\,kHz$
sample rotation to an ensemble of dipolar coupled two spin systems with
$_D/2 = 10\,kHz$. The plots in Fig. 2(a) are appropriate for an ensemble of
identical two-spin systems with $= /2$ and $= 0$ while the plots in Fig. 2(b)
describe a somewhat more realistic case involving an isotropic distribution of
and values. In both of these plots the feedback due to angular momentum
conservation requires that the alternating magnetization will modulate the
$_r/2 = 10\,kHz$ applied sample spin rate. In addition the phase of the periodicity introduced into the spin rate is 180 out of phase with the periodicity in
J_z, consistent with the negative sign in (6). The same two spin system used
in Fig. 2(a) was used to determine the peak z magnetization as a function
of applied rotation rate in Fig. 2(c). This plot demonstrates that the largest
Barnett magnetization can be obtained in zero field when $_r \quad _D\,.$

It is natural to ask if an analogue of the above experiments can be obtained by causing an ensemble of two-spin systems at thermal equilibrium
in zero field to rotate by instantaneously jumping a real DC magnetic field.
Introduction of a real DC magnetic field H_0 along the $+ z$ direction in the
laboratory frame where $= 0$ adds a Zeeman term $H_z = H_0(I_{z,1}+I_{z,2})$
to the dipolar coupling Hamiltonian H_D yielding the full laboratory frame
Hamiltonian as $H_{lab} = H_z + H_D$. One consequence of the field H_0 is to add
an additional term to the zero field angular momentum conservation relation
shown in (5) since $d\,L/dt = i\,[H_{lab},L] = i\,[H_D,L] = -i\,[H_D,J]$ and
$d\,J/dt = i\,[H_{lab},J] = i\,[H_z,J] +i\,[H_D,J]$. Rearranging these equations,
noting that L commutes with H_z, and using the fact that $L = I_r$ in the
classical limit recasts (5) as

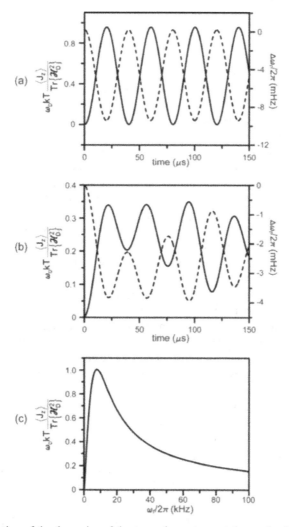

Fig. 2. Simulation of the dynamics of the two spin system rotating at the frequency $\omega_r/2\pi$ with $\omega_D/2\pi = 10\,\text{kHz}$ and $\omega_0/2\pi = 0$. In each case (a)–(c) the $t < 0$ thermal equilibrium condition corresponds to an ensemble of two spin systems in zero magnetic field. At the time $t = 0$ the spin rate is instantaneously increased from zero to $\omega_r/2\pi = 10\,\text{kHz}$ in (a) and (b) and a variable level in (c). The effects of the sample rotation are to create magnetization (solid line) as shown by the left ordinate in (a) and (b) for the $\theta = \pi/2$ orientation and the isotropic powder respectively. The polarization $J_z = I_{z,1} + I_{z,2}$ is scaled by the initial density operator so that a left ordinate value of 1 corresponds to a polarization of $\omega_D/2kT$. The right ordinate shows how the magnetization feeds back via angular momentum conservation to provide only mHz changes in the spin rate (dashed line) because the ratio $\hbar N_s/I$ is only $10^{-3}\,\text{s}^{-1}$ for a real sample containing N_s spins. The plot in (c) suggests that the maximum transfer of rotational angular momentum into Zeeman polarization occurs when $\omega_r = \omega_D$

$$\frac{d}{dt}(I_r) = -\frac{d}{dt} J + i [H_z, J] = -i [H_D, J] . \tag{8}$$

The expectation values are the same regardless of reference frame since the trace is independent of unitary transformation. Comparison of (8) to (6) shows that the interaction of the spin angular momentum J with the applied DC field H_0 represents an additional source of angular momentum that indirectly influences the momentum conservation. The rotor only exchanges angular momentum with the spins through H_D, the dipolar interaction – not through the direct coupling with H_z. The external H_z or any other applied field represents an energy-momentum source from a solenoid or an oscillator not directly coupled to the rotor. When initial angular momentum flows into J_z by jumping the rotation frequency from zero to $_r$, the built in feedback in (6) and demonstrated by simulation in Fig. 2 gives rise to the ghost field $H_{ghost} = _r/$. This field can not couple to the solenoid. On the other hand, when an applied field is jumped from zero to H_0 the source of angular momentum flowing into J_z is the solenoid. The solenoid plays the role now of the rotor because of the application of an electromotive force which launches a current and therefore momentum into the applied H_0 field. The resulting exchange of order between the dipolar and Zeeman reservoirs is that of the Strombotne and Hahn experiment [14]. Therefore, H_z has only an indirect e ect on the rotor but remains to define the Zeeman Bloch equations when $_r = 0$. Equation (8) also explains why neither the Barnett nor the Einstein-de Haas e ects have been noticed in routine NMR experiments in solids performed at high field. At high magnetic field the nuclear Zeeman interaction scales the non-secular $q = 0$ part of the dipole-dipole coupling to roughly $_D^2/_0 \quad _D \quad _0$. To zeroth order in perturbation theory the NMR spectrum is governed by the Zeeman interaction H_z and the right hand side of (8) is identically zero thus removing any coupling between the nuclear spin and mechanical angular momenta. Extending this argument to first order in perturbation theory also does not yield any useful coupling between $_r$ and J when the applied DC field is parallel to the sample spinning axis because the spatial term $T_0^{(2)}(I_1, I_2) R_0^{(2)}(\ ,\ )$ does not have any $_r$ rotational dependence. It is the time dependence imparted on the $q = 0$ parts of $H_D(t)$ that translate into the generation of magnetization in zero magnetic field or the onset of sample rotation in a magnetic field. This can be appreciated by considering the Hamiltonian for the two-spin system with a magnetic field applied parallel to the sample spinning axis along the $+z$ direction. This time dependent Hamiltonian $H_{lab} = H_z + H_D(t)$ is best considered in the rotating frame at the frequency $_r$. Since this transformation involves a rotation about the z axis, the nuclear Zeeman interaction H_z is simply added to the Hamiltonian in (7) to give

$$H_{rot} = (_r + _0)(I_{z,1} + I_{z,2}) + _D \quad (-1)^q T_q^{(2)}(I_1, I_2) R_{-q}^{(2)}(\ ,\ ), \tag{9}$$

a result suggesting that sample rotation can either add to or subtract from the apparent Larmor frequency $_0$. Judging from (9) the largest e ect of magnetic field on sample rotation would be in low-field where ω_r is comparable to ω_0, and given (8) combined with the zero field results mentioned above, the largest effect should be observed when ω_r is comparable to ω_D.

In direct analogy to the zero field case mentioned in Fig. 2, a combination of the solution to the Liouville-von Neumann equation in the rotating frame using H_{rot} in (9) with the angular momentum conservation rule in (8) permits the dynamics of the magnetization and the rotor frequency to be determined starting from a stationary sample in zero magnetic field. Figure 3 demonstrates the effect of instantaneously switching on an $H_0 = 2.35$ G DC magnetic field or equivalently an $\omega_0/2\pi = 10$ kHz ^{1}H Larmor frequency to an ensemble of dipole-dipole coupled two spin systems with $\omega_D/2\pi = 10$ kHz. The plots in Fig. 3(a) correspond to the same ensemble of identical two-spin systems with $\theta = \pi/2$ and $\phi = 0$ shown in Fig. 2(a) while the plots in Fig. 3(b) show results for the same isotropic distribution of θ and ϕ values used in Fig. 2(b). In both of these plots the feedback due to angular momentum conservation requires that the magnetization will generate a periodic sample rotation. Since the commutator in (8) is always zero for the case of a field applied parallel to the maximum moment of inertia, the negative sign in (8) causes the phase of the periodicity in the spin rate to be 180° out of phase with the periodicity in J_z. The same two-spin system used in Fig. 3(a) was used to determine the peak spin rate ω_r as a function of applied DC field and hence Larmor frequency ω_0 in Fig. 3(c). This plot demonstrates that the largest induced sample rotation rate starting in zero field can be obtained when the applied magnetic field H_0 is comparable to the inherent dipolar field i.e. when $H_0 \approx \omega_D/\gamma$ or $\omega_0 \approx \omega_D$.

The effects predicted by the admittedly crude two-spin system in Figs. 2 and 3 continue to manifest themselves in larger more realistic cases [14]. Figure 4 shows the Barnett initially induced magnetization in (a) and the DC field jump induced rotation in (b) for an eight spin system. In these examples the eight spins are positioned on the corners of a stationary cube in zero magnetic field that occupies the normal x, y, and z cartesian axis system while the gated sample rotation and magnetic field directions are along the $+z$ axis. The side length of the cube is 2.3 Å giving an $\omega_D/2\pi = 10$ kHz dipolar coupling for protons with the smallest separation. In this example thermal equilibrium is appropriate for an ensemble of identical eight spin cubes held at the temperature T. To remain consistent with Figs. 2 and 3, the rotation speed in Fig. 4(a) is jumped from zero to $\omega_r/2\pi = 10$ kHz while the field is jumped from zero to $H_0 = \omega_0/\gamma = 2.35$ G in Fig. 4(b). Comparison of Fig. 4 to Figs. 2 and 3 suggest that the dynamics of the eight spin system are similar to that for the two spin system. The major difference is that the presence of three different dipolar coupling values corresponding to spins separated on the edge, face diagonal, and body diagonal of the cube, have the net effect of smearing the Zeeman-dipolar oscillations in Figs. 2 and 3, an averaging that makes the eight spin simulation closely resemble oscillations observed in real field cycled experiments [14]. Admittedly, Figs. 2(a) and 3(a), Figs. 2(b) and 3(b), Figs. 2(c) and 3(c), and Fig. 4(a) and Fig. 4(b) look

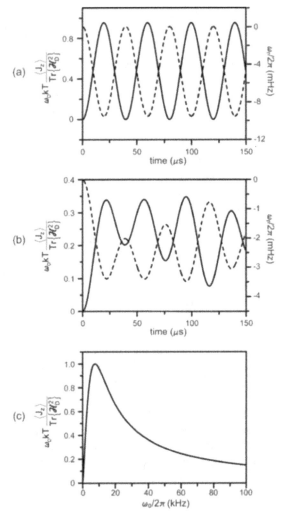

Fig. 3. Simulation of the dynamics of the two spin system with applied DC field at frequency $\omega_0/2\pi$ with $\omega_D/2\pi = 10\,\text{kHz}$ and $\omega_r/2\pi = 0$. In each case (a)–(c) the $t < 0$ thermal equilibrium condition corresponds to an ensemble of two spin systems in zero magnetic field. At the time $t = 0$ the DC magnetic field is instantaneously increased from zero to $H_0 = \omega_0/2\pi = 10\,\text{kHz}$ in (a) and (b) and a variable level in (c). The effects of this field are to create magnetization (solid line) as shown by the left ordinate in (a) and (b) for the $\theta = \pi/2$ orientation and the isotropic powder respectively. The polarization $\langle J_z \rangle$ is scaled by the initial density operator so that a left ordinate value corresponds to a polarization of $\omega_D/2kT$. The right ordinate shows how the magnetization feeds back via angular momentum conservation so that a small fraction of N spins in the sample becomes polarized, where the acquired sample rotation rate ω_r is typically $N\hbar/I = 10^{-3}\,\text{s}^{-1}$. The plot in (c) suggests that the maximum production of Zeeman polarization occurs when $\omega_0 = \omega_D$

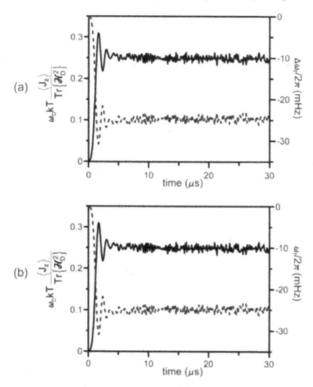

Fig. 4.   Simulation of the dynamics of an eight spin system with the spins centered on the corners of a cube. Taking the spins as protons, the $\omega_D/2\pi = 10\,\text{kHz}$ dipolar coupling corresponds to the strongest coupling or the shortest distance between nuclei on the 2.3 Å long cube side. The plot in (a) corresponds to jumping of an $\omega_r/2 = 10\,\text{kHz}$ sample rotation rate while in (b) similar results are explored by turning on an $\omega_0/2 = 10\,\text{kHz}$ Larmor frequency. In both cases the $t < 0$ thermal equilibrium situation reflects an ensemble of dipolar coupled eight spin systems. Here the value of the magnetization on the left ordinate is scaled to the thermal equilibrium density operator for the eight spin system while the right ordinate pertains to changes in the sample rotation frequency due to angular momentum conservation

identical. This similarity is intended as the ghost field due to sample rotation in Figs. 2 and 4(a) yield the same dynamics by independent calculations as the real field in Figs. 3 and 4(b). In Figs. 2– 4 one may interpret in the case of jumping the rotor frequency from zero to ω_r, that when momentum flows from J_z to the rotor, the process is Einstein-de Haas. If the momentum flows from the rotor to J_z the process is Barnett. Hence there is an oscillatory display of both effects. These oscillations due to angular momentum conservation can be connected with the oscillations in energy due to population exchange between Zeeman and dipolar order.

6 Adiabatic Demagnetization and Remagnetization

Because of spin entropy conservation in a homonuclear dipolar coupled spin system [15, 16] in the absence of relaxation, the adiabatic demagnetization of a spin system converts Zeeman order to dipolar order after an initial polarizing Zeeman field H_i is reduced adiabatically to zero. During this process, the initial spin temperature is reduced from T_i to a final dipolar spin temperature $T_f = (H_{loc}/H_i) T_i$ assuming $H_i \quad H_{loc}$, where $H_{loc} \quad _D/2$ is the local average dipolar field. Suppose that the thermodynamic relation d $U = -M dH$ governing the adiabatic process of demagnetization in the presence of a real field H applies also to the ghost field H_{ghost}. More accurately, this means that the spin temperature will be a function of the ghost field H_{ghost} as if it were a real field. On this basis any polarization induced by sample spinning will be Barnett in character. When a real field H_i is reduced during demagnetization, the adiabatic reduction in Zeeman energy $-M H$ requires that the spins must do work on the solenoid. Conversely the solenoid must do work on the spins to restore $M H$ in the process of remagnetization. With the case of a Barnett polarization mediated by sample spinning, it is necessary to show experimentally that the ghost field H_{ghost} can bring about an analagous adiabatic response. This would seem to follow by virtue of the identity dU $= -(M/)(dH_{ghost}) = - d_r$, if in this case positive and negative work must be done by the rotor instead of the solenoid. If this picture is true, the outside work by the rotor must be different and not related to the microscopic changes in rotor energy and momentum mentioned in connection with spin lattice relaxation. There is an ambiguity here because the same terms that apply to either situation have been invoked by assuming that the same spin temperature applies in both cases. A negative or positive H_{ghost} added parallel to H by sample spinning in opposite directions would require both of these mechanisms to operate simultaneously.

In the limit that the ghost field behaves just like a real field, a sizeable portion of initially polarized magnetization $M_i = C_{spin} H_i/T_i$ may be recovered and sustained after adiabatic demagnetization by turning on a small ghost field H_{ghost}. Here the initial field H_i is sizeable and C_{spin} is the nuclear spin Curie constant. After adiabatic remagnetization in the ghost field H_{ghost}, a good fraction of M_i given by

$$M_f = M_i \frac{H_{ghost}}{H_{ghost}^2 + H_{loc}^2} = M_i \frac{_r}{_r^2 + _D^2} \qquad (10)$$

should be recovered due to the sample spinning. Since the adiabatically recovered magnetization in (10) is not directly proportional to H_{ghost} or equivalently $_r$ as it is in the genuine Barnett effect, the final magnetization M_f should be considered as a re-magnetizéd "pseudo–Barnett" polarization suspended in the absence of real fields. Beginning at room temperature without demagnetization one would obtain a much smaller magnetization $M_i = C_{spin} H_{ghost}/T_i$.

The dynamics of the "pseudo-barnett" polarization in a small ghost field can be easily tested using the ensemble of eight spin systems described above. Here the initial thermal equilibrium polarization is taken to be in the high field high temperature limit $_{eq} = 1 - H_z/kT - H_D/kT$ where both the Zeeman H_z and dipolar coupling H_D Hamiltonians are appropriate for eight spins situated at the corners of a cube. The Zeeman and dipolar temperatures are taken to be the same, and the field H_i is parallel to the sample spinning direction in addition to defining the $+$ z axis. Figure 5(a) shows the timing of the real magnetic field and sample rotation ramps during the adiabatic demagnetization and remagnetization process while Figs. 5(b)–(d) show how the total magnetization J_z, dipolar order H_D, and spin rate $_r$ evolve in time respectively. Comparison of Figs. 5(b) and (c) indicate that the demagnetization process from a field of $H_i = 2,350\,G$ to zero completely transfers proton Zeeman order into dipolar order at the end of the ramp at $t = 100\,\mu s$. After persisting as dipolar order for $t = 250\,\mu s$, the adiabatically ramped sample spinning from zero to a final value of $_r/2 = 30\,kHz$ causes the "pseudo-Barnett" magnetization to appear. The dashed line in Fig. 5(b) corresponds to the ratio M_f/M_i anticipated on the basis of (10) with $_r/2 = 30\,kHz$ and $_D^2{}^{1/2} = 16\,kHz$ for the eight spin problem. Figures 5(b)–(d) all display what appears to be a large amount of noise at low field. Closer inspection of these plots reveals that the noise is periodic and that the source of the periodicity is the angular momentum feedback due to the Barnett e ect.

7 Sample Spinning Non-axial with DC Field

In all of the previous cases with the DC field and sample rotation applied in the $+$ z direction, $J_x = J_y = 0$ and any feedback contributions due to o axis sample rotation are discarded in (6) and (8). The more useful case applying to narrow NMR lines in solids involves rapid sample spinning at some angle with respect to an applied DC field. New time dependent terms in the Hamiltonian are generated from rotation transformations of only the secular terms. In a small magnetic field comparable to the dipolar field, inclusion of the transformation of the non-secular terms show that any static field component normal to the axis of spin rotation results in saturation of any polarization M_0 that may be acquired by the Barnett e ect. The coupling of the sample spinning to the magnetization in the case when a real DC field is applied at some initial angle that is not parallel to the direction of sample rotation can be understood by writing the dipolar coupling interaction in the principal axis frame of the moment of inertia tensor I. In this frame the z-axis corresponds to the direction of the maximum moment of inertia. Realizing that the applied static field can be at any orientation with respect to this frame recasts the Zeeman interaction as $H_z = _x(t)J_x + _y(t)J_y + _z(t)J_z$ where it is understood that the J_x, J_y, and J_z operators pertain to the total spin angular momentum in the x, y, and z directions in the inertial frame

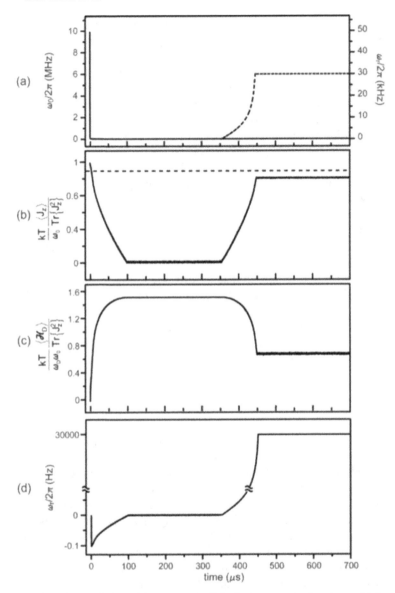

Fig. 5.   Eight spin simulation of the e ects of adiabatic demagnetization from a high
H$_i$ = $_0$/ = 2 , 350 G magnetic field to zero field and subsequent remagnetization
in an e ective spinning field of   H$_{ghost}$ = $_r$/ = 7 .05 G. The timing of the magnetic
field and sample spinning is shown in (a) while the change in magnetization   J$_z$,
dipolar order   H$_D$, and sample spin rate $_r$ are included in (b)–(d) respectively

while $\omega_x(t)$, $\omega_y(t)$, and $\omega_z(t)$ correspond to the time dependent orientation of the field $H = \mu(t)/\gamma = (\omega_x(t)/\gamma)i + (\omega_y(t)/\gamma)j + (\omega_z(t)/\gamma)k$ in the inertia frame as the sample rotates. The three principal components of the inertia tensor I_{xx}, I_{yy}, and I_{zz} can be used along with the angular momentum conservation relation in (8) to develop a similar conservation rule for any general orientation of the DC field with respect to the sample rotation direction as

$$\frac{d}{dt}\begin{vmatrix} I_{xx}\,\omega_{r,x}(t) \\ I_{yy}\,\omega_{r,y}(t) \\ I_{zz}\,\omega_{r,z}(t) \end{vmatrix} = -\frac{d}{dt}\begin{vmatrix} J_x \\ J_y \\ J_z \end{vmatrix} + \begin{vmatrix} 0 & -\omega_z(t) & \omega_y(t) \\ \omega_z(t) & 0 & -\omega_x(t) \\ -\omega_y(t) & \omega_x(t) & 0 \end{vmatrix}\begin{vmatrix} J_x \\ J_y \\ J_z \end{vmatrix} ,$$

(11)

where $\omega_i(t) = \omega_i(t) + \omega_{r,i}(t)$, $i = x,y,z$ and $\omega_{r,i}(t)$ corresponds to the frequency of rotation of the object along the principal axes of I in the inertial frame. The expectation values of the total magnetization in (11) in the presence of a DC field are most easily obtained by expressing the Zeeman and homonuclear dipolar interactions in a rotating frame at the sample rotation frequency as

$$H_{rot} = \sum_{i\{x,y,z\}}\sum_{j=1}^{N}(\omega_{r,i} + \omega_i(t))I_{i,j} + \sum_{i=1}^{N}\sum_{j=i}^{N}\omega_D(i,j)$$

(12)

$$\times \sum_{q=-2}^{2}(-1)^q T_q^{(2)}(I_i,I_j)R_{-q}^{(2)}(\theta_{ij},\phi_{ij}),$$

where the sum over all of the spins considered in the sample is included for clarity, $\omega_D(i,j) = \gamma_i\gamma_j\hbar/r_{i,j}^3$, and the polar angles $\theta_{i,j}$ and $\phi_{i,j}$ are included in the final term to specify that the internuclear direction $r_{i,j}$ between each spin pair in the sample might have a different orientation with respect to the moment of inertia frame. In the special case of Barnett induced magnetization, the discussion in the previous sections suggests that any changes due to the feedback predicted in (11) in practical spin rates of several kHz along the $+z$ direction are small and most likely negligible. Taking $\omega_{r,x}(t) = \omega_{r,y}(t) \approx 0$ and $\omega_{r,z}(t) = \omega_r$ suggests that the only time dependence remaining in (12) in the rotating inertia frame is due to the precession of the applied DC field around the sample. The transverse $\omega_x(t)$ and $\omega_y(t)$ components will modulate at ω_r or equivalently, are at exact resonance with M_0 and will thus cause saturation transitions while the ω_z term is time independent and will add a detuning effect and thus a resonance offset. It is important to note that a similar effect occurs in standard magnetic resonance experiments when a DC field perpendicular to the Zeeman polarizing field is turned on in the lab frame. An initially polarized value of M_0 would disappear as it precesses around the effective field at a rate depending on the size of the perpendicular field.

8 Lattice Structure Dependence of the Barnett E ect

In a liquid the dipole-dipole interaction between spins is cut o because of
random fluctuations of the lattice coordinates. The competition between mutually interacting spins precessing in the local dipolar field and the apparent
precession caused by a small ghost field H_{ghost} disappears if the local field
averages to zero. To understand this averaging assume first that the dipolar
coupling between two spins is zero. The two spins of course would precess
about any applied DC field, but for the moment let them remain pointed
in fixed directions in space like gyroscopes. One can easily see for example
that rotation of the internuclear vector r by some precession angle about a
perpendicular axis will alter the orientation angles of both dipoles relative to
one another. Now turn on the dipole-dipole interaction. Comparison of the
interaction energy at the two di erent orientations reveals that it is in fact
di erent. If instead of two discrete angles a continuous set of angles is explored
through sample rotation one finds that if the rotation rate $_r$ is comparable
to or exceeds the dipole-dipole coupling strength, the spins appear to be precessing coherently about the rotation axis perpendicular to r in the rotating
frame. It is in this case that an apparent torque can be attributed to the
fictitious field H_{ghost} = $_r/$. As long as r retains the integrity of the lattice
structure, the spins can sense Fourier components characteristic of the lattice
rotation appearing in that frame and ultimately polarize by spin-lattice relaxation. But as stated above, if the lattice structure disappears because of
coordinate fluctuations as in a liquid due to motional narrowing, the Barnett
e ect will be quenched. Basically the lattice environment becomes isotropic,
looking virtually the same to the spins regardless of the rotational aspects of
the liquid. If a torque is to be attributed to H_{ghost} in a rotating sample, the
spins must mutually interact over many cycles of rotation in a reference frame
that connects them to a fixed lattice structure.

9 Conclusion

One motivation for this work was the lure of polarizing nuclear spins in solids
in conventional high field applications by transferring the massive angular momentum I $_r$ from a rotating macroscopic object. The arguments provided in
the above sections suggest that a small fraction of this massive sample angular
momentum can indeed be transferred but only in low field situations where
the sample rotation rate $_r$ and Larmor frequency $_0$ is comparable to the
dipolar coupling strength $_D$. At present conventional high field applications
of this method to homonuclear dipole-dipole coupled spin systems are limited
because the high magnetic field quenches the spin-lattice coupling manifest
in the non-secular dipole-dipole terms that drive the e ect. Before identifying
possible uses of the Barnett e ect in nuclear spin systems, experiments must
be completed to verify whether or not the ghost field does in fact behave like

a real DC field. Instead of relying only on conservation laws to account for the transfer of angular momentum between the spin system and the rotor, the challenge remains to devise a more rigorous two reservoir formalism similar to that applied to cross relaxation between two spin species.

Acknowledgements

In particular, ELH is grateful to Dietmar Stehlik for stimulating discussions and initiating interest in the Einstein-de Haas and Barnett e ects. We also gratefully acknowledge useful discussions with Alex Pines, Maurice Goldman, John Waugh, Jean Jeener, Eugene Commins, Carlos Meriles, Demitrius Sakellari, Andreas Trabesinger and Jamie Walls. MPA is a David and Lucile Packard and Alfred P. Sloan foundation fellow.

References

1. R. Mc Dermott, A.D. Trabesinger, M. Muck, E.L. Hahn, A. Pines, J. Clarke: Science 295 , 2247–2249 (2002)
2. D. Budker, D.F. Kimball, V.V. Yashchuk, M. Zolotorev: Phys. Rev. A   65 , 55403 (2002)
3. L.F. Bates: Modern Magnetism   (University Press, Cambridge 1951)
4. (a) S.J. Barnett: Phys. Rev.   6, 239–270 (1915)
   (b) S.J. Barnett: Rev. Mod. Phys.   7, 129–166 (1935)
5. (a) A. Einstein, W.J. de Haas: Verhandl. Deut. Phsik. Ges.   17, 152–170 (1915)
   (b) A. Einstein, W.J. de Haas: Verhandl. Deut. Phsik. Ges.   18, 173–177 (1916)
6. J.H. Van Vleck: The Theory of Electric and Magnetic Susceptibility   (Clarendon, Oxford 1932) pp 94–97
7. J.M.B. Kellogg, I.I. Rabi, N.F. Ramsey, J.R. Zacharias: Phys. Rev.   56 , 728–743 (1939)
8. (a) F. Bloch, W.W. Hansen, M. Packard: Phys. Rev.   70 , 474–485 (1946)
   (b) E.M. Purcell, H.C. Torrey, R.V. Pound: Phys. Rev.   69 , 37–38 (1946)
9. A. Abragam, B. Bleaney: Electron Paramagnetic Resonance of Transition Ions (Clarendon, Oxford 1970)
10. E.M. Purcell: Astrophys. J.   231 , 404–416 (1979)
11. A. Lazarian, B.T. Draine: Astrophys. J.   520 , L67–70 (1999)
12. D.K. Sodickson, J.H. Waugh: Phys. Rev. B   52 , 6467–6469 (1995)
13. B. Black, B. Majer, A. Pines: Chem. Phys. Lett.   201 , 550–554 (1993)
14. R.L. Strombotne, E.L. Hahn: Phys. Rev.   133 , A1616–A1629 (1964)
15. A. Abragam: Principles of Nuclear Magnetism   (Clarendon, Oxford 1961)
16. M. Goldman: Spin Temperature and Nuclear Magnetic Resonance in Solids (Clarendon, Oxford 1970)

Distance Measurements in Solid-State NMR and EPR Spectroscopy

G. Jeschke and H.W. Spiess

Max Planck Institute for Polymer Research, Postfach 3148, 55021 Mainz, Germany
jeschke@mpip-mainz.mpg.de
spiess@mpip-mainz.mpg.de

Abstract.   Magnetic resonance techniques for the measurement of dipole-dipole couplings between spins are discussed with special emphasis on the underlying concepts and on their relation to site-specific distance determination in complex materials. Special care is taken to reveal the approximations involved in data interpretation and to examine the range of their validity. Recent advances in the understanding of measurements on multi-spin systems and in the extraction of spin-to-spin pair correlation functions from dipolar evolution functions are highlighted and demonstrated by selected experimental examples from the literature.

1 Introduction

The majority of our knowledge on the geometric structure of matter has been obtained by scattering techniques; mainly by X-ray, electron, and neutron diffraction. Diffraction methods are the natural choice for systems with translational symmetry and are very useful for all repetitive structures, in particular, structures with at least some degree of long-range order. Scattering techniques in general can reveal the size and shape of certain objects in a system, provided that the distribution of sizes and shapes is reasonably narrow and that sufficient contrast between the objects and their environment can be achieved. A limitation of scattering techniques results from the fact that destructive interference cancels any signals due to   non-repetitive   features of the structure. This limits the complexity of the structures that can be understood on the basis of scattering data.

The structural picture of a system obtained by magnetic resonance techniques, such as nuclear magnetic resonance (NMR) and electron paramagnetic resonance (EPR), is to a large extent complementary to the one obtained by scattering techniques. Nuclear and electron spins can be considered as   local probes of the structure. Signals due to diverse environments of the observed spins may overlap, but they do not cancel. Indeed, analysis of NMR spectra provides detailed information about the degree of disorder, e.g., in incommensurate systems or spin glasses [1, 2]. Moreover, interactions of spins with

G. Jeschke and H.W. Spiess: Distance Measurements in Solid-State NMR and EPR Spectroscopy , Lect. Notes Phys.   684 , 21–63 (2006)
www.springerlink.com

their environment are mostly so weak that external perturbations can compete with them. It is therefore possible to manipulate the Hamiltonian during the experiment almost at will [3, 4, 5]. By such manipulations it is possible to separate interactions and thus to reduce signal overlap, for instance by disentangling the spectrum into two or more dimensions. Complexity of the spectra can also be reduced by intentionally and selectively suppressing signal contributions of spins in more mobile or more rigid environments [4]. Finally, only the structural features of interest for a given problem can be addressed by site-selective isotopic labelling in NMR or site-selective spin-labelling in EPR. With this arsenal of approaches for obtaining just the information relevant in a certain context, magnetic resonance techniques are well suited for studying complex structures.

Limitations arise mainly from the comparatively low sensitivity of magnetic resonance experiments, which is caused by the low energy of spin transitions. These limitations often dictate the choice between NMR and EPR. Except for systems featuring native paramagnetic centers, NMR is the first choice, as it can be applied to the system as it is or after a mere isotope substitution that usually does not influence structure and dynamics significantly. However, if the structural features of interest correspond to only a small fraction of the material, EPR spin labelling techniques may be required to obtain su ciently strong signals. Examples are chain ends in polymers or single residues in proteins with molecular weights larger than 20 kDa. As the paramagnetic moiety of the common nitroxide spin probes has a size of approximately 0 .5 nm, such spin-labelling approaches are restricted to structural features that are larger than 1 nm, and it must be ascertained that the labelling does not change the structure or function of the system.

In the past, most method development and applications work in NMR and EPR has been devoted to elucidation of the   chemical structure of diamagnetic and paramagnetic molecules, respectively. Starting with proteins in solution [6, 7, 8], this has changed recently, and obtaining information on   geometric structure and on structural dynamics has now become the main goal of method development. Geometric structure is derived by a molecular modelling approach that takes into account general knowledge on bond lengths, bond angles, and dihedral angles in molecules [9] as well as constraints on spin-to-spin distances (for application of such an approach to a weakly ordered system, see [10]). The distance constraints in turn result from the magnetic resonance spectra. Constraints on the relative orientation of groups (angular constraints) may also be useful. As the experience with protein structures has shown, such an approach works quite well even if only part of the structure is well defined. In fact, for many problems in biomolecular and materials science knowledge about part of the structure already provides valuable insight into the relationships between structure, properties, and function.

Determination of the full geometric structure requires a large number of constraints. Moreover, this number increases strongly with increasing size and complexity of the system. Nevertheless, protein structure determination by high-resolution NMR in solution successfully implements such an approach [6, 7]. Usually precision of the constraints derived from high-resolution NMR spectra is rather limited, but for proteins with molecular weights up to approximately 20 kDa it is often possible to obtain significantly more constraints than would be strictly required to solve the structure. Overdetermination of the problem then compensates for lack of precision of the individual measurements. In this chapter we explore the basics of the alternative approach in which only part of the structure is solved by a smaller number of more precise constraints. Such precise information may be required for understanding self-organization phenomena in supramolecular systems, for fine-tuning the properties of materials, and for understanding how the function of biomacromolecules is optimized.

In the approach discussed here, which is applicable mainly to solid materials and soft matter, information on spin-to-spin distances is obtained by measurements of the dipole-dipole coupling, which scales with the inverse cube of the distance. The strength of this coupling depends only on the distance, fundamental constants, and the angle between the spin-to-spin vector and the quantization axis of the two spins. It is not influenced by the medium in between the spins, so that such measurements are potentially very precise. Furthermore, information on the spin-to-spin pair correlation function can be obtained even in cases where the distances of the spins under consideration are broadly distributed. To fully utilize these advantages, the most appropriate experiment for a given problem has to be selected. This requires an overview of the existing techniques and, in particular, of the concepts on which they are based. In these Notes we attempt for the first time to provide such an overview. Our concentration on concepts implies that only a limited selection of experiments can be treated. Further useful experiments are discussed in NMR and EPR monographs [4, 5, 11] and in a number of reviews [12, 13, 14, 15, 16, 17].

This chapter is organized as follows. In Sect. 2 we introduce the dipole-dipole Hamiltonian and discuss its truncation for several cases of interest as well as its spectrum. We also examine under which conditions isotropic spin-spin couplings, such as J couplings, are significant and how they influence the spectra. E ects of molecular motion and spin delocalization on the average dipole-dipole interaction are considered. In Sect. 3 we introduce the principal approaches for measuring the spectrum of the dipole-dipole Hamiltonian and discuss them with special emphasis on the range of their applicability. Complications in the measurement of the distance between two spins that are caused by couplings to further spins of the same kind are considered in Sect. 4. In this Section we also examine how dipolar time evolution functions are related to the spin-spin pair correlation function and under which conditions the pair correlation function can be determined. Where applicable, the approaches based on sample spinning are compared with experiments on static samples.

2 Dipole-Dipole Interaction in a Two-Spin System

Coupling between magnetic dipoles is a pairwise interaction. We may thus first discuss a two-spin system and consider later the complications that arise in multi-spin systems. The dipole-dipole coupling is the energy of the magnetic dipole moment μ_1 in the field induced by dipole moment μ_2 and vice versa:

$$E = -\mu_1 \cdot B_2(r_{12},\mu_2) = -\mu_2 \cdot B_1(r_{12},\mu_1) , \tag{1}$$

where r_{12} is the distance vector connecting the two moments. Hence, the coupling depends only on the distance and the relative orientation of the two dipole moments with respect to each other. With the expression for the field induced by the dipole,

$$B(r,\mu) = -\frac{\mu_0}{4}\frac{1}{r^3}\left[\mu - \frac{3}{r^2}(\mu \cdot r)r \right] , \tag{2}$$

we have

$$E = -\frac{\mu_0}{4}\frac{1}{r_{12}^3}\left[\mu_1 \cdot \mu_2 - \frac{3}{r_{12}^2}(\mu_1 \cdot r_{12})(\mu_2 \cdot r_{12}) \right] . \tag{3}$$

2.1 Hamiltonian of the Dipole-Dipole Interaction

For two magnetic moments associated with spins S and I, the Hamiltonian for the dipole-dipole coupling can be derived from (3) by the correspondence principle. Writing the Hamiltonian in units of angular frequencies we find

$$\hat{H}_{dd} = \frac{1}{r_{SI}^3}\frac{\mu_0}{4}\gamma_S \gamma_I \left[\hat{S} \cdot \hat{I} - 3\frac{1}{r_{SI}^2} \left(\hat{S} \cdot r_{SI} \right)\left(\hat{I} \cdot r_{SI} \right) \right] , \tag{4}$$

where the

$$\gamma_{S,I} = g_{S,I}\, \mu_{S,I}/\hbar \tag{5}$$

are the magnetogyric ratios of the spins with the appropriate magnetons $\mu_{S,I}$ (Bohr magneton μ_B for electron spins, nuclear magneton μ_n for nuclear spins) and the appropriate g values. For historical reasons, the electron g value is taken as positive in most literature. In these notes we conform to the convention in which the electron g value is negative [18] to ensure that (5) is valid for both nuclear and electron spins.

In all the experimental situations that we shall discuss the dipole-dipole coupling is small compared to the Zeeman interaction of at least one of the two spins. Indeed, in magnetic fields above 1 T the coupling is small compared to the Zeeman interactions of both spins, except for hyperfine couplings in EPR. With the magnetic field axis chosen as the z axis, it is therefore convenient to consider the dipole-dipole Hamiltonian in a basis spanned by the eigenstates $|s_S s_I\rangle, |s_S\bar{s}_I\rangle, |\bar{s}_S s_I\rangle$, and $|\bar{s}_S\bar{s}_I\rangle$ of the operator $\hat{F}_z = \hat{S}_z + \hat{I}_z$. The energy level schemes for the cases of comparable and strongly different Zeeman

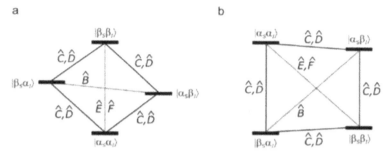

Fig. 1.   Energy level schemes of two-spins systems for the cases where the Zeeman interactions of both spins (a) or of spin S (b) are much larger than the dipole-dipole coupling. Labelling of the states corresponds to the NMR case (positive magneto-gyric ratio) in (a) and to the EPR case (negative magnetogyric ratio) in (b).   Solid lines correspond to allowed transitions,   dotted lines   to forbidden transitions. Operators designate the assignment of terms in the dipolar alphabet to the transitions

energies are shown in Fig. 1. Considerable simplifications are possible if the quantization axes of the two spins are parallel to each other. Experiments are usually performed at magnetic fields B_0 that are sufficient to align the quantization axes with the magnetic field axis. Exceptions are nuclear spins $S, I > 1/2$ with substantial quadrupole couplings, where the required fields may be technically inaccessible, and electron spins   S of transition metal or rare earth metal ions with large   g anisotropy, where alignment of the quantization axis with the external field generally cannot be achieved. For all other cases we may define a common frame for the two spins. As the   z axis we choose the quantization axis, the   x and y axes are for the moment left unspecified. Introducing polar coordinates and the shift operators

$$\hat{S}^+ = \hat{S}_x + i\,\hat{S}_y \, ,$$
$$\hat{S}^- = \hat{S}_x - i\hat{S}_y \, ,$$
$$\hat{I}^+ = \hat{I}_x + i\,\hat{I}_y \, ,$$
$$\hat{I}^- = \hat{I}_x - i\hat{I}_y \, , \tag{6}$$

we thus obtain a representation in terms of the   dipolar alphabet :

$$\hat{H}_{dd} = \frac{1}{r_{SI}^3}\frac{\mu_0}{4}\, S \, I \, \left[\hat{A} + \hat{B} + \hat{C} + \hat{D} + \hat{E} + \hat{F}\right] \tag{7}$$

with

$$\hat{A} = \hat{S}_z\hat{I}_z \left[1 - 3\cos^2\right] \, , \tag{8}$$

$$\hat{B} = -\frac{1}{4}\left[\hat{S}^+\hat{I}^- + \hat{S}^-\hat{I}^+\right]\left[1 - 3\cos^2\right] \, , \tag{9}$$

$$\hat{C} = -\frac{3}{2}\left[\hat{S}^+\hat{I}_z + \hat{S}_z\hat{I}^+\right]\sin\,\cos\,e^{-i} \, , \tag{10}$$

$$\hat{D} = -\frac{3}{2}\ \hat{S}^-\hat{I}_z + \hat{S}_z\hat{I}^-\ \sin\ \cos\ e^{i}\ , \tag{11}$$

$$\hat{E} = -\frac{3}{4}\hat{S}^+\hat{I}^+\ \sin^2\ e^{-2i}\ , \tag{12}$$

$$\hat{F} = -\frac{3}{4}\hat{S}^-\hat{I}^-\ \sin^2\ e^{-2i}\ . \tag{13}$$

Which terms are relevant depends on the relative magnitude of the dipolar frequency

$$_{dd}(r_{SI}) = \frac{1}{r_{SI}^3}\frac{\mu_0}{4}\quad S\ I\ , \tag{14}$$

given here in angular frequency units, with respect to the splittings between the levels in the absence of dipole-dipole coupling. These splittings can be expressed in terms of the Larmor frequencies $_S$ and $_I$ in the absence of coupling. The secular term $\hat{A}$ is significant in all experimental situations and corresponds to a first-order correction of the Zeeman energies. It causes a splitting of both the S and I spin transitions in doublets. The double-quantum terms $\hat{E}$ and $\hat{F}$ do not significantly influence the eigenvalues and eigenvectors of the Hamiltonian if the Larmor frequency of at least one spin is much larger than the dipolar frequency, which is the case for all experiments discussed in these Notes. Hence these terms can safely be neglected. For like spins,

$$_{dd}\quad _{SI} = |\ _S - \ _I|\ , \tag{15}$$

the term $\hat{B}$ causes significant mixing of the $|\ _S\ _I$ and $|\ _S\ _I$ states. This situation is usually encountered in homonuclear NMR experiments, where S and I are spins corresponding to the same isotope, and may be encountered in pulse EPR experiments when both the S and I spins are excited by pulses at the same microwave freq uency. The $\hat{B}$ term is significant in such EPR experiments unless both the width of the EPR spectrum and the excitation bandwidth of the microwave pulses are much larger than $_{dd}$. In pulse electron electron double resonance (ELDOR) experiments this term can generally be neglected. In the following we denote experiments where both the S and I spins are excited by the same irradiation frequency – and the $\hat{B}$ term is thus significant – as experiments on like spins, and experiments where the $\hat{B}$ term can be neglected as experiments on unlike spins.

The terms $\hat{C}$ and $\hat{D}$ are negligible if

$$_{dd}\quad _{S},\ _I\ . \tag{16}$$

This inequality is fulfilled except for dipolar hyperfine couplings in electron nuclear double resonance (ENDOR) and electron spin echo envelope modulation (ESEEM) experiments, where the dipole-dipole coupling may be comparable to the nuclear Zeeman frequency. In the latter situation only the terms $\hat{S}^+\hat{I}_z$ and $\hat{S}^-\hat{I}_z$ in $\hat{C}$ and $\hat{D}$ are non-secular and thus negligible. Neglecting these terms and now choosing the x axis so that $= 0$, we find the truncated dipole-dipole Hamiltonian for the electron-nuclear two-spin system

$$\hat{H}_{dd} = \omega_{dd} \left[\left(1 - 3\cos^2\theta\right) \hat{S}_z\hat{I}_z - 3\sin\theta\cos\theta \, \hat{S}_z\hat{I}_x \right] . \tag{17}$$

For like spins (most homonuclear NMR experiments and single-frequency pulse EPR experiments under certain conditions) the truncated dipole-dipole Hamiltonian is given by

$$\hat{H}_{dd} = \omega_{dd} \left(1 - 3\cos^2\theta\right) \left[\hat{S}_z\hat{I}_z - \frac{1}{4} \left(\hat{S}^+\hat{I}^- + \hat{S}^-\hat{I}^+ \right) \right] , \tag{18}$$

and, finally, for unlike spins (ELDOR and heteronuclear NMR experiments), it is given by

$$\hat{H}_{dd} = \omega_{dd} \left(1 - 3\cos^2\theta\right) \hat{S}_z\hat{I}_z . \tag{19}$$

In either of the three cases, the dipolar frequency ω_{dd} and thus the interspin distance r_{SI} is uniquely determined by the spectrum of $\hat{H}_{dd}$ if the distribution of angle θ between the spin-spin vector and the common quantization axis is known. In particular, for macroscopically disordered systems orientations with a given angle θ are realized with probability $\sin\theta$. Note also that the dipole-dipole coupling is purely anisotropic – if we express it in tensorial form as $\hat{H}_{dd} = \hat{S}\mathbf{D}\hat{I}$, we find that the tensor $\mathbf{D}$ is traceless. Usually the spectrum of $\hat{H}_{dd}$ is not directly accessible by analyzing the lineshape of NMR or EPR spectra, as ω_S and ω_I are broadly distributed due to anisotropy of the Zeeman interaction or due to interactions with other spins in the sample. Separation of interactions as described in Sect. 3 is then required to measure the spectrum of $\hat{H}_{dd}$. As we shall see, it is usually possible to eliminate the contributions of all other interactions by applying appropriate external perturbations and it is often even possible to factor out relaxational broadening. However, J coupling between the S and I spins is described by the same product operators and is thus inseparable from the dipole-dipole coupling for fundamental reasons. It is therefore necessary to discuss in which situations J coupling may interfer with the measurement of ω_{dd} and how reliable interspin distances can be obtained in such a situation.

2.2 J Coupling and Isotropic Hyperfine Coupling

Coupling between two spins may arise not only due to the dipole-dipole inter-action through space, but may also be mediated by the electron cloud. Overlap of the singly occupied molecular orbitals of two unpaired electrons leads to Heisenberg exchange between the two electron spins. Exchange coupling be-tween electron spins may also proceed through orbitals of neighbor molecules, as for instance solvent molecules. This superexchange relies on a correlation of the spin states of the unpaired electron and electrons of the solvent molecules in spatial regions where the corresponding orbitals overlap. Similarly, the nu-clear spin state is correlated to the spin states of electrons in s orbitals at this nucleus, and this correlation can be transported through a chain of overlap-ping orbitals to another nucleus. With the exception of the hyperfine coupling between the nucleus and an unpaired electron in its s orbitals (Fermi contact coupling), all these coupling mechanisms are usually denoted as   J couplings and are often referred to as through-bond couplings. Note that the latter term may be misleading, as through-solvent or through-space   J couplings can be substantial in certain situations [19].

In general, J couplings may have both an isotropic and an anisotropic contribution [20, 21]. However, the anisotropic contribution, which is also called a pseudodipolar contribution, is usually negligible in NMR experiments on elements in the first and second row of the periodic system and in EPR experiments on spin pairs with a distance exceeding 0   .5 nm. In most cases of interest, we are thus left with a situation where the   J coupling is purely isotropic and the dipole-dipole coupling is purely anisotropic. The two interactions can then in principle be separated by sample rotation. For macroscopically oriented systems such as crystals or liquid-crystals such a separation is feasible in both NMR and EPR spectroscopy by studying the orientation dependence of the spectra. For macroscopically disordered systems fast sample reorientation during an NMR experiment (see Sect. 2.5) can be used to average the dipole-dipole coupling, so that the   J coupling can be measured separately. In the next Section we shall see that analysis of the lineshape corresponding to the total spin-spin coupling (dipole-dipole and isotropic   J coupling) also provides unique values for $_{dd}$ and J . In many cases, J coupling can be neglected altogether, as it is much smaller than dipole-dipole coupling. This situation is usually encountered in solid-state NMR if only first-row and second-row elements are involved and in EPR when r_{SI} is longer than 1 .5 nm and the two spins are separated by an insulating matrix. Note however that   J coupling in solid-state NMR may also be useful for detecting through-bond correlations by the two-dimensional INADEQUATE experiment [22, 23, 24].

Where necessary, we shall use the Hamiltonian

$$\hat{H}_J = J \hat{S} \hat{I} = J \left(\hat{S}_x \hat{I}_x + \hat{S}_y \hat{I}_y + \hat{S}_z \hat{I}_z \right) = J \left(\hat{S}_z \hat{I}_z + \frac{1}{2} \left(\hat{S}^+ \hat{I}^- + \hat{S}^- \hat{I}^+ \right) \right) ,$$

(20)

for isotropic J coupling in both the NMR and EPR cases. For the EPR case, this definition of J is the one adopted in most modern textbooks [5, 25, 26]. Note that two other definitions $\hat{H}_J = -2J \hat{S} \hat{I}$ [27] and $\hat{H}_J = -J \hat{S} \hat{I}$ [28] are also widely used in EPR literature, so that care must be taken when comparing reported values of J .

2.3 The Pake Pattern

We are now in a position to discuss spectral patterns that result from coupling of two spins. For negligible J coupling, the Hamiltonians in (18) and (19) both give rise to the Pake pattern [29] shown in Fig. 2a. Note that for like spins the splitting   is 1.5 times as large as for unlike spins, since in the former

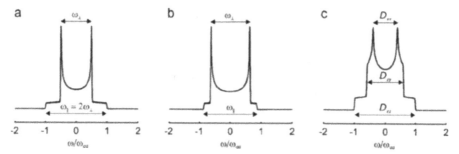

Fig. 2.   Spectral patterns arising from two-spin coupling (simulations). (a) Pake pattern. For like spins, $= 3 \;_{dd} / 2$, for unlike spins, $= \;_{dd}$. (b) Pattern in the presence of J coupling for unlike spins, $J = 0.275 \;_{dd}$. (c) Orthorhombic dipolar pattern for unlike spins as it may be observed when the point-dipole approximation is violated, here $= 0.2$

case the pseudo-secular zero-quantum term $\hat{B}$ contributes to the splitting. Furthermore, for like spins,   J coupling does not influence the spectral pattern. For unlike spins in the presence of significant   J coupling, the singularities and outer inflection points are found at

$$= \pm| \;_{dd} + J | ,$$
$$= \pm|- 2 \;_{dd} + J | , \tag{21}$$

respectively (Fig. 2b). If the whole pattern can be measured the dipolar frequency can thus always be extracted.

2.4 The Point-Dipole Approximation

So far we have assumed that both spins are strictly localized in space, i.e., that they can be considered as point dipoles. This   point-dipole approximation is certainly well justified for nuclear spins, as nuclear radii are of the order of only a few femtometers. For electron spins of some paramagnetic species, no- tably nitroxide radicals and many transition metal complexes, the point-dipole approximation is valid at distances of 2 nm and longer, as the distribution of distances implied by the conformational freedom of the molecules is much broader than implied by the spatial distribution of the electron spin. If the unpaired electron is delocalized on the length scale of the measured distance, this delocalization has to be taken into account explicitly. This case is usually encountered when a distance between an electron spin   S and a nuclear spin I_N is determined from the dipolar contribution to the hyperfine coupling. As- suming that we know all significant spin densities $_k$ at the other nuclei as well as the distances $R_k > 0.25$ nm of these nuclei from the nucleus with index N , we may compute a dipole-dipole coupling tensor   D by the electron-nuclear point-dipole formula

$$D = \frac{\mu_0}{4} \sum_{k=N}^{k} S_I \frac{3 n_k n_k^T - 1}{R_k^3} , \tag{22}$$

where the n_k are unit vectors denoting the direction cosines of R_k and super-script T denotes the transpose. If the spatial distribution of the electron spin $S(r)$ is known from a quantum-chemical computation, we have

$$D = \frac{\mu_0}{4} S_I \int S(r) \left[3 \frac{r_{SI} r_{SI}^T}{r_{SI}^5} - \frac{1}{r_{SI}^3} \right] dr , \tag{23}$$

where r_{SI} depends on integration variable r. For a system of two distributed electron spins we find

$$D = \frac{\mu_0}{4} S_I \int \int S(r_S) I(r_I) \left[3 \frac{r_{SI} r_{SI}^T}{r_{SI}^5} - \frac{1}{r_{SI}^3} \right] dr_I dr_S , \tag{24}$$

where r_{SI} depends on both integration variables r_S and r_I. Using these for-mulas it is possible to check whether a quantum-chemical computation is consistent with experimental findings or to obtain estimates for the devia-tion between a computed and an experimental geometry of the system. It has been demonstrated that such approaches can distinguish between possi-ble alternatives for the inner electronic structure of strongly coupled clusters of paramagnetic ions [30, 31].

If the point-dipole approximation is not valid, the dipole-dipole coupling tensor does not in general have axial symmetry. It is thus characterized by three principal values, which can be expressed as $D_{xx} = (1 + \eta) \overline{\omega}_{dd}$, $D_{yy} = (1 - \eta) \overline{\omega}_{dd}$, and $D_{zz} = -2 \overline{\omega}_{dd}$ with the average dipolar frequency $\overline{\omega}_{dd}$ and the asymmetry η. The approximate spin-to-spin distance obtained by substituting the average dipolar frequency for the dipolar frequency in (14) is usually shorter than the distance between the centers of gravity of the spatial distributions of the two spins, as the averaging is over $1/r_{SI}^3$. Note how-ever that this is only a rule of thumb – for strong delocalization contributions from different spatial regions may cancel each other due to the orientation dependence of the sign of the coupling. This may lead to a smaller average dipolar frequency than expected and thus to an overestimate of the distance.

For such an orthorhombic dipole-dipole coupling tensor the orientation dependence of the Hamiltonian can be written as

$$\hat{H}_{dd} (\theta, \phi) = \overline{\omega}_{dd} \left[P_{sec} \hat{S}_z \hat{I}_z + P_{ZQ} \left(\hat{S}^+ \hat{I}^- + \hat{S}^- \hat{I}^+ \right) + P_{SQ} \hat{S}_z \hat{I}_x \right] , \tag{25}$$

where the orientation dependence of the secular, zero-quantum, and single-quantum contributions is given by

$$P_{sec} (\theta, \phi) = 1 - 3 \cos^2 \theta + \eta \sin^2 \theta \cos 2\phi ,$$

$$P_{ZQ} (\theta, \phi) = -\frac{1}{4} P_{sec} ,$$

$$P_{SQ} (\theta, \phi) = -\sin \theta \cos \theta (3 + \eta \cos 2\phi) . \tag{26}$$

The corresponding spectral pattern is shown in Fig. 2c.

2.5 Averaging of the Dipole-Dipole Interaction by Internal or External Motion

The dependence of the dipole-dipole interaction on the relative orientation of the spin-to-spin vector r_{SI} with respect to the external field (angle θ) leads to partial or complete averaging when the spin pair reorients on the time scale of the experiment. In soft matter such reorientation occurs due to local dynamics. For fast reorientation of a spin-to-spin vector r_{SI} of constant length on a cone, the Pake pattern is simply scaled by a local dynamic order parameter. Such an effect can be recognized if the distance r_{SI} is known a priori or if a second measurement can be performed at temperatures that are sufficiently low to obtain the static dipolar spectrum. With $\omega_{dd,stat}$ and $\omega_{dd,dyn}$ being the dipolar frequencies observed under static and dynamic conditions, respectively, the local dynamic order parameter is defined by [32]

$$ S_{SI} = \frac{\omega_{dd,stat}}{\frac{1}{2}(3\cos^2\theta - 1)\,\omega_{dd,dyn}} \,, \tag{27} $$

where 2θ is the opening angle of the cone.

Note that for fast anisotropic motion of a spin-to-spin vector of constant length, the average dipole-dipole coupling is in general not axially symmetric [4, 33].

For the case of isotropic dynamics during which both length and orientation of the spin-to-spin vector change, the Pake pattern is preserved if and only if the dynamics is much faster than the time scale of the experiment and the changes in length and orientation of the vector are uncorrelated. In this situation an average dipolar frequency can be computed by averaging (14) over $r_{SI}(t)$ and the pattern is furthermore scaled by a local dynamic order parameter analogous to the one defined in (25). If the changes of length and orientation are correlated, the spectral pattern has to be computed by averaging over $P_2(\theta)/r_{SI}^3$, where $P_2(\theta) = (3\cos^2\theta - 1)/2$ is the second Legendre polynomial. If dynamics proceeds on the time scale of the experiment spectral patterns can be computed numerically by the general approach introduced in [34].

Intentional reorientation of the sample with respect to the external magnetic field can be used to separate isotropic from anisotropic interactions. Such an approach is most easily realized by sample rotation. As magnetic resonance linewidths in solids are usually dominated by the anisotropy of interactions, sample rotation leads to motional narrowing [35, 36]. For a spin-spin coupling with both an isotropic component J and a purely anisotropic dipolar component with axial symmetry, we find for the time dependence of the secular part of the coupling Hamiltonian

$$ \hat{H}_{SI} = \hat{H}_J + \hat{H}_{dd}(t) \tag{28} $$
$$ = \hat{S}_z\hat{I}_z \left[C_0 + C_1 \cos(\omega_{rot} t + \psi) + C_2 \cos(2\omega_{rot} t + 2\psi) \right], $$

where ω_{rot} is the angular frequency of sample rotation and β is an Euler angle relating the principal axis frame of the dipolar tensor D to the rotor-fixed frame. The coefficients are

$$C_0 = J - \frac{\omega_{dd}}{4}\left(3\cos^2\theta_{rot} - 1\right)(1 + 3\cos 2\beta),$$

$$C_1 = \frac{3\omega_{dd}}{2}\sin 2\theta_{rot}\sin 2\beta,$$

$$C_2 = -\frac{3\omega_{dd}}{2}\sin^2\theta_{rot}\left(1 - \cos^2\beta\right), \tag{29}$$

where β is another Euler angle and θ_{rot} is the angle between the rotation axis and the magnetic field axis.

The time dependence simplifies for two choices of θ_{rot}, the magic angle $\theta_{rot} = 54.74°$, and the right angle, $\theta_{rot} = 90°$. At the magic angle the time-independent part (coefficient C_0) is purely isotropic. The spectrum then consists of a series of sidebands that are spaced by the rotation frequency and whose amplitudes roughly trace the Pake pattern (Fig. 3). For significant J coupling, the centerband and each sideband become doublets with splitting J. At moderate rotation frequencies, $\omega_{rot} < \omega_{dd}/5$ the dipolar frequency ω_{dd} can be determined by fitting the sideband pattern. Generally, the effect of magic angle spinning (MAS) can be considered as a refocusing of the anisotropy of interactions. As a result, rotational echoes are observed at integer multiples of the rotor period. At high rotation frequencies, $\omega_{rot} \gg \omega_{dd}$, the dipole-dipole coupling of an isolated spin pair is averaged completely, at least if chemical shift anisotropy is negligible. For homonuclear spin pairs with coinciding chemical shift, the dipole-dipole coupling is not fully averaged unless the chemical shift anisotropy is also fully averaged [37, 38].

While MAS is widely applied in solid-state NMR spectroscopy [12, 39], it is not directly applicable to EPR spectroscopy as technically feasible sample rotation frequencies are much smaller than the anisotropy of the electron Zeeman and hyperfine interaction. This limitation can in principle be overcome

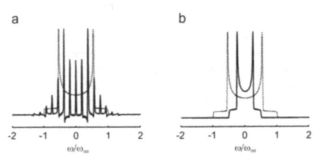

Fig. 3.   Dipolar patterns during sample rotation.   Dashed lines are the static Pake patterns. (a) Moderately fast rotation at the magic angle $\omega_{rot} = 54.74°$, $J = 0$, $\omega_{rot} = 0.1875\,\omega_{dd}$. (b) Fast rotation at the right angle $\theta_{rot} = 90°$, $J = 0$, $\omega_{rot} = 5\,\omega_{dd}$

by applying the magic angle turning experiment [40]. In this experiment, the Larmor frequency is averaged over only three discrete orientations, which is sufficient to fully suppress anisotropic broadening. Using fast sample rotation with a frequency of 20 kHz or higher the required 240° turn can be completed in a time shorter than the longitudinal relaxation time of the electron spins [41]. The experiment is, however, still limited by the requirement that the excitation bandwidth of the pulses is at least comparable to the total width of the EPR spectrum, which cannot yet be achieved for most samples [42].

At the right angle the term with coefficient C_1 vanishes. Due to the term with coefficient C_2 there is still anisotropic broadening, however, the width of the dipolar pattern is scaled by a factor of 1/2. Such right-angle spinning has been demonstrated to provide resolution enhancement in EPR spectroscopy for a broad range of samples [43, 44]. Routine use of this technique is hampered by the problem that sufficiently stable sample rotation is hard to achieve at temperatures below 200 K, where many EPR experiments have to be performed. An additional wiggling magnetic field perpendicular to the static magnetic field provides a similar relative motion of sample and magnetic field as sample rotation and can thus be used to overcome this technical problem [45].

3 Measurement Techniques for Isolated Spin Pairs in Solids

As we have pointed out in Sect. 2 the spin-spin distance r_{SI} in a spin pair can be determined or at least estimated from dipolar patterns or dipolar MAS sideband patterns of one of the spins. In practice, magnetic resonance spectra of solids are always spectra of multi-spin systems. Dipolar spectra of such multi-spin systems are more complicated to analyse and will be discussed in Sect. 4. However, in many cases an appropriate combination of experimental techniques and data analysis procedures can provide dipolar spectra that are dominated by the interaction of a single pair of spins. Such approaches are discussed in the following.

3.1 Lineshape Analysis

If the dipole-dipole interaction in the spin pair under consideration is of a similar magnitude as all the other interactions of spin S or even larger, distance information can be extracted from ordinary NMR or EPR spectra. This situation may occur when the distance r_{SI} is much shorter than all distances of spin S to other spins that are of the same kind as spin I. In NMR spectroscopy, this is the usual case for rare spins I or if spin I has been introduced by isotope labelling. In EPR spectroscopy, the situation is commonplace for distances up to approximately 2 nm between nitroxide radicals introduced by site-directed spin-labelling [46, 47]. Usually the other interactions in the spin

Hamiltonian cannot be neglected, thus lineshape analysis requires that the spectrum of spin S in the absence of dipole-dipole coupling is known. If S and I are like spins both spectra have to be known.

Spectrum Simulation

Dipolar splittings are resolved in the spectra if the distance is well defined and the dipole-dipole interaction is larger than the width of the most narrow peaks in the NMR or EPR spectrum. In this situation the spectrum may contain su cient information to extract not only the spin-spin distance r_{SI} but also the orientation of vector r_{SI} with respect to the molecular frame defined by the chemical shift anisotropy tensor or g tensor [48]. The reliability of this kind of lineshape analysis is much enhanced when a global fit of several spectra obtained at di erent magnetic fields is performed, as this corresponds to a variation of the ratio of dipolar to Zeeman anisotropy [49].

Deconvolution and Convolution Approaches

In some applications the orientation of vector r_{SI} with respect to the molecular frames of spins S and I is random or it can be assumed that e ects of orientation correlation on the lineshape are negligible. Thus, the absorption lineshape $A(\ )$ is the convolution of the absorption lineshape $A_0(\ )$ in the absence of dipole-dipole coupling with the dipolar pattern $S(\ )$. The dipolar pattern can then be extracted even for cases where the dipole-dipole interaction merely causes broadening of the lineshape rather than resolved splittings. Furthermore, a precise analysis is possible even if the spectrum in the absence of dipole-dipole coupling cannot be simulated, provided that this spectrum is experimentally accessible. According to the convolution theorem of Fourier transformation, a convolution of the spectrum with the dipolar pattern corresponds to the product of the Fourier transforms F of the spectrum and the dipolar pattern

$$F\{A(\ )\} = F\{A_0(\ )\}F\{S(\ )\} , \qquad (30)$$

so that the dipolar pattern can be obtained by [50]

$$S(\ ) = F^{-1}\ \frac{F\{A(\ )\}}{F\{A_0(\ )\}} , \qquad (31)$$

where F^{-1} denotes the inverse Fourier transformation.

Alternatively, the distance or a distribution of distances can be fitted by computing the corresponding dipolar pattern and simulating the spectrum $A(\ )$ by inverse Fourier transformation of (30) [46].

3.2 Double-Resonance Techniques

In many cases dipolar frequencies are significantly smaller than the width of the most narrow peaks in the inhomogeneously broadened static NMR or EPR spectrum, but still larger than or comparable to the homogeneous linewidth $_{\text{hom}} = 2/T_2$, where T_2 is the transverse relaxation time. Dipolar broadening of the ordinary NMR or EPR lineshape is then negligible, however, the dipolar frequency can still be measured if the influence of all the other broadening mechanisms is suppressed. Such suppression can be achieved by refocusing the other anisotropic interactions in echo experiments or by averaging them by MAS. In both cases the dipole-dipole interaction is usually also suppressed, exceptions being the coupling between like spins in the Hahn echo experiment [51] and the coupling between unlike spins in the solid-echo experiment [52]. In the following Sections we shall discuss ways to recouple exclusively the desired dipole-dipole interaction. For the case of unlike spins this can be achieved by double resonance experiments.

Static Solid-State NMR

In a two-pulse echo sequence $90_x - - 180_x - -$ echo applied to spin S the 180 - pulse with phase x refocuses all contributions to the spin Hamiltonian that are linear in the $\hat{S}_z$ operator. In particular, this includes the Zeeman interaction of spin S and the coupling to spin I, provided that the pulse does not excite spin I. Refocusing is based on inversion of the spin magnetic moment with respect to the external field and to the local dipolar field generated by spin I. The coupling to spin I can thus be selectively reintroduced by inverting only the local dipolar field, which can in turn be achieved by applying a 180 -pulse to spin I (Fig. 4). In this spin echo double resonance (SEDOR) experiment [53] the variation of the echo amplitude with time is described by

$$V_{\text{dip}} = \cos \quad_{\text{dd}} \quad 1 - 3\cos^2 \quad + J \quad . \tag{32}$$

For a macroscopically disordered system, the dipolar time evolution function $V_{\text{dip}}( )$ is the Fourier transform of the dipolar pattern. In the usual case where J coupling can be neglected, it is the Fourier transform of the Pake pattern (Fig. 4c). Due to tranverse relaxation, the dipolar time evolution function is damped, $V( ) = V_{\text{dip}}( )\exp(-2/T_2)$.

In another double-resonance approach, the heteronuclear dipole-dipole coupling drives the equilibration of spin temperatures between baths of S and I spins, i.e., it determines the dynamics of cross polarization. Precise measurements are possible if the spin temperature of the I spins is inverted halfway through the polarization transfer and homonuclear couplings among the I spins are suppressed by Lee-Goldburg decoupling. In such Lee-Goldburg decoupling the I spins are spin-locked along the magic angle. This polarization inversion spin-exchange at the magic angle (PISEMA) experiment is a

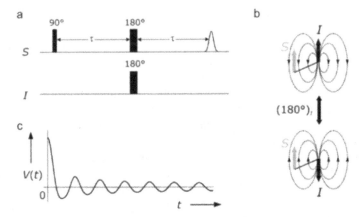

Fig. 4.   SEDOR experiment. (a) Pulse sequence. (b) E ect of the 180   -pulse applied
to the I spin on the local field at the   S spin. (c) Variation of the echo amplitude   S ( )
for a macroscopically disordered system with well defined distance   r$_{SI}$ (simulation)

two-dimensional experiment with dipolar evolution in the indirect dimension
and acquisition of a free induction decay corresponding to the ordinary decoupled NMR spectrum in the direct dimension [54]. Compared to its more
simple predecessor, separated local field spectroscopy [55], the introduction
of Lee-Goldburg decoupling in PISEMA strongly improves resolution in the
indirect dimension and thus the precision of distance measurements.

Solid-State MAS NMR

Sample rotation at the magic angle with   rot   dd averages dipole-dipole
coupling to zero. This is because the local dipolar field imposed by spin   I at
spin S changes its sign during rotation and the average over a full rotor period
vanishes. Recoupling is again possible by inverting the local field by 180   -
pulses applied to the I spin. In the basic version of the experiment [56, 57],
two 180 -pulses are applied during each rotor period, one after the first half
and one at the end of the rotor period. To obtain a pure dipolar evolution,
at least one 180 -pulse must be applied to the   S spins in the center of the
total evolution period, as MAS does not refocus resonance o sets due to the
isotropic chemical shifts. The dipolar evolution is traced by the amplitudes of
rotational echoes. Usually the basic unit of the experiment consists of several
rotor periods t$_{rot}$.

   In this rotational echo double resonance (REDOR) experiment, the dipolar
evolution is also damped by transverse relaxation. For longer distances, the
dipolar oscillation may be overdamped, so that it becomes necessary to factor
out relaxational decay. This can be done by measuring the rotational echo
amplitudes   S$_0$ (nt$_{rot}$) in the absence of the recoupling pulses for the   I spins
and the amplitudes   S (nt$_{rot}$) in their presence. The di erence   S = S$_0$ − S

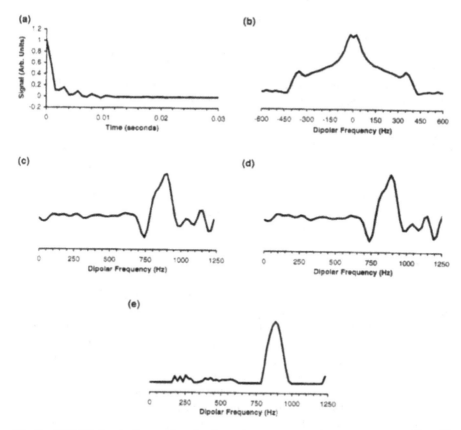

Fig. 5. REDOR data in di erent domains (experiment on 10% selectively 2   – 13 C – 15
N- labelled glycine diluted into unlabelled gylcine). (a) Time-domain data (dipolar
evolution function). (b) Frequency-domain data obtained by complex Fourier transform. (c) Distribution of dipolar frequencies obtained by the REDOR transform.
(d) Distribution of dipolar frequencies obtained by REDOR asymptotic rescaling.
(e) Distribution of dipolar frequencies obtained by Tikhonov regularization. Reproduced with permission from [87]

corresponds to the variation of the rotational echo amplitude caused by both
dipole-dipole interaction and transverse relaxation, while the normalized signal $V_{REDOR} = 1 - S/S_0$ corresponds to the REDOR dipolar evolution function (Fig. 5a) [58]:

$$V_{REDOR}\ (\omega_{dd}, t) = \frac{\sqrt{2}}{4} J_{1/4}\left(\sqrt{2}\,\omega_{dd}\,t\right) J_{-1/4}\left(\sqrt{2}\,\omega_{dd}\,t\right), \tag{33}$$

where $J_{1/4}$ and $J_{-1/4}$ are Bessel functions of the first kind. The di erence with
respect to the static dipolar evolution function, as observed in the SEDOR
experiment (32), results since rotational averaging is only partially o set by

the recoupling pulses. Accordingly, the Fourier transform of this function does not correspond to a Pake pattern (Fig. 5b).

Numerous variants of the basic REDOR experiment have been proposed that correct for imperfections encountered in certain situations [13, 14]. For instance, it is usually impossible to fully invert quadrupole spins $I > 1/2$ by a 180°-pulse. The rotational echo adiabatic passage double resonance (REAPDOR) experiment circumvents this problem by applying a prolonged pulse to the I spins that inverts these spins with higher efficiency than a 180°-pulse [59]. The prolonged pulse achieves inversion since during rotation the resonance frequency of the I spins adiabatically passes the frequency of the applied pulse.

The PISEMA experiment introduced in the previous section can also be applied under MAS conditions [60]. In contrast to REDOR there is no need to synchronize the pulses with sample rotation.

Pulse EPR

In pulse EPR spectroscopy, excitation bandwidths are often significantly smaller than the width of the whole spectrum of a single paramagnetic species. For example, spectra of nitroxide spin labels have a width of 180 MHz at X-band frequencies of 9.6 GHz, which is mainly due to nitrogen hyperfine anisotropy. As experiments with good sensitivity can be performed with pulse lengths of 32 ns, corresponding to excitation bandwidths of 30 MHz, two microwave frequencies can be placed in the spectrum that excite nitroxide radicals with different orientations or with different magnetic spin quantum numbers of the nitrogen nucleus. As pulses at one frequency excite exclusively S spins and pulses at the other frequency excite exclusively I spins, such ELDOR experiments can be considered as experiments on unlike spins, although both spins are electron spins of nitroxide radicals. The situation thus corresponds to a heteronuclear experiment in NMR.

Accordingly, the principle of SEDOR can also be applied in EPR spectroscopy. Because of a more unfavourable ratio between typical dipolar frequencies (0.1–10 MHz) and typical transverse relaxation times (1 μs), it is necessary to factor out relaxation. This can be done by using a constant interpulse delay , and varying the delay of the 180°-pulse for the I spins with respect to the 90°-pulse for the S spins, which results in the pulse sequence $(90°)_S - t - (180°)_I - (\tau - t) - (180°)_S - \tau - $ echo. The local field at the S spin is thus changed at time t during the defocusing period of length τ, which corresponds to an interchange of magnetization between the two transitions of the dipolar doublet. Hence, the magnetization vector precesses with frequency $\omega_S \pm \omega_{dd} (1 - 3\cos^2\theta)/2$ during time t and with frequency $\omega_S - \omega_{dd} (1 - 3\cos^2\theta)/2$ during time $\tau - t$ as well as in the refocusing period of duration τ. The echo amplitude as a function of t is then modulated with frequency $\omega_{dd} (1 - 3\cos^2\theta) + J$. It does not vary due to relaxation as the total duration of the experiment is constant. The choice of the fixed interpulse delay   in this pulse ELDOR (PELDOR) or double electron electron resonance (DEER) experiment [61, 62] thus involves a tradeo between sensitivity and resolution, as the signal amplitude decreases with increasing   while resolution of the dipolar frequency increases with increasing maximum observation time t_{max} .

In EPR typical instrumental deadtimes t_d of a few ten nanoseconds after a pulse compare unfavourably to the period of dipolar oscillations, at least for distances up to 2 .5 nm. With standard equipment $t > t_d$ has to be chosen to avoid signal distortion. This problem can be circumvented either by using a bimodal microwave cavity and separate high-power amplifiers for the two frequencies [62] or by inserting another 180   -pulse for refocusing the two-pulse echo. In this four-pulse DEER experiment [63, 64] $t = 0$ for the dipolar decay corresponds to the unobserved first echo (see Fig. 6a), so that the complete time evolution can be observed when choosing $_1$ t_d.

Selective excitation by the micr owave pulses implies that for a given   S spin the corresponding I spin is flipped by the 180   -pulse with probability < 1. Thus, for an isolated spin pair only a fraction $( )$ of the echo is modulated:

$$V_{DEER} (_{dd}, t) = 1 - \int_0^{/2} ( ) 1 - \cos _{dd} t 1 - 3\cos^2 \sin d . \quad (34)$$

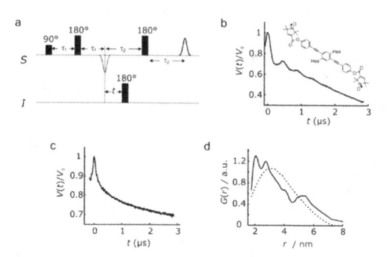

Fig. 6.   Four pulse DEER experiment. (a) Pulse sequence. (b) Experimental data set for a well defined distance r_{SI} 2.8 nm The inset shows the structure of the biradical. (c) Experimental data set for a broad distribution of distances in a [2]catenane [82]. (d) Pair correlation functions obtained from the data in (c) by direct transformation (solid line) and by fitting a simplified geometric model of the structure (dotted line)

If correlation between the molecular frames of the two spins and the orientation of vector r_{SI} is negligible, does not depend on angle and can be pulled in front of the integral. A fraction $1 -$ of the echo is then unmodulated.

In practice, spin pairs are never completely isolated and the signal is thus also a ected by coupling of spin S to remote spins. For a homogeneous spatial distribution of remote spins with concentration c, the e ect is an exponential damping of the signal by a factor

$$V_{hom}(c,t) = \exp\left[-\frac{2}{9}\frac{g_S g_I \mu_B^2 \mu_0 N_A}{3}\,ct\right], \tag{35}$$

where N_A is the Avogadro constant, μ_B is the Bohr magneton, and the concentration is given in units of mmol L^{-1}. A typical experimental data set for a rigid biradical with $r_{SI} = 2.8$ nm diluted into a matrix at a concentration of 2 mmol L^{-1} is shown in Fig. 6b.

3.3 Rotational Resonance

Per definition double resonance methods are not applicable to like spins. However, in solid-state MAS NMR homonuclear recoupling can be achieved by a technique that is similar in spirit to the PISEMA experiment. In the PISEMA experiment, double resonance irradiation establishes a degeneracy of levels of the two coupled spins in the doubly rotating frame, so that the dipole-dipole coupling can drive a magnetization exchange. In other words, an external perturbation o sets the di erence in the resonance frequencies, and as a result, dipole-dipole coupling causes strong mixing.

In homonuclear MAS NMR, di erences in resonance frequencies of the two spins are usually of the same order of magnitude as the frequency of sample rotation $_{rot}$. By matching an integer multiple of the rotation frequency to this di erence [65, 66],

$$n_{rot} = _S - _I, \tag{36}$$

rotational resonance occurs, at which a mixing of spatial and spin-dependent contributions to the Hamiltonian takes place. The phenomenon can best be discussed by Floquet theory, as it corresponds to a degeneracy of levels in Floquet space [67]. Such Floquet states correspond to spin states in the presence of a periodic external perturbation that couples to the spins. In the case at hand this perturbation is sample rotation. Thus, degeneracy of states in Floquet space means that, in the presence of this perturbation, the states can be mixed by a small o -diagonal element of the Hamiltonian. Here this o -diagonal element is the flip-flop term $\hat{B}$ in the dipolar alphabet, (9). State mixing in the vicinity of the level anti-crossing, (36), causes line broadening from which the dipolar frequency can be estimated. If the MAS NMR spectrum contains lines from multiple spins, this rotational resonance condition can usually be established selectively for each pair of spins. Site-selective distance measurements by this method thus do not require acquisition of a complete two-dimensional data set.

3.4 Pulse-Induced Recoupling During MAS
in Homonuclear Spin Systems

In Sect. 3.2 we have seen that radiofrequency pulses resonant with one of
the spins in the spin pair can partially o set MAS averaging of the dipole-
dipole coupling. For homonuclear spin systems pulses excite both spins simultaneously. To achieve similar recoupling, the 180   -pulse applied in REDOR
halfway through the rotor period has to be replaced by a pair of 90   -pulses
with phases x and $-$x that are separated by an interpulse delay   [69]. The
pulse pair is placed symmetrically with respect to the centre of the rotor period. To ensure averaging of the chemical shift anisotropy and of resonance
o sets, 180   -pulses with alternating phases   x and $-$x are applied at the end
of each second rotor period, leading to a basic cycle that extends over four
rotor periods. In this dipolar recovery at the magic angle (DRAMA) experiment, the recoupling e ciency and shape of the dipolar pattern depend on
the ratio   /t $_{rot}$, where t_{rot} $= 2$ / $_{rot}$ is the duration of the rotor period. The
recoupling is achieved as a 90 $_x$ -pulse applied to both spins interconverts $\hat{S}_z \hat{I}_z$
and $\hat{S}_y \hat{I}_y$ terms. Several alternative experiments for homonuclear recoupling
during MAS have been developed, some of which can be performed with narrowband excitation and combined with the rotational resonance experiment
[70]. A systematic treatment of the combined influence of radiofrequency fields
and sample rotation based on symmetry considerations can provide optimized
pulse schemes [71].

3.5 Build-Up of Double-Quantum Coherence

In the low-temperature limit, $k_B T$/   s, I, only the lowest level of the
four-level system (see Fig. 1a) of a spin pair is populated in thermal equilibrium. A 90 -pulse applied to this initial state excites coherence at all six
transitions. However, usually magnetic resonance experiments are performed
in the high-temperature limit, $k_B T$/   s, I, where population di erences between the levels are small and all four single-quantum transitions
are equally polarized in thermal equilibrium. In this situation a 90   -pulse excites coherence exclusively at the four single-quantum transitions. The zero-
quantum and double-quantum (DQ) transition are therefore sometimes called
forbidden transitions. Indeed, for irradiation in the linear regime their excitation is forbidden by selection rules at any temperature. However, even
in the high-temperature limit, coherence can be excited on forbidden transitions by a combination of pulses and free evolution [72, 73, 74]. The basic
building block of such experiments consists of a 90   -pulse that excites coherence on all single-quantum transitions, an interpulse delay   during which
the coherences of the two transitions of each spin acquire a phase di erence
    = $_{dd} t$ 1$-$ 3 cos^2 / 2, and another 90   -pulse that converts a fraction
sin   of the single-quantum coherence to DQ coherence. At a later stage of

the experiment, a third 90°-pulse transfers the DQ coherence back to single-quantum coherence in antiphase ($\hat{S}_y\hat{I}_z$ and $\hat{S}_z\hat{I}_y$). Single-quantum coherence in antiphase does not give rise to a signal, but is reconverted to observable single-quantum coherence by evolution under $\hat{H}_{dd}$. By appropriate phase cycling all signals can be eliminated that stem from other coherence transfer pathways [3], so that the build-up of DQ coherence can be monitored as a function of time τ.

Homonuclear MAS NMR

This basic scheme for excitation of DQ coherence cannot simply be applied under MAS conditions, where $\hat{H}_{dd}$ is averaged. However, the scheme can be combined with DRAMA recoupling or with other recoupling techniques that use rotor-synchronized pulses [75]. In a DRAMA-type pulse cycle $90_y - \tau_1 - 90_x - \tau_2 - 90_{-x} - \tau_1 - 90_{-y}$ with $2\tau_1 + \tau_2 = t_{rot}$, the double-quantum contribution to the dipolar Hamiltonian under MAS [76] is effective only during interpulse delay τ_2. This contribution is thus not averaged over the rotor cycle and DQ coherence is generated. The double-quantum contribution of $\hat{H}_{dd}$ can also be reintroduced by the back-to-back (BABA) pulse cycle $90_x - t_{rot} - 90_{-x} 90_y - t_{rot} - 90_{-y}$ [77].

Information on the dipolar frequency can be obtained either from the DQ build-up curves or from DQ sideband patterns. To obtain the patterns, a variable delay t_1 with increment $\Delta t_1 < t_{rot}$, corresponding to the indirect dimension of a two-dimensional experiment, is introduced in between the DQ excitation and reconversion subsequences. Each subsequence consists of one or several of the recoupling pulse cycles introduced above. Fourier transformation along the t_1 dimension yields DQ spinning-sideband patterns whose frequency dispersion results from rotor-encoding of the DQ Hamiltonian. In other words, sample reorientation between excitation and reconversion rather than the evolution of DQ coherence determines the width of the patterns and the intensity distribution among the sidebands. This has two consequences. First, sidebands are also observed for rotation frequencies that significantly exceed the dipolar frequency ω_{dd}. This results from recoupling. Second, the pattern depends on the number of rotor cycles (pulse cycles) used for excitation and reconversion of the DQ coherence. For a given rotation frequency, the experiment can thus be adapted to the expected magnitude of the dipolar frequency by applying an appropriate number of recoupling pulse cycles. This approach is restricted by relaxation which imposes an upper limit on the number of recoupling pulse cycles that can be used. For small couplings, $\omega_{dd} < \omega_{rot}/10$, one may thus be confined to a regime where only the two first-order sidebands can be observed and where only the total intensity rather than the shape of the pattern contains information on ω_{dd}. In this situation, the information can be extracted from build-up curves, i.e., from the dependence of the total intensity of the DQ sideband pattern on the number of recoupling pulse cycles. The fact that DQ NMR can be efficiently applied at high MAS frequencies makes this experiment particularly suitable for the measurement of proton-proton distances [12].

Pulse EPR

In EPR spectroscopy, the basic scheme for excitation and reconversion of DQ coherences can be applied, but must be supplemented with 180°-pulses for refocusing the dispersion of frequencies ω_s and ω_I caused by g and hyperfine anisotropy and by unresolved isotropic hyperfine couplings. The resulting excitation subsequence, $90 - \tau_1/2 - 180 - \tau_1/2 - 90$, and reconversion subsequence, $90 - \tau_2/2 - 180 - \tau_2/2 -$ echo, sandwich a period $t_1/2 - 180 - t_1/2$ during which the DQ coherence evolves. To factor out relaxation, this DQ EPR experiment is performed with fixed t_1 and a constant sum $\tau_1 + \tau_2$ of the other interpulse delays, varying the difference $\tau_1 - \tau_2$ [78, 79]. Assuming excitation of the whole spectrum, the signal for an isolated spin pair is then given by [5]

$$V_{DQ}(\tau_1, \tau_2) = \frac{1}{2}\cos\left[\omega_{dd}(\tau_1 - \tau_2)\left(1 - 3\cos^2\theta\right)/2\right]$$
$$+ \frac{1}{2}\cos\left[\omega_{dd}(\tau_1 + \tau_2)\left(1 - 3\cos^2\theta\right)/2\right], \qquad (37)$$

where we have assumed that the total width of the EPR spectrum is much larger than ω_{dd}, so that the $\hat{B}$ term in the dipolar alphabet can be neglected. Note that the second term on the right-hand side of (37) is constant as $\tau_1 + \tau_2$ is kept constant. Fourier transformation of the first term gives the Pake pattern.

3.6 Solid-Echo Techniques

The solid-echo sequence $90_x - \tau - 90_y - \tau -$ echo refocuses the dipole-dipole coupling of like spins but does not refocus the one of unlike spins [80]. In solid-state NMR, this sequence can thus be used to separate heteronuclear from homonuclear couplings [52]. An echo decay that is purely due to heteronuclear couplings and transverse relaxation is obtained if resonance offsets, chemical shift anisotropies and homonuclear multi-spin effects can be neglected. By supplementing the experiment with additional 180°-pulses halfway through the evolution periods, an echo experiment is obtained that refocuses resonance offsets, couplings between unlike spins, and couplings in isolated pairs of like spins. By introducing a difference $\tau_1 - \tau_2$ between the defocusing and refocusing time, $90_x - \tau_1/2 - 180_x - \tau_1/2 - 90_y - \tau_2/2 - 180_x - \tau_2/2 -$ echo, we obtain an experiment in which variation of the echo amplitude as a function of $\tau_1 - \tau_2$ for constant $\tau_1 + \tau_2$ is solely due to couplings between like spins [81]. In EPR spectroscopy, this single-frequency technique for refocusing dipole-dipole couplings (SIFTER) allows for observing dipolar evolution for longer times than alternative experiments. This is because the contribution of remote like spins to the echo decay is smaller for the solid echo than for the Hahn echo.

4 Complications in Multi-Spin Systems

Considering isolated spin pairs is a useful approximation for a measurement on spin S when the distance $r_{SI(k)}$ to spin I_k under consideration is much shorter than the distance to any other I spin. The signal contribution due to the latter, remote spins can then be neglected at short dipolar evolution times or can be accounted for in a summary way as indicated in Sect. 3.2 for the DEER experiment. For many problems of interest such a situation can be generated by choosing the appropriate experiment, e.g., an experiment with selective perturbation of I spins, or by isotope or spin labelling. However, in a sizeable number of cases the intrinsic structure of the material precludes this approximation or a selective labelling approach would be too tedious. Experiments and data analysis procedures for multi-spin systems are thus required.

While experiments on isolated spin pairs correspond to two-body problems, which can in principle be solved exactly, any experiment on a multi-spin system corresponds to a many-body problem for which there is no general solution. Depending on the topology of the system exact solutions may or may not exist. If no exact solution exists, it is often possible to find a regime where approximations are su ciently precise, or the ambiguities of the many-body problem can be overcome by analyzing the data in terms of a preconceived structural model. Finally, one may resort to calibration, i.e., to systematic comparison of the experimental data to data obtained on a similar system with known structure.

4.1 Dipolar Broadening and Moment Analysis

Consider the resonance line of S spins that is broadened by dipole-dipole coupling to a large number of like spins, in this case also denoted as S spins, or to a large number of unlike spins, denoted as I spins. As the number of coupled spins roughly scales with r^2, the $r_{S(j)S(k)}$ (or $r_{S(j)I(k)}$) are in e ect continuously distributed at long distances, so that no a priori limit can be put on the number of spins that have to be considered in computation of the lineshape. Thus, the problem cannot be solved by explicit computation of resonance frequencies and amplitudes of all transitions of the multi- spin system. However, the lineshape can be analyzed by the method of moments , in which characteristics of the absorption line A( ) are defined that can be computed analytically for any given spatial distribution of spins and can be obtained easily from the experimental lineshape [83, 84]. The first moment of the lineshape,

$$= \frac{\int_0 A(\ )d\ }{\int_0 A(\ )d\ },$$
(38)

is the average frequency of the resonance line, which is not influenced by the dipole-dipole coupling. Higher moments are defined by

$$\langle\omega^n\rangle = \frac{\int_0^\infty (\omega - \langle\omega\rangle)^n A(\omega)d\omega}{\int_0^\infty A(\omega)d\omega}.$$ (39)

Analysis is often restricted to the second moment, $\langle\omega^2\rangle$, which is of the order of the square of the linewidth. In general, experimental precision decreases with increasing order n of the moment, as the low-amplitude wings of the lineshape with lower signal-to-noise ratio contribute increasingly. The second moment can be computed without diagonalizing the Hamiltonian of the system [83]. For the case of like spins, where the $\hat{A}$ and $\hat{B}$ terms of the dipolar alphabet have to be included, it is given by

$$\langle\omega^2\rangle_{SS} = \frac{3}{4}\left(\frac{\mu_0}{4\pi}\right)^2 \gamma_S^4 \hbar^2 S(S+1) \frac{1}{N}\sum_{j,k}\frac{1-3\cos^2\theta_{jk}}{r_{jk}^6},$$ (40)

where N is the number of spins included in the summation. This sum can be computed for a given crystal lattice. For an isotropic system where all spins S are similarly situated, angular correlations can be neglected, and we have [83]

$$\langle\omega^2\rangle_{SS} = \frac{3}{5}\left(\frac{\mu_0}{4\pi}\right)^2 \gamma_S^4 \hbar^2 S(S+1) \frac{1}{N}\sum_k r_{jk}^{-6},$$ (41)

where the sum does no longer depend on j, as it is the same for any given spin S.

For the case of unlike spins, the $\hat{B}$ term has to be neglected, which leads to scaling of the second moment by a factor of $(2/3)^2$:

$$\langle\omega^2\rangle_{SI} = \frac{1}{3}\left(\frac{\mu_0}{4\pi}\right)^2 \gamma_S^2 \gamma_I^2 \hbar^2 I(I+1) \frac{1}{N}\sum_{j,k}\frac{1-3\cos^2\theta_{jk}}{r_{jk}^6}.$$ (42)

Equation (41) changes accordingly. Note that the width of the Pake pattern for an isolated spin pair scales by a factor $2/3$ on neglecting the $\hat{B}$ term, but that formulas for the nth moment cannot generally be converted by scaling by $(2/3)^n$.

4.2 Relation of the Dipolar Pattern to the Spin-Spin Pair Correlation Function

For an isotropic system with similarly situated observer spins S_j, the second moment is fully defined by the distribution of distances r_{jk} to coupled spins S_k or I_k as can be seen in (41). It can thus be expressed in terms of the spin-spin pair correlation function $G(r)$, which gives the probability to find a coupled spin at distance r from the observer spin. In fact, if angular correlations between spin pairs can be neglected, this applies to any moment, and as the lineshape (or dipolar pattern of the multi-spin system) is fully determined by a series expansion into its moments, the dipolar pattern itself is fully determined by $G(r)$. Because of the unique mapping between distances r_{SI} and dipolar frequencies   dd the dipolar pattern or the dipolar evolution function in time domain can be converted to the pair correlation function   G (r). Direct integral transformations of experimental data to the pair correlation function or the distribution of dipolar frequencies have been proposed for the REDOR [58], see Fig. 5c,d, and DEER [85], see Fig. 6d, experiments. Tikhonov regularization with an adaptive choice of the regularization parameter [86] is generally applicable for this task if an analytical expression for the dipolar evolution function is available. As demonstrated in Fig. 5, Tikhonov regularization is advantageous for noisy data in cases where the pair correlation function consists of narrow peaks [87]. Distance distributions with broad peaks are harder to analyze, as transformation of the dipolar evolution function to a distance distribution corresponds to an ill-posed problem. A systematic comparison of di erent procedures for data analysis in this situation can be found in [88].

Whether or not   G (r) can be obtained from magnetic resonance data by one of the procedures mentioned above depends on the availability of an analytical expression for the dipolar time evolution function. For isolated spin pairs such expressions are available for most experiments. In the following we shall see that under certain conditions, the expression for a multi-spin system can be obtained from the expression for an isolated spin pair.

4.3 E ective Topology of Spin Systems

Consider a system consisting of an observer spin   S with dipole-dipole couplings to several spins I_k . In general, the I spins will also be coupled among themselves, so that a true multi-body problem arises (Fig. 7a). In that situation the dipolar evolution function for the multi-body system cannot be expressed in terms of the functions of the pairs SI_k . On the other hand, if couplings among the I spins can be neglected due to the design of the experiment or if they are much smaller than the couplings in pairs SI_k (Fig. 7b), the

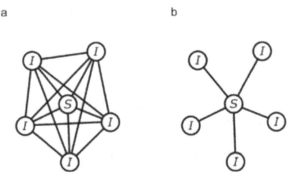

a   b

Fig. 7.   E ective topology of multi-spin systems. (a) All couplings are signifcant during the experiment. Factorization into pair contributions is impossible. (b) Only couplings between spin   S and spins I_k are signifcant. The signal is a product of pair contributions

Hamiltonian is block diagonal with the blocks corresponding to the spins I_k. The Hamiltonian and density operator of the spin system can then be factorized into pair contributions, which can be treated separately (see, for instance, [89]). As a result, the signal of the multi-spin system can be expressed as a product of the signals of the pairs

$$V_{multi} = \prod_k V_{SI\,(k)} . \tag{43}$$

This situation is typical for electron-nuclear spin systems, in which usually one electron spin S is hyperfine-coupled to several nuclear spins I_k with the hyperfine coupling being by several orders of magnitude larger than the couplings among the nuclear spins.

For all the other cases, couplings among I spins are generally comparable to couplings between S and I spins. Nevertheless, an effective topology allowing for factorization can be achieved when the I spins are excited by only one pulse with a duration that is short compared to the inverse of the dipolar linewidth of these spins. The couplings among I spins can then be neglected during the pulse. They do influence the evolution of I spin coherence after the pulse, but as there is no subsequent mixing between S and I spins, this influence does not extend to the signal observed on S spin transitions. Thus, (43) applies to SEDOR and DEER experiments. For experiments with prolonged irradiation of I spins or experiments in which several pulses are applied to these spins, it has to be checked explicitly whether or not (43) is a good approximation for the signal of the multi-spin system. Note that simplification of the effective topology can also be achieved by homonuclear decoupling of the I spins during the dipolar evolution period of the experiment. In the PISEMA experiment [54] this is done by Lee-Goldburg decoupling.

Effective topology of the spin system is also the key to understanding why the build-up of I spin multiple-quantum coherences in heteronuclear NMR experiments can be used to count I spins in the vicinity of the S spin [90]. Such spin counting relies on the fact that the maximum coherence order which can be excited in a given evolution time is limited by the number of spins in a cluster. The cluster consists of spins coupled among themselves by dipole-dipole interactions that are comparable to the inverse of the evolution time [74]. In the heteronuclear case, the build-up of I spin multiple-quantum coherence is due to the coupling of all the I spins to a single S spin. As the number of I spins increases only as r_{SI}^2 while the coupling decreases with r_{SI}^{-3}, the build-up curve converges with time [90, 91].

4.4 Multi-Spin Effects in MAS NMR

As long as the rotation frequency ω_{rot} does not strongly exceed the total anisotropy of the static NMR spectrum, the extent of line narrowing under MAS depends on the commutation properties of the Hamiltonian at different times t_1 and t_2 during the rotor cycle [37]. Consider the Hamiltonian $\hat{H}_0$ that includes all anisotropic interactions of the S spins. If $[\hat{H}_0(t_1), \hat{H}_0(t_2)] = 0$ for all times t_1 and t_2, $\hat{H}_0$ is inhomogeneous in the sense of Maricq and Waugh [37]. For this case MAS leads to complete refocusing of the transverse S spin magnetization to a rotational echo after a full rotor cycle. Already at low MAS frequencies, the anisotropic spectrum of the S spins is then resolved into a sideband pattern, with the widths of the centerband and all individual sidebands being determined by the transverse relaxation time T_2. If, on the other hand, $[\hat{H}_0(t_1), \hat{H}_0(t_2)] = 0$ at least for some combinations of t_1 and t_2, such complete refocusing does not occur and $\hat{H}_0$ is called homogeneous in the sense of Maricq and Waugh. For a homogeneous $\hat{H}_0$, the width of the sidebands depends on sideband order and on $_{rot}$. Complete narrowing to the limit of the relaxational linewidth is only achieved for $_{rot}$ that are much larger than the total anisotropy of the static spectrum. In many cases, such high rotation frequencies may be technically inaccessible.

Any combination of chemical shift anisotropy terms and $\hat{A}$ terms of the dipolar alphabet is inhomogeneous, as $[\hat{S}_z, \hat{S}_z \hat{I}_{kz}] = 0$. Thus, high resolution can be achieved at moderate $_{rot}$ as long as only heteronuclear dipole-dipole couplings are involved. This is true irrespective of whether an isolated spin pair or a multi-spin system is considered, since $\hat{A}$ terms of different spins also commute. For the homonuclear case, where the $\hat{B}$ terms of the dipolar alphabet are significant, $\hat{H}_0$ is still inhomogeneous for an isolated spin pair in the absence of significant chemical shift anisotropy, as $[\hat{S}_z \hat{I}_z, \hat{S}_x \hat{I}_x] = [\hat{S}_z \hat{I}_z, \hat{S}_y \hat{I}_y] = 0$. In the presence of chemical shift anisotropy, $\hat{H}_0$ becomes homogeneous, as $\hat{S}_z$ does not commute with the $\hat{B}$ term. For an isolated homonuclear spin pair with coinciding isotropic shifts [37] or for small clusters of nuclei of the same isotope [38] the centreband and each sideband are then broadened into characteristic dipolar patterns. More significantly for the problem at hand, $\hat{H}_0$ is also homogeneous for multi-spin systems of homonuclei, as $\hat{B}$ terms of different spins do not commute and furthermore the $\hat{B}$ term of one spin does not commute with the $\hat{A}$ term of another spin.

Detailed insight into sideband broadening for homogeneous $\hat{H}_0$ can be obtained by applying Floquet theory to MAS NMR on multi-spin systems [92]. Although quantitative analysis requires numerical computations for systems consisting of a small finite number of nuclei and moment analysis for a large or infinite number of nuclei, analytical expressions reveal how multi-spin correlations influence the patterns [12, 93]. It is found that the influence of coupling terms that involve correlations of k spins scales with $_{rot}^{-k}$. Thus, the multi-spin character of the signal decreases with increasing rotation frequency.

4.5 Multi-Spin Effects in Double-Quantum Sideband Patterns

The DQ sideband pattern of an isolated spin pair with distance r_{SI} consists of only odd-order sidebands, spaced by $(2n+1)$ $_{rot}$ from the center of the pattern, where n is an integer number. The effect of a third spin on such a pattern can be described in terms of a dimensionless perturbation parameter [12]

$$\kappa = \frac{2\,\delta_{dd}^{pert}}{\omega_{rot}}.\qquad(44)$$

where the perturbing dipole-dipole coupling δ_{dd}^{pert} is defined as the coupling to the spin of the original pair that is closer to it. Generally, in such a multi-spin system even-order sidebands spaced by $2\,n\,\omega_{rot}$ from the center of the pattern appear and their intensity increases with increasing κ. While the intensity of the additional sidebands can provide an estimate of the overall influence of perturbing spins, the detailed intensity pattern and the occurence of broadening of the sidebands depends on the spatial arrangement of the three spins. In a linear configuration, no additional broadening is observed irrespective of the magnitude of κ, i.e., the linear three-spin system behaves inhomogeneously. Furthermore, the centreband is the most intense of the new bands for this geometry. In a arrangement, where the perturbing spin has the same distance from both spins of the original pair, the second-order sideband is the most intense of the new bands, and significant broadening of the original odd-order sidebands is observed. The three-spin system in a arrangement thus behaves homogeneously. In contrast to DQ build-up curves, DQ sideband patterns thus exhibit a dependence on the geometrical arrangment of the spins in the multi-spin system that can provide additional information in favorable cases where the total number of spins is small or further constraints are available.

5 Application Examples

How many distances have to be measured and which precision has to be achieved strongly depends on the system of interest and on the information that is required to understand its properties or function. In highly disordered systems it may be su cient to prove spatial proximity of certain substructures. On the other hand, full determination of the well-defined structure of a biomacromolecule may require a sizeable number of constraints on distances and angles. For soft matter, a static structure often has to be supplemented by some information on dynamics to understand the function of the system. In the following, we illustrate these issues on a number of application examples and model studies.

5.1 Detecting ^{31}P-^{31}P-Spatial Proximity in Phosphate Glasses

Glasses are characterized by a rather low degree of order beyond the trivial constraints on bond lengths, bond angles, and dihedral angles. Nevertheless, it has been found that some heterogeneity, and hence some order, is present over a wider range of length scales. For any particular glass it is of interest what types of order do persist, as such insight may show ways of tailoring the material to certain applications.

Inorganic silicate and phosphate glasses can be described in terms of elementary building blocks in which the central silicon or phosphorous atom is coordinated by four oxygen atoms in a roughly tetrahedral geometry. Such a building block with n 4 oxygen atoms that bridge to a neighbour building block is called a $Q^{(n)}$ group. The distribution of $Q^{(n)}$ groups, and thus the structure of the glass, can be influenced by varying the ratio between the network formers SiO_2 or P_4O_6 and network modifiers, which may be akali or earth alkali oxides. This distribution can be determined by lineshape analysis of ^{29}Si or ^{31}P MAS NMR spectra, as the different $Q^{(n)}$ groups are characterized by different chemical shift ranges. However, the connectivities among the $Q^{(n)}$ groups, which are a characteristic for the homogeneity or heterogeneity of the structure, are not accessible from such spectra.

These connectivities can be detected by homonuclear ^{31}P DQ NMR spectroscopy [94]. Two-dimensional DQ spectra of two phosphate glasses containing 58 and 35 mol% of the network modifier Na_2O are displayed in Fig. 8. The glass with the higher content of Na_2O is expected to form chain-like networks consisting mainly of Q^1 and Q^2 groups. As seen by the crosspeaks $1-2$ and $2-1$ in Fig. 8a, significant fractions of the Q^1 and Q^2 groups are close enough for build-up of DQ coherence within the excitation time of 640   µs used in the experiments. This excludes structures consisting mainly of rings of Q^2 groups and isolated Q^1 groups and is consistent with the presence of relatively short chains of Q^2 groups that are end-capped by Q^1 groups. The larger mean

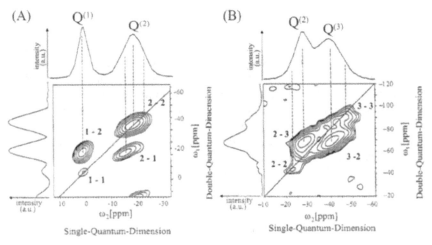

Fig. 8. ^{31}P-^{31}P-double quantum NMR on two phosphate glasses. Connectivities between structural units $Q^{(n)}$ and $Q^{(n')}$ are indicated by $n-n'$ (a) Glass containing 58 mol% Na_2O. (b) Glass containing 35 mol% Na_2O. Reproduced with permission from [94]

isotropic chemical shift of -16 ppm for the $2-1$ crosspeak compared to the shift of -18 ppm for the $2-2$ autopeak strongly indicates that the type of the adjacent building blocks has a stronger influence on chemical shift than varaiations of bond lengths and bond angles caused by strain in the glass.

Such strain is stronger in the glass with lower content of the network modifier (Fig. 8b), which is expected to form a three-dimensional network consisting mainly of Q^2 and Q^3 groups. Connectivity between Q^2 and Q^3 groups is again manifest by crosspeaks $2-3$ and $3-2$, and again the dominating contribution to chemical shift variation appears to come from the type of neighboring $Q^{(n)}$ group.

5.2 Measurement of ^1H-^1H Distances in Bilirubin

One of the principal limitations of x-ray crystallography lies in the difficulty to detect the position of hydrogen atoms, which is caused by the small electron density on these atoms. Hydrogen atoms involved only in conventional chemical bonds do not present too much of a problem, as their position can usually be predicted from the known positions of the heavier atoms by reyling on the known potentials for bond lengths, bond angles, and dihedral angles. However, the strength of hydrogen bonds and hence the position of hydrogen atoms involved in a hydrogen bond cannot be predicted easily and precisely with the currently available computational approaches. Furthermore, structures that are dominated or strongly influenced by hydrogen bonding often have to be studied in the solid state, as the hydrogen bonds are broken on dissolving the material. Recently, it has been demonstrated in a number of cases that hydrogen-bonded structures can be conveniently studied by DQ MAS NMR.

As an example, consider the yellow-orange pigment bilirubin (for the structure, see Fig. 9e), which is a product in the metabolism of hemoglobin. Strong hydrogen bonding renders this compound insoluble under physiological conditions and thus unexcretable, unless it is enzymatically conjugated with glucoronic acid. This process is usually performed in the liver and its failure causes the yellow discolouration of the skin associated with hepatitis. The hydrogen bonds involve three protons in each of the two pseudo-symmetry-related moieties of the molecules. The triangle made up by these protons is fully characterized by two distances and one angle, which can be determined from DQ MAS measurements [95]. As discussed in Sect. 4.5, intensities in the sideband pattern of such a three-spin system are dominated by the largest dipole-dipole coupling corresponding to the shortest proton-proton distance and modified by the perturbation due to the third proton. Hence, in the case at hand the shortest distance of 0.186 nm between the lactam and pyrrole NH protons can be determined most precisely (error of 0.002 nm), while the distance of 0.230 nm between the lactam NH and carboxylic acid OH protons is less certain (error of 0.008 nm). Likewise, the H-H-H angle of 122 is uncertain by 4. Note however, that even the precision of these less well defined values

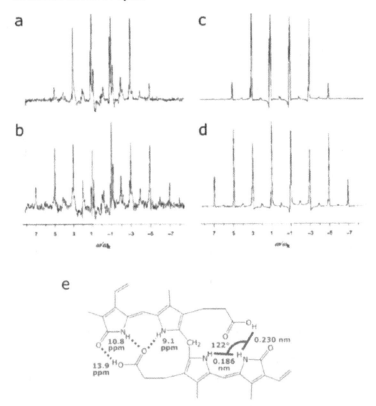

Fig. 9. ¹H (700.1 MHz) DQ MAS NMR on bilirubin. (a) Spinning- sideband pattern for a BABA excitation period 2 t_{rot}. (b) Best-fit simulation of the pattern in (a) assuming an isolated three- spin system. (c) Spinning-sideband pattern for a BABA excitation period 3 t_{rot}. (d) Best-fit simulation of the pattern in (c) assuming an isolated three-spin system. (e) Structure of bilirubin. Hydrogen bonds and proton chemical shifts (in ppm) are indicated in the left-hand moiety, distances and angles determined from the best fits are indicated in the right-hand moiety. Adapted with permission from [95]

is still much better than the precision of proton positions derived from x-ray data.

5.3 Site-Selective Measurement of Distances Between Paramagnetic Centers

Supramolecular assemblies that extend over several nanometers can be designed on the basis of metal ions, multidentate ligands, which provide a well defined coordination geometry at the metal, and rigid spacers between these ligands [96]. The known structures of such assemblies have been derived by x-ray crystallography. However, crystallizing these materials may become more and more di cult if one increases the complexity of the structure. An alternative way of characterizing the assemblies in frozen solution would be the measurement of distances between the metal centers, between a metal center and selected sites on the linkers, or between two sites on the linkers. As typical distances between such sites exceed 2 nm, pulse EPR is the techniques of choice. This applies particularly to assemblies that contain paramagnetic transition metals such as copper(II).

The potential of this kind of structure determination has been demonstrated on a model complex of copper(II) with two ligands, each of which consists of a terpyridine coordinating unit, a rigid spacer, and a nitroxide spin label as an endgroup (Fig. 10a) [97]. As the EPR spectra of the copper centre and the nitroxide labels overlap only slightly (Fig. 10b), it is possible to measure the end-to-centre and end-to-end distance separately by two DEER experiments with di erent choices of the observer and pump frequencies. Pumping at position A in the nitroxide spectrum and observing at position B in the copper spectrum provides a dipolar evolution function that is solely due to the copper-nitroxide pair (Fig. 10c). The distance of nm obtained by fitting this function is in nice agreement with the distance of 2 .43 nm predicted by molecular modelling. Pumping again at position A but observing now at position C also in the nitroxide spectrum provides a dipolar evolution function that is solely due to the nitroxide- nitroxide pair (Fig. 10d). In this case the fit value of 5 .2 nm overestimates the expected distance by 0 .34 nm. A more precise measurement of this long distance would require observation of the dipolar evolution function for a longer time, which is precluded here by enhanced relaxation of the nitroxide due to the nearby copper centre. However, the precision that could be achieved should be su cient to elucidate the principal structure of a supramolecular assembly.

5.4 Averaging of Dipole-Dipole Interactions by Dynamics in the Discotic Phase of a Hexabenzocoronene

Some materials are applied not in the solid state but as a liquid or in a liquid-crystalline state. In cases like this, a static structure is insu cient to explain the properties. However, once a static structure at lower temperature is known, it is often possible to derive information on structural dynamics at higher temperatures by an analysis of motional averaging of the dipole-dipole interaction (see Sect. 2.5).

The interest in polycyclic aromatic materials, which form liquid-crystalline columnar mesophases derives from their ability to form vectorial charge transport layers. Such transport layers may be suitable for applications in xerography, electrophotography, or in molecular electronic devices. Hexaalkyl-substituted hexa- peri -benzocoronenes (for a structure, see Fig. 11c) are a class of such materials with exceptionally high one-dimensional charge carrier mobility. The solid-state structure of the columns (stacks) of HBC-C$_{12}$ molecules could be elucidated by analyzing ring-current e ects on the chemical shift of

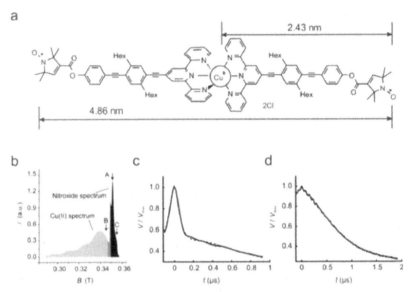

Fig. 10. Spectral selection in DEER measurements on a model compound for coordination polymers. (a) Structure of the model complex with end-to-centre and end-to-
end distances predicted by molecular modelling. (b) EPR spectrum of the complex
and excitation positions for DEER. (c) DEER data obtained for observer (S spin)
frequency B and pump (I spin) frequency A, corresponding to the copper-nitroxide
distance. The fit (dashed line) corresponds to a distance of 2   .43 nm. (d) DEER data
obtained for observer (S spin) frequency C and pump (I spin) frequency A, corresponding to the nitroxide-nitroxide distance. The fit (dashed) line corresponds to a
distance of 5 .20 nm

a given polycycle due to the adjacent polycycles in the stack. Furthermore,
spatial proximities of certain types of aromatic protons were determined from
cross-peaks in two-dimensional DQ MAS NMR spectra [98] in the same way
as discussed above for the phosphate glasses. The structure of the columns
is characterized by a herring-bone packing as it was also found in an x-ray
structure of unsubstituted hexabenzocorenene. This regular packing is lost in
the discotic phase as the polycycles begin to rotate independently about an
axis parallel to the long axis of the stack and perpendicular to the aromatic
plane. This leads to reorientation of the spin-to-spin vector in the proton
pairs marked by ellipses Fig. 11c. As a result, the dipole-dipole interaction
is partially averaged. By analysing the DQ MAS NMR sideband pattern, a
dipole-dipole coupling of 15 .0 kHz is found for these pairs at a temperature of
333 K in the solid state. In the liquid crystalline state (T = 386 K), the same
analysis yields a coupling of only 6 kHz, corresponding to a reduction by a factor of 0.4. Fast axial rotation about an axis perpendicular to the polycycle and
passing through its centre of symmetry would result in a reduction by a factor
of 0.5. This indicates that there is either a significant out-of-plane motion of

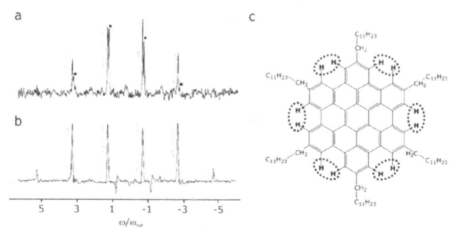

Fig. 11.   Motional averaging of the proton-proton dipole-dipole coupling in the hexabenzocoronene HBC-C $_{12}$. (a) DQ MAS NMR sideband pattern of a proton pair in the solid state (333 K) measured at an MAS frequency of 35 kHz. (b) Sideband pattern measured in the discotic phase (386 K) at an MAS frequency of 10 kHz. (c) Structure of HBC-C $_{12}$. Isolated proton pairs are marked by   dashed ellipses . Reproduced with permission from [98]

the C-H bonds for these protons or, maybe more likely, an out-of-plane motion of the polycycles themselves.

5.5 Structure Determination of a Peptide by Solid-State NMR

Full structure determination of proteins or peptides by solid-state NMR relies on sequence-specific assignment of backbone carbon and nitrogen resonances in ^{13}C and ^{15}N NMR spectra. Furthermore, distances have to be measured for a su cient number of pairs of backbone nuclei and sidechain nuclei. The number of required distance constraints can be drastically reduced if also torsion angles can be determined. By correlating spectra of two dipole-dipole coupled spin pairs in a three-dimensional NMR experiment, the orientation dependence of the dipole-dipole coupling can be utilized to obtain such constraints [99]. This method is based on the fact that the bond length and hence the magnitude of the dipole-dipole coupling for pairs of directly bonded nuclei are known. In a three-dimensional experiment with two chemical shift dimensions and one dimension corresponding to dipolar evolution, the frequency in the dipolar dimension thus depends on the relative orientation of the two dipolar tensors.

For sensitivity reasons, application of such approaches to proteins requires ^{13}C and ^{15}N isotope labelling. In many model studies, selective labelling has been used for the distance measurements, so that the approximation of isolated spin pairs could be applied. However, broad application of such methodology may depend on techniques that can solve the structure of uniformly labelled proteins, as this significantly reduces the e ort required in the labelling process.

For the chemotactic peptide   N -formyl-L-Met–L- Leu–L-Phe-OH a full structure determination based on uniform isotope labelling was demonstrated by applying a frequency-selective version of the REDOR experiment [99]. In this frequency-selective REDOR technique, broadband recoupling is combined with chemical shift refocusing by weak Gaussian-shaped pulses that are resonant only with one ^{13}C and one ^{15}N nucleus [100]. As in site-selective DEER it is thus possible to obtain dipolar evolution functions that are solely due to one selected spin pair (Fig. 12b-d). With this technique, 16 long-range ^{13}C-^{15}N-distances between 0.3 and 0 .6 nm were measured. Furthermore, 18 torsion angle constraints on 10 angles could be obtained with four di erent three-dimensional experiments. Sequence-specific shift assignment was achieved from a three-dimensional shift-correlation experiment that is also based on spatial proximity. The complete data set allowed for constructing a structural model of the peptide (Fig. 12a) by simulated annealing techniques or full search of the conformational space. In the latter procedure, conforma-

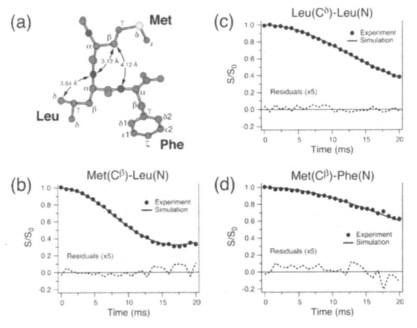

Fig. 12.   Frequency-selective REDOR measurements of carbon-nitrogen distances in a uniformly ^{13}C-^{15}N-labelled peptide diluted into the unlabelled peptide. (a) Structural model of the peptide. (b-d) Experimental REDOR curves for selected spin pairs together with fits and fit residuals. Reproduced with permission from [99]

tional space is factored into subspaces and then reduced by excluding all those subspaces in which at least one experimental constraint is violated. In the case at hand, the structure could be fully determined except for the orientations of the phenyl group and of the C terminus, for which no constraints were available.

5.6 Constraints on Long Distances in a Protein by EPR

Structure determination by high-resolution NMR is restricted to soluble proteins, while structure determination by x-ray crystallography requires that the protein can be crystallized. For most membrane proteins, which are notoriously hard to crystallize and may not fold into their functional structures in solution, neither of the two approaches is applicable. Solid-state NMR spectroscopy as well as EPR spectroscopy on singly and doubly spin-labeled mutants can provide at least partial information on structure and structural dynamics for this important class of proteins.

Determining the fold of a protein or at least recognizing that a protein belongs to a known class of folds may be possible even if only a few distances can be measured. For this purpose, long distances exceeding 2 nm provide particularly valuable constraints as they contain information on the relative arrangement of secondary structure elements such as   -helices and   -sheets. Pulse EPR methods such as the DEER experiment or DQ EPR can be used to measure distances between spin labels in the range between 2 and 5 nm, and in favourable cases up to 8 nm [11, 17, 64, 79]. That such methods can be applied to protein structure determination in the solid state has been demonstrated on the soluble protein T4 lysozyme consisting of 164 amino acid residues [101].

Distances ranging between 2.1 and 4   .7 nm could be measured by DQ EPR for eight selected pairs of spin-labelled residues in shock-frozen solutions of this protein. Experimental data for three pairs are shown in the left column of Fig. 13a together with fits by distance distribution consisting of a single Gaussian peak. Broadening due to a distribution of distances with a typical widths of 0 .2 nm is also apparent in the dipolar spectra shown in the left column of this figure. The determined distances are by 0   .3–1.0 nm larger than the distances between the respective   - or   -protons of the residues in the crystal structure. Without relying on this structure, it is possible to construct a rough three-dimensional model from the measured distances by the triangulation approach that is illustrated in Fig. 13b. It turned out that the number of eight distance constraints was too small to fully specify the relative positions of all the residues which had been spin-labelled. However, as is also shown in Fig. 13b a suggestion for an additional label site could be derived which should complete the triangulation.

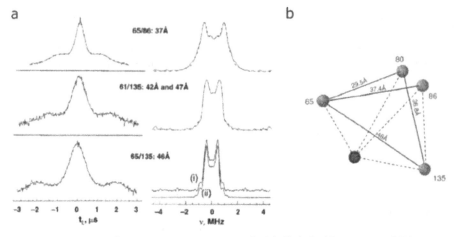

Fig. 13. DQ EPR distance measurements on spin-labelled double mutants of T4 lysozyme. (a) DQ coherence temporal envelopes and their fits (dashed lines) for three double mutants (left column) and dipolar spectra obtained by Fourier transformation (right column). Simulations of dipolar spectra for double mutants 65/86 und 61/135 are shown as dashed lines, the simulation for double mutant 65/135 (ii) is shifted downward with respect to the experimental spectrum (i). (b) Experimental distances for several double mutants showning "triangulation" in progress. Grey spheres correspond to average widths of distributions obtained by fitting the experimental data. The position of residue 86 cannot be fixed for lack of a su cient number of distance constraints. The black sphere depicts an additional label site which would complete the triangulation. Reproduced with permission from [101]

6 Conclusion

Precision and sensitivity of distance measurements by magnetic resonance methods depend substantially on the choice of technique, in particular, on the elimination of signal contributions due to other interactions and due to relaxation. Adapting spectral selectivity of the measurement to the problem at hand is also important. Generally, better defined structures are studied with higher site selectivity than less well defined structures. Both precision and sensitivity may also be strongly influenced by the data analysis procedures used in the interpretation of experimental raw data. Consideration of experimental imperfections, in particular of imperfect suppression of other interactions or of the unavoidable influence of isotropic spin-spin couplings (J couplings) may be crucial. It should be borne in mind that deriving spin-spin pair correlation functions from dipolar evolution functions is an ill-posed mathematical problem, so that noise may influence the results in a di erent way than in the more familiar Fourier transformation techniques. Choosing a numerically stable procedure and cross-checking the results by model computations is necessary to ensure reliability of the results [88]. In general, the reliability of structural models derived from NMR and EPR distance measurements can be better estimated if they are compared to the results of molecular modelling techniques. NMR crystallography is based on such an approach [102]. Quantum-chemical computation of magnetic resonance parameters may help to resolve ambiguities and keep the number of adjustable parameters and their ranges in the fitting process to an absolute minimum.

A study of a solid-state structure by magnetic resonance techniques thus amounts to a complex task, for which no simple set of rules can be given. This is not expected to change in the future, as the complexity of the process derives from the complexity of the structures that are studied by magnetic resonance techniques. It is exactly the strong variability of magnetic resonance experiments and the possibility to adapt them to the problem under investigation that allows for a study of such structures. The question which technique is optimum for given types of spins thus cannot be answered in general. The history of method development and subsequent application or non-application of the methods suggests that Einstein's dictum is valid also in this field: Make things as simple as possible – but no simpler.

Acknowledgment

We thank I. Schnell for helpful discussions. Financial support from the Deutsche Forschungsgemeinschaft is gratefully acknowledged.

A Appendix

BABA back-to-back: an NMR recoupling pulse sequence for broadband excitation of multiple-quantum coherences during fast MAS

DEER double electron electron resonance: acronym for a spin-echo double resonance experiment in EPR spectroscopy; used synonymously with PELDOR

DQ double-quantum: designates transitions of spin S that involve a change $m_S = 2$; build-up of coherence on such transitions depends on the dipole-dipole coupling

DRAMA dipolar recovery at the magic angle: an NMR recoupling pulse sequence that reintroduces homonuclear dipole-dipole coupling during MAS

ELDOR electron electron double resonance: a collective term including continuous-wave and pulse EPR experiments in which two microwave frequencies or a magnetic field step are applied to obtain information on dynamics or on couplings between electron spins

ENDOR electron nuclear double resonance: indirect detection of the NMR spectrum of nuclei that are hyperfine coupled to an electron spin by observation on electron spin transitions to increase sensitivity and bandwidth

ESEEM electron spin echo envelope modulation: a modulation in the decay of the primary (Hahn) or stimulated echo that is caused by coherence transfer echoes; can be used for indirect detection of an NMR spectrum of hyperfine coupled nuclei if the normally forbidden electron-nuclear zero- and double-quantum transitions are slightly allowed

INADEQUATE incredible natural abundance double quantum transfer experiment: signals from pairs of rare S spins are selectively detected in the presence of much more abundant isolated S spins by applying a double-quantum filter

MAS magic angle spinning: fast rotation of the sample about an axis that includes an angle of 54.74 with the magnetic field axis; increases resolution if broadening in the spectra is caused by anisotropy of interactions

PELDOR pulse ELDOR : an abbreviation used mostly for spin echo double resonance experiments in EPR; synonymous with DEER

PISEMA polarization inversion exchange at the magic angle: an NMR double resonance technique in which I spin magnetization is spin-locked at an angle of 54.74 with respect to the magnetic field axis and inverted halfway through the evolution period; polarization transfer between I and S spins driven by the heteronuclear dipole-dipole couplings can then be observed over a longer time, thus enhancing resolution

REAPDOR rotational echo adiabatic passage double resonance: broadband recoupling technique for heteronuclear couplings of spins $I > 1/2$ that relies on long pulses and on the change of the resonance frequency during MAS

REDOR rotational echo double resonance: recoupling of heteronuclear dipole-dipole coupling during MAS by applying 180 pulses to one of the spins, so that rotational averaging is disturbed

SEDOR spin echo double resonance: the heteronuclear dipole-dipole coupling is reintroduced into the nuclear primary (Hahn) echo decay of S spins by applying an additional 180 pulse to the I spins

SIFTER single frequency technique for refocusing: by using a combination of primary (Hahn) echo and solid-echo refocusing, the coupling of like spins is separated from Zeeman anisotropy and hyperfine couplings

References

1. R. Blinc, T. Apih: Progr. Nucl. Magn. Reson. 41, 49 (2002)
2. R. Blinc, J. Dolinsek, A. Gregorovic, B. Zalar, C. Filipic, Z. Kutnjak, A. Levstik, R. Pirc: Phys. Rev. Lett. 83, 424 (1999)
3. R.R. Ernst, G. Bodenhausen, A. Wokaun: Principles of Nuclear Magnetic Resonance in One and Two Dimensions (Clarendon Press, Oxford 1987)
4. K. Schmidt-Rohr, H.W. Spiess: Multi-Dimensional NMR and Polymers (Academic Press, London 1994)
5. A. Schweiger, G. Jeschke: Principles of pulse electron paramagnetic resonance (Oxford University Press, Oxford 2001)
6. K. W¨uthrich: J. Biol. Chem. 265, 22059 (1990)
7. K. W¨uthrich: Acta Crystallogr. D 51, 249 (1995)

8. I. Bertini, C. Luchinat, G. Parigi: Concepts Magn. Reson.   14, 259 (2002)
9. A.R. Leach: Molecular Modelling: Principles and Applications, 2nd ed.   (Prentice Hall, Harlow 2001)
10. A. Martinez-Richa, R. Vera-Graziano, D. Likhatchev: ACS Sym. Ser.   834, 242 (2003)
11. L.J. Berliner, S.S. Eaton, G.R. Eaton (Ed.):   Biological magnetic resonance, Vol. 19 (Plenum, New York 2000)
12. I. Schnell, H.W. Spiess: J. Magn. Reson.   151, 153 (2002)
13. K. Saalw"achter, I. Schnell: Solid State Nucl. Mag.   22, 154 (2002)
14. L. Frydman: Annu. Rev. Phys. Chem.   52, 463 (2001)
15. D.D. Laws, H.M.L. Bitter, A. Jerschow: Angew. Chem. Int. Ed.   41, 3096 (2002)
16. S.P. Brown, H.W. Spiess: Chem. Rev.   101, 4125 (2001)
17. G. Jeschke: Macromol. Rapid Commun.   23, 227 (2002)
18. J.M. Brown, R.J. Buenker, A. Carrington, C. Di Lauro, R.N. Dixon, R.W. Field, J.T. Hougen, W. Huttner, K. Kuchitsu, M. Mehring, A.J. Merer, T.A. Miller, M. Quack, D.A. Ramsay, L. Veseth, R.N. Zare: Mol. Phys.   98, 1597 (2000)
19. J.E. Peralta, V. Barone, R.H. Contreras, D.G. Zaccari, J.P. Snyder: J. Am. Chem. Soc.   123, 9162 (2001)
20. N. Bloembergen, T.J. Rowland: Phys. Rev.   97, 1679 (1955)
21. M. Tanaka: J. Phys. Soc. Jpn.   27, 784 (1969)
22. A. Lesage, C. Auger, S. Caldarelli, L. Emsley: J. Am. Chem. Soc.   119, 7867 (1997)
23. S.P. Brown, M. Perez-Torralba, D. Sanz, R.M. Claramunt, L. Emsley: J. Am. Chem. Soc.   124, 1152 (2001)
24. S. Hediger, A. Lesage, L. Emsley: Macromolecules   35, 5078 (2002)
25. J.A. Weil, J.R. Bolton, J.E. Wertz: Electron paramagnetic resonance   (Wiley, New York 1994)
26. N.M. Atherton: Principles of electron spin resonance   (Ellis Horwood, New York 1993)
27. J.E. Harriman: Theoretical Foundations of Electron Spin Resonance   (Academic Press, New York 1978)
28. J.R. Pilbrow: Transition ion electron paramagnetic resonance   (Clarendon, Oxford 1990)
29. G.E. Pake: J. Chem. Phys.   16, 327 (1948)
30. P. Betrand, C. More, B. Guigliarelli, A. Fournel, B. Bennet, B. Howes: J. Am. Chem. Soc.   116, 3078 (1994)
31. C. Elsasser, M. Brecht, R. Bittl: J. Am. Chem. Soc.   124, 12606 (2002)
32. M. Wind, K. Saalw" achter, U.M. Wiesler, K. M" ullen, H.W. Spiess: Macromolecules 35, 10071 (2002)
33. V. Macho, L. Brombacher, H.W. Spiess: Appl. Magn. Reson.   20, 405 (2001)
34. H.W. Spiess: Chem. Phys.   6, 217 (1974)
35. E.R. Andrew, A. Bradbury, R.G. Eades: Nature   183, 1802 (1959)
36. I. Lowe: Phys. Rev. Lett.   2, 285 (1959)
37. M.M. Maricq, J.S. Waugh: J. Chem. Phys.   70, 3300 (1979)
38. G. Jeschke, W. Ho bauer, M. Jansen: Chem. Eur. J.   4, 1755 (1998)
39. D.D. Laws, H.M.L. Bitter, A. Jerschow: Angew. Chem. Int. Ed.   41, 3096 (2002)
40. Z. Gan: J. Am. Chem. Soc.   114, 8307 (1992)
41. M. Hubrich, C. Bauer, H.W. Spiess: Chem. Phys. Lett.   273, 259 (1997)

42. D. Hessinger, C. Bauer, M. Hubrich, G. Jeschke, H.W. Spiess: J. Magn. Reson. 147 , 217 (2000)
43. G. Sierra, A. Schweiger: Mol. Phys.   95 , 973 (1998)
44. D. Hessinger, C. Bauer, G. Jeschke, H.W. Spiess: Appl. Magn. Reson.   20 , 17 (2001)
45. R.A. Eichel, A. Schweiger: J. Chem. Phys.   115 , 9126 (2001)
46. W.L. Hubbell, D.S. Cafiso, C. Altenbach: Nature Struct. Biol.   7, 735 (2000)
47. L.J. Berliner (Ed.):   Biological magnetic resonance, Vol. 14   (Plenum, New York 1998)
48. E.J. Hustedt, A.H. Beth: 'Structural Information from CW-EPR Spectra of Dipolar Coupled Nitroxide Spin Labels'. In: Biological Magnetic Resonance , Vol. 19, ed. by L.J. Berliner, G.R. Eaton, S.S. Eaton (Kluwer, New York 2000) pp. 155–184
49. E.J. Hustedt, A.I. Smirnov, C.F. Laub, C.E. Cobb, A.H. Beth: Biophys. J.   72 , 1861 (1997)
50. W. Xiao, Y.-K. Shin: 'EPR Spectroscopic Ruler: the Method and its Applications'. In: Biological Magnetic Resonance , Vol. 19, ed. by L.J. Berliner, G.R. Eaton, S.S. Eaton (Kluwer, New York 2000) pp. 249–276
51. M. Engelsberg, R.E. Norberg: Phys. Rev. B   5, 3395 (1972)
52. N. Boden, M. Gibb, Y.K. Levine, M. Mortimer: J. Magn. Reson.   16 , 471 (1974)
53. M. Emshwiller, E.L. Hahn, D. Kaplan: Phys. Rev.   118 , 414 (1960)
54. C.H. Wu, A. Ramamoorthy, S.J. Opella: J. Magn. Reson. A   246 , 325 (1995)
55. J.S. Waugh: Proc. Natl. Acad. Sci. USA   73 , 1394 (1976)
56. T. Gullion, J. Schaefer: J. Magn. Reson.   81 , 196 (1989)
57. T. Gullion, J. Schaefer: Adv. Magn. Reson.   13 , 57 (1989)
58. K.T. Mueller, T.P. Jarvie, D.J. Aurentz, B.W. Roberts: Chem. Phys. Lett. 242 , 535 (1995)
59. T. Gullion: Chem. Phys. Lett.   246 , 325 (1995)
60. A. Ramamoorthy, S.J. Opella: Solid State Nucl. Mag.   4, 387 (1995)
61. A.D. Milov, K.M. Salikhov, M.D. Shirov: Fiz. Tverd. Tela (Leningrad)   23 , 957 (1981)
62. A.D. Milov, A.G. Maryasov, Yu.D. Tsvetkov: Appl. Magn. Reson.   15 , 107 (1998)
63. M. Pannier, S. Veit, A. Godt, G. Jeschke, H.W. Spiess: J. Magn. Reson.   142 , 331 (2000)
64. G. Jeschke: ChemPhysChem   3, 927 (2002)
65. D.P. Raleigh, G.S. Harbison, T.G. Neiss, J.E. Roberts, R.G. Gri n: Chem. Phys. Lett.   138 , 285 (1987)
66. B.H. Meier, W.L. Earl: J. Am. Chem. Soc.   109 , 7937 (1987)
67. A. Schmidt, S. Vega: J. Chem. Phys.   96 , 2655 (1992)
68. T. Nakai, C.A. McDowell: J. Chem. Phys.   96 , 3452 (1992)
69. R. Tycko, G. Dabbagh: Chem. Phys. Lett.   173 , 461 (1990)
70. G. Goobes, S. Vega: J. Magn. Reson.   154 , 236 (2002)
71. M.H. Levitt: 'Symmetry-Based Pulse Sequences in Magic-Angle Spinning Solid-State NMR'. In: Encyclopedia of Nuclear Magnetic Resonance, Vol. 9   , ed. by D.M. Grant, R.K. Harris (John Wiley & Sons, Chichester 2002) pp. 165–196
72. H. Hatanaka, T. Terao, T. Hashi: J. Phys. Soc. Jpn.   39 , 835 (1975)
73. H. Hatanaka, T. Hashi: J. Phys. Soc. Jpn.   39 , 1139 (1975)
74. M. Munowitz, A. Pines: Adv. Chem. Phys.   66 , 1 (1987)

75. R. Graf, D.E. Demco, J. Gottwald, S. Hafner, H.W. Spiess: J. Chem. Phys. 106 , 885 (1997)
76. D.E. Demco, S. Hafner, H.W. Spiess: J. Magn. Reson. A   116 , 36 (1995)
77. M. Feike, D.E. Demco, R. Graf, J. Gottwald, S. Hafner, H.W. Spiess: J. Magn. Reson. A 122 , 214 (1996)
78. P.P. Borbat, J.H. Freed: Chem. Phys. Lett.   313 , 145 (1999)
79. P.P. Borbat, J.H. Freed: 'Double-Quantum ESR and Distance Measurements'. In: Biological Magnetic Resonance   , Vol. 19, ed. by L.J. Berliner, G.R. Eaton, S.S. Eaton (Kluwer, New York 2000) pp. 383–459
80. J.G. Powles, P. Mansfield: Phys. Lett.   2, 58 (1962)
81. G. Jeschke, M. Pannier, A. Godt, H.W. Spiess: Chem. Phys. Lett.   331 , 243 (2000)
82. G. Jeschke, A. Godt: ChemPhysChem   4, 100 (2003)
83. J.H. van Vleck: Phys. Rev.   74 , 1168 (1948)
84. C.P. Slichter: Principles of Magnetic Resonance   (Springer, Berlin 1990)
85. G. Jeschke, A. Koch, U. Jonas, A. Godt: J. Magn. Reson.   155 , 72 (2001)
86. J. Weese: Comput. Phys. Commun.   69 , 99 (1992)
87. F.G. Vogt, D.J. Aurentz, K.T. Mueller: Mol. Phys.   95 , 907 (1998)
88. G. Jeschke, G. Panek, A. Godt, A. Bender, H. Paulsen: Appl. Magn. Reson. 26 , 223 (2004)
89. W.B. Mims: Phys. Rev. B   5, 2409 (1972)
90. K. Saalw"achter, H.W. Spiess: J. Chem. Phys.   114 , 5707 (2001)
91. H.W. Spiess: 'Double-Quantum NMR Spectroscopy of Dipolar Coupled Spins Under Fast Magic-Angle Spinning'. In: Encyclopedia of Nuclear Magnetic Resonance , Vol. 9, ed. by D.M. Grant, R.K. Harris (John Wiley & Sons, Chichester 2002) pp. 44-58
92. C. Filip, X. Filip, D.E. Demco, S. Hafner: Mol. Phys.   92 , 757 (1997)
93. C. Filip, S. Hafner, I. Schnell, D.E. Demco, H.W. Spiess: J. Chem. Phys.   110 , 423 (1999)
94. M. Feike, C. J" ager, H.W. Spiess: J. Non-Cryst. Solids.   223 , 200 (1998)
95. S.P. Brown, X.X. Zhu, K. Saalw" achter, H.W. Spiess: J. Am. Chem. Soc.   123 , 4275 (2001)
96. J.M. Lehn: Angew. Chem. Int. Ed. Engl.   29 , 1304 (1990)
97. E. Narr, A. Godt, G. Jeschke: Angew. Chem. Int. Ed.   41 , 3907 (2002)
98. S.P. Brown, I. Schnell, J.D. Brand, K. M"   ullen, H.W. Spiess: J. Am. Chem. Soc. 121 , 6712 (1999)
99. C.M. Rienstra, L. Tucker-Kellog, C.P. Jaroniec, M. Hohwy, B. Reif, M.T. McMahon, B. Tidor, T. Lozano-P´   erez, R.G. Gri n: Proc. Natl. Acad. Sci. USA 99 , 10260 (2002)
100. C.P. Jaroniec, B.A. Tounge, J. Herzfeld, R.G. Gri n: J. Am. Chem. Soc.   123 , 3507 (2001)
101. P.P. Borbat, H.S. Mchaourab, J.H. Freed: J. Am. Chem. Soc.   124 5304 (2002)
102. F. Taulelle: Curr. Opin. Sol. State Mater. Sci.   5 397 (2001)

NMR Studies of Disordered Solids

J. Villanueva-Garibay and K. M" uller

Institut f" ur Physikalische Chemie, Universit" at Stuttgart, Pfa enwaldring 55
70569 Stuttgart, Germany
k.mueller@ipc.uni-stuttgart.de

Abstract.   In this contribution an introduction to dynamic solid state NMR spectroscopy is presented. The main emphasis is given to dynamic ^{2}H NMR techniques,
since these methods – in combination with selectively or partially deuterated compounds – have demonstrated a particular suitability for studying the molecular properties (i.e. order and dynamics) of solid, semisolid materials as well as anisotropic
liquids. A general overview about the theoretical background of dynamic NMR spectroscopy is provided in the first part, which also includes the description of the main
experimental methods in dynamic ^{2}H NMR spectroscopy. In the second part representative results from model simulations are given, considering various types of
motional processes which are frequently discussed in disordered materials. Applications of dynamic ^{2}H NMR techniques during the study of inclusion compounds are
shown in the last section.

1 Introduction

Dynamic NMR spectroscopy is a well established technique for the evaluation of molecular dynamics in condensed media. Apart from the frequent
application of such techniques in the field of high-resolution (liquid) NMR
spectroscopy [1, 2], dynamic NMR methods were also applied successfully
on quite di erent types of anisotropic materials [3], such as polymers [4, 5],
thermotropic liquid crystals [6, 7], (lyotropic) lipid bilayers and biological
membranes [8, 9], guest-host systems (clathrates, zeolites) [10], molecular and
plastic crystals [11], etc. These latter studies have clearly demonstrated that
dynamic solid state NMR spectroscopy represents a powerful method for the
determination of the dynamic and structural features of even complex systems. Dynamic NMR spectroscopy thus can be used to probe the ordering
characteristics in terms of conformational, orientational and positional order.
Likewise, a comprehensive analysis of such experiments gives access to the
inherent motional contributions, comprising conformational, reorientational
and lateral motions. In favourable cases, dynamic NMR spectroscopy is able

J. Villanueva-Garibay and K. M"   uller: NMR Studies of Disordered Solids   , Lect. Notes Phys.   684 ,
65–86 (2006)
www.springerlink.com   c Springer-Verlag Berlin Heidelberg 2006

to follow such motions over a very broad time-scale, ranging from the sub-kHz to the GHz region [12, 13].

In this contribution we will provide a brief introduction to dynamic solid state NMR methods. In particular, we will focus on dynamic ^{2}H NMR techniques, since these methods in combination with selectively or partially deuterated compounds have demonstrated their particular suitability for studying the aforementioned solid and semisolid materials as well as anisotropic liquids [13, 14, 15]. In the first part of this contribution we briefly describe the theoretical background of dynamic NMR which also includes the description of the main experimental methods. In the second part some representative model simulations are provided. Results from the application of dynamic NMR techniques during the study of guest-host systems are shown in the last section.

2 Theoretical Background

2.1 General Theory

The description of NMR experiments in general is done by considering the time evolution of the spin-density operator (t). In the absence of molecular motions the time evolution of the spin-density operator (t) is given by the Liouville-von Neumann equation [16]

$$\frac{d\,(t)}{dt} = -\frac{i}{}[\,(t), H]\,. \tag{1}$$

H is the time-independent Hamiltonian of dimension n (n: dimension of Hilbert space) which includes various terms for magnetic interactions of the nuclei with their local surrounding as well as terms for the r.f. pulses

$$H = H_{rf} + H_{CS} + H_D + H_Q + \ldots\,. \tag{2}$$

The formal solution of (1) is given by

$$(t) = U(t)\,(0)\,U(t)^{-1}\,, \tag{3}$$

where the propagator $U(t)$ is defined as

$$U(t) = e^{-(i/\,)Ht}\,. \tag{4}$$

The NMR experiment is furthermore subdivided in time intervals $_1, _2 \cdots _n$ that possess a constant Hamiltonian, e.g. with and without r.f. pulses and/or particular magnetic interactions. The density matrix operator (t) at a particular time t can then be easily calculated via the summation of the intervals with constant Hamiltonian [11]

$$(t) = \qquad _i\,. \tag{5}$$

In the presence of molecular motions   the starting point is the Stochastic Liouville equation, which in general is given by [16, 17]

$$\frac{d\,\rho(t)}{dt} = -\frac{i}{\hbar}[\rho(t), H] + \frac{d\rho}{dt}\bigg|_{dyn}. \tag{6}$$

Here, the second term on the right side accounts for the contribution due to dynamic processes, such as molecular motion or chemical exchange. Equation (6) can be solved after rewriting in the form

$$\frac{d\,\rho(t)}{dt} = -\frac{i}{\hbar}L\,\rho(t) \tag{7}$$

to yield $\rho(t)$ [11]

$$\rho(t) = \rho(0)\,e^{-\frac{i}{\hbar}Lt}. \tag{8}$$

As before, the full time evolution of the density matrix is then described by dividing the experiment in intervals with constant Hamiltonian. The Liouville superoperator L in (7) and (8) is derived from the Hamiltonian by the prescription [12]

$$L = H \otimes E - E \otimes H, \tag{9}$$

where E is the identity operator in Hilbert space. The matrix L is further expanded by the part that accounts for molecular motion to give the matrix $\tilde{L}$. The final dimension of $\tilde{L}$ is given to n^2N, where N is the number of exchanging sites [11, 18].

In the most general case the following terms are included in the spin Hamiltonian [19]

$$H = H_{rf} + H_{CS} + H_D + H_Q. \tag{10}$$

They refer to the radio frequency part H_{rf}

$$H_{rf} = \sum_i \omega_{rf}^i(t)(I_{ix}\cos\varphi_i + I_{iy}\sin\varphi_i) \tag{11}$$

and the contributions from several magnetic interactions, namely

(i) chemical shift interaction (H_{CS})

$$H_{CS} = \sum_i \omega_{CS,0}^i(t)\,I_{iz}, \tag{12}$$

(ii) dipole-dipole interaction (H_D) between spins i and k

$$H_D = \sum_{i,k} \omega_{D,0}^i(t)\,\frac{1}{6}(3I_{iz}\,I_{kz} - I_i \cdot I_k), \tag{13}$$

(iii) and quadrupolar interaction (H_Q) with first and second order contributions

$$H_Q = \sum_i \omega_{Q,0}^i(t) \frac{1}{6} \left[3I_{iz}^2 - I_i^2 \right]$$

$$+ \frac{1}{2\omega_0} \sum_i \begin{bmatrix} \omega_{Q,-2}^i(t) \, \omega_{Q,+2}^i(t) \left[2I_i^2 - 2I_{iz}^2 - 1 \right] I_{iz} + \\ \omega_{Q,-1}^i(t) \, \omega_{Q,+1}^i(t)(4I_i^2 - 8I_{iz}^2 - 1) I_{iz} \end{bmatrix} .$$

(14)

In (11) to (14) ω_0 and ω_i denote the Larmor frequency and the corresponding
interaction constants, which can be found elsewhere [19].

The angular dependence (anisotropy) of the various magnetic interactions
is obtained by the transformation of the respective magnetic interaction from
its own principle axis system (PAS) into the laboratory (LAB) frame using
second-rank rotation matrices [20]. In the presence of molecular motions a
minimum of two transformations is required, namely (i) from the PAS to an
intermediate axis system (IAS), defined by the symmetry axis of the motional
process, and from the IAS to the LAB frame. If several motions are superimposed, additional transformations (through intermediate axis systems IAS-1
to IAS-n , n = number of superimposed motions) are required [11, 14, 21].

The spin-part of the Hamiltonian, i.e., the spin operators, is given in its
matrix representation. For a general spin system the basis of the Hamiltonian
is obtained via the direct product of the relevant single spin operators I_i^μ (i =
1...n, μ = x, y, z), e.g.

$$I_{1x}I_{2y} = I_x \otimes I_y \otimes 1 \otimes ... \otimes 1_{(n)} ; \quad I_{3x} = 1 \otimes 1 \otimes I_x \otimes ... \otimes 1_{(n)} ; \quad \text{etc.} \quad (15)$$

1 is the single spin unity operator. The single spin operators are obtained by
the relations

$$\langle I,m |I_z|I,m \rangle = m , \tag{16}$$

$$\langle I,m \pm 1|I_\pm |I,m \rangle = \sqrt{I(I+1) - m(m \pm 1)} . \tag{17}$$

On this basis, the time evolution of the density matrix, as given by (3) or (8),
can then be calculated. In the most general case this requires diagonalization
of the spin Hamiltonian or the matrix L within the various time intervals.
The dimension of these matrices is given by Π^n or by n^2N in the absence and
presence of molecular motions, respectively [11, 18].

2.2 Dynamic ^{2}H NMR Spectroscopy

In the following we will restrict ourselves to the case of dynamic ^{2}H NMR
spectroscopy on static samples, i.e., broadline NMR conditions. In ^{2}H NMR
spectroscopy we have a very particular situation, since the spin Hamiltonian is dominated by the quadrupolar interaction with a coupling constant
$\omega_Q/2\pi = e^2qQ/h$ between about 165 kHz (aliphatic deuterons) and 185 kHz

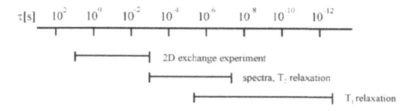

Fig. 1.   Sensitive time-scales of different types of dynamic NMR experiments

(aromatic deuterons). In dynamic NMR spectroscopy various sensitive time-scales (see Fig. 1) and NMR experiments can be distinguished, where – instead of the most general formalism (see Sect. 2.1) – suitable approaches can be employed for the theoretical description of these experiments. In the following we thus distinguish among (i) the fast (rate constant $k \gg \delta_Q$, $k \gg \delta_Q$), (ii) intermediate ($k \approx \delta_Q$), and (iii) ultraslow motional region ($k \ll \delta_Q$) [13, 22].

From the experimental point of view (see Fig. 2), the quadrupole echo sequence is used for the detection of ^{2}H NMR line shapes and spin-spin (T_2) relaxation. The time-scale of such NMR line shape studies and T_2 effects is given by the strength of the (motionally) modulated magnetic interaction – here the quadrupolar interaction – which is in the MHz range [11, 22, 23, 24, 25, 26]. Spin-lattice (T_1) relaxation is probed with a modified inversion recovery experiment from which fast molecular motions in the vicinity of the Larmor frequency, i.e., in the MHz to GHz-region, are accessible [11, 14, 24]. Finally, 2D exchange and related NMR experiments can be used to study ultra-slow motional processes in the Hz- and sub-Hz range. Here, the accessible

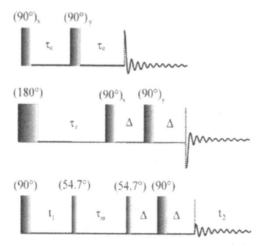

Fig. 2.   Basic pulse experiments in dynamic ^{2}H NMR spectroscopy, top: quadrupole echo sequence, middle: inversion recovery sequence, bottom: 2D exchange sequence

time-scale is determined by the length of the exchange interval τ_m, and is limited by spin-lattice (lower limit) and spin-spin relaxation (upper limit) [12, 13, 27].

^{2}H NMR Spectra in the Rigid Limit

If the motional processes are slow on the NMR time-scale, and the sample refers to a polycrystalline material, then the observed ^{2}H NMR powder spectrum ("Pake" pattern) is the sum over all (static) orientations of the crystallites with respect to the external magnetic field (see Fig. 3). Thus, the powder spectrum is the weighted sum of individual pairs of lines whose frequencies ω_q are given by [13]

$$\omega_q = \pm \frac{3}{8} \omega_Q \left[3\cos^2\theta - 1 + \eta \sin^2\theta \cos 2\phi \right] . \tag{18}$$

θ and ϕ are the spherical polar angles which specify each crystallite orientation, and η is the asymmetry parameter (which for aliphatic deuterons normally is close to 0). The powder spectrum is obtained via Fourier transformation of the free induction decay (FID) signal

$$S(t) = N \int_0^\pi \sin\theta \, d\theta \int_0^{2\pi} d\phi \, e^{-i\omega_q(\theta, \phi)t} , \tag{19}$$

where N is a normalization constant. It should be noted that the singularities in the ^{2}H NMR powder spectra of rigid samples directly reflect the three main components of the quadrupolar interaction tensor (Q_{ii}^{PAS}) in its principle axis system (see Fig. 3) [4, 22, 24].

For the case of an axially symmetric quadrupolar interaction tensor ($\eta = 0$), a splitting of $3\omega/4 \omega_Q$ between the perpendicular singularities is registered.

^{2}H NMR Spectra in the Fast Exchange Region

If the molecular motions are in the fast exchange region, then the inspection of the experimental line shapes already gives an indication about the symmetry (or type) of the underlying motional process (see Fig. 4). In fact, the description of the fast exchange NMR line shapes does not require a complex line shape simulation. Rather, it is just necessary to transform the quadrupolar interaction tensor Q^{PAS} from its principle axis system to the IAS coordinate system, which is defined by the particular motional process (i.e., motional symmetry axis), using appropriate transformation matrices $R(\alpha, \beta, \gamma)$ (α, β, γ = Euler angles specifying the coordinate transformation)

$$Q^{IAS}(\Omega) = R(\Omega)Q^{PAS} R^{-1}(\Omega) \tag{20}$$

with

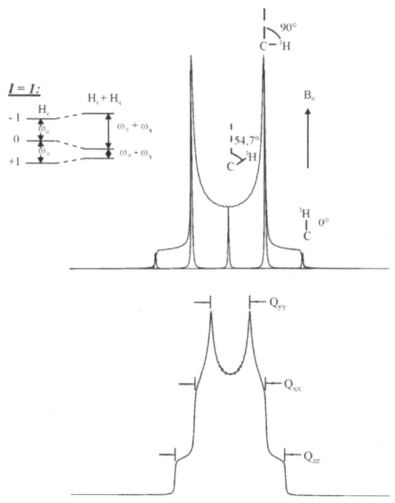

Fig. 3.   Relationship between crystal orientation and ^{2}H NMR powder pattern (upper spectrum: = 0; lower spectrum: = 0)

$$Q^{PAS} = \frac{3}{4} \frac{e^2 qQ}{h} \begin{pmatrix} 1+ & 0 & 0 \\ 0 & 1- & 0 \\ 0 & 0 & -2 \end{pmatrix} . \qquad (21)$$

The averaged tensor components are then calculated by using the equilibrium
population of the relevant molecular orientations $P_{eq}(_k, _k, _k)$ which are
necessary for the particular motion under consideration.

$$\bar{Q}_{ij} = \sum_k P_{eq}(_k, _k, _k) Q_{ij}^{IAS}(_k, _k, _k) . \qquad (22)$$

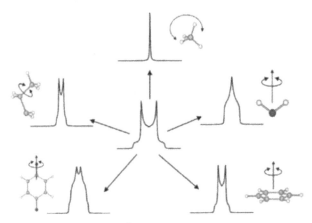

Fig. 4.   Theoretical fast exchange ^{2}H NMR spectra based on di erent motional models

The fast exchange ^{2}H NMR line shapes are determined by the residual principle tensor components $\bar{Q}_{ii}^{PAS}$. These quantities, which again can be taken directly from the spectral singularities, are obtained via diagonalization of the averaged tensor matrix, according to [24]

$$\bar{Q}_{ij} - {}_{ij} \bar{Q}_{ij}^{PAS} = 0 . \qquad (23)$$
$$_{j}$$

The procedure for the analysis of the fast exchange spectra is thus very simple. The most convenient way is the implementation of the above procedure into standard mathematical software packages [28], such as Mathcad ^{TM} [29], Matlab ^{TM} [30], etc.

^{2}H NMR Line Shape and T$_2$ E ects

If the molecular motions occur in the   intermediate time-scale   (k $_Q$), then ^{2}H NMR line shapes from the   quadrupolar echo experiment   can be adequately described via (24), where infinitesimally sharp   -pulses are assumed, and finite pulse e ects thus are neglected [22, 23, 24, 25, 26].

$$S(t, 2_e) = 1\,e^{\,At}\,e^{A\,_e}\,e^{A\,_e}\,P_{eq}(0) . \qquad (24)$$

Here, the evolution of the magnetization (i.e., spin-spin relaxation) is explicitly considered during the intervals $_e$ between the r.f. pulses. S(t, 2 $_e$) is the FID starting at the top of the quadrupole echo, and the vector   P$_{eq}$ (0) denotes the fractional populations of the N exchanging sites in thermal equilibrium.   A is a complex matrix of size N with

$$A = i \quad + K . \qquad (25)$$

The imaginary part of A is given by the diagonal matrix Ω whose elements $\Omega_{ii} (= \omega_{q,i})$ describe the frequencies of the exchanging sites. The real part corresponds to a kinetic matrix K. Here, the non-diagonal elements k_{ij} are the jump rates from site j to site i, while the diagonal elements k_{ii} represent the sums of the jump rates for leaving site i; they also contain the residual line widths in terms of $1/T_2^0$.

$$\Omega = \begin{pmatrix} \omega_{q,1} & & & \\ & \omega_{q,2} & & \\ & & \ddots & \\ & & & \omega_{q,N} \end{pmatrix} ;$$

$$K = \begin{pmatrix} -\sum_i k_{i1} + 1/T_{2,1}^0 & k_{12} & \cdots & k_{1N} \\ k_{21} & -\sum_i k_{i2} + 1/T_{2,2}^0 & \cdots & k_{2N} \\ \vdots & \vdots & \ddots & \vdots \\ k_{N1} & k_{N2} & \cdots & -\sum_i k_{iN} + 1/T_{2,N}^0 \end{pmatrix} .$$

$$(26)$$

It should be noted that (26) also accounts for the general case that – depending on the complexity of the system studied – several (superimposed) internal and intermolecular processes might be present at the same time. Equation (24) is solved numerically using standard diagonalization routines, from which ^2H NMR line shapes, partially relaxed spectra and spin-spin relaxation times T_2 are derived. The influence of finite pulse effects can be taken into account after calculation of the FID by making use of analytical expressions, as shown in [31].

^2H NMR Line Shape and T_1 Effects

The simulation of partially relaxed ^2H NMR spectra from the modified inversion recovery experiment is feasible with the help of (27) [24]

$$S(t, 2\tau_e, \tau_r) = [1 - 2e^{-\tau_r/T_1}] \, S(t, 2\tau_e) . \tag{27}$$

$S(t, 2\tau_e)$ is the FID signal obtained by (24) or by tensor averaging in the fast motional limit, as also described earlier. The delay τ_r refers to the relaxation interval between the inversion pulse and the quadrupole echo sequence used for signal detection.

In order to calculate spin-lattice relaxation effects, second order perturbation theory [32, 33, 34] is employed which, however, is not only restricted to the fast motional limit. Rather, the condition must hold that the spin-lattice relaxation rate is slower than the motional rate k responsible for spin relaxation $(1/T_1 \ll k)$ [11]. Again, if only the quadrupolar interaction determines spin-lattice relaxation, then the ^2H spin-lattice relaxation time T_1 is given by

$$\frac{1}{T_1} = \frac{3}{16} \frac{e^2 qQ}{h}^2 [J_1(\omega) + 4 J_2(2\omega)] . \tag{28}$$

The spectral densities J_m can be derived by solving the following equation for a general N-site exchange [35, 36]

$$J_m(\omega) = 2 \sum_{a,a'=-2}^{2} d_{ma}^{(2)}(\beta) d_{ma'}^{(2)}(\beta) \sum_{n,l,j=1}^{N} X_l^{(0)} X_l^{(n)} X_j^{(0)} X_j^{(n)} d_{0a}^{(2)}(\theta_l) d_{0a'}^{(2)}(\theta_j)$$

$$\times \cos(a\phi_l - a'\phi_j) \frac{\lambda_n}{\lambda_n^2 + \omega^2} \tag{29}$$

with $\quad \omega_i = \omega_i - \omega . \tag{30}$

Here, $X^{(n)}$ and λ_n are the corresponding eigenvectors and eigenvalues of the symmetrized rate matrix $\tilde{K}$ (see (27) without $1/T_2^0$ terms). The angles θ and ϕ are the spherical polar angles between the PAS and an IAS (determined by the motional process), whereas the angles β and γ connect the IAS and the LAB system. $d_{ab}^{(2)}(\beta)$ are elements of the reduced Wigner rotation matrix. If there is a superposition of several motional modes, then the transformation from the PAS to the LAB frame is subdivided into several steps according to the number of motional contributions (see above).

2D Exchange ^{2}H NMR Spectra

In order to describe the 2D exchange NMR experiments for the detection of ultraslow motions , motional e ects normally are only considered during the exchange interval τ_m. With the assumption of infinitesimally sharp r.f. pulses, the quadrupolar order (S_Q) and Zeeman order (S_Z) signals are calculated using the following equations [13, 27, 37]

$$S_Q(t_1, t_2; \tau_m) = C \sum_{ij} \sin(\omega_q^i t_1) e^{-t_1/T_2^i} P_{ij}(\tau_m) \sin(\omega_q^j t_2) e^{-t_2/T_2^j},$$

$$S_Z(t_1, t_2; \tau_m) = C \sum_{ij} \cos(\omega_q^i t_1) e^{-t_1/T_2^i} P_{ij}(\tau_m) \cos(\omega_q^j t_2) e^{-t_2/T_2^j} \tag{31}$$

with $\quad P_{ij}(\tau_m) = P_i(0) e^{K\tau_m}{}_{ij} .$

P_{ij} and $P_i(0)$ denote the conditional probability that a nucleus jumps from site j to site i during τ_m and the equilibrium population of site i, respectively. K is the exchange matrix, already introduced above (26). In (31) spin-lattice relaxation contributions during τ_m were neglected. The processing of the data sets $S_z(t_1, t_2; \tau_m)$ and $S_Q(t_1, t_2; \tau_m)$ in order to obtain the pure absorption 2D exchange NMR spectrum is described elsewhere [27, 38].

3 Simulation Programs

The theoretical NMR spectra and relaxation effects for $I = 1$ spin systems were obtained by employing appropriate FORTRAN programs [39, 40, 41] that are based on the theoretical approaches and assumptions, as outlined in Sect. 2.2. In general, a numerical diagonalization is required in order to calculate the theoretical line shapes and relaxation times, which is achieved by employing appropriate routines [42]. All simulations were performed on personal computers (Windows and LINUX platforms) or SUN workstations (UNIX platform) [43]. The parameters which enter in the simulation programs are the quadrupolar coupling constant, the transformation angles between the PAS of the quadrupolar interaction, i.e., the C-^{2}H bond direction, and an internal coordinate system that is defined by the motional process. In the case of superimposed motions further transformation angles are necessary. Furthermore, pulse intervals, motional correlation times, equilibrium populations of the various jump sites are required along with a residual line width, the sweep width, the number of acquired data points, and the number of crystallite orientations in order to calculate the NMR powder spectrum.

4 Model Simulations

In the following, a few representative results from model simulations are provided that demonstrate the impact of various molecular or simulation parameters on the ^{2}H NMR line shapes and relaxation data. It should be emphasized that these examples are closely related to cases that are frequently encountered during the study of various types of disordered solids.

To begin with, we recall the effect of different molecular motions on the fast exchange ^{2}H NMR line shapes, which is demonstrated in Fig. 4. As can be seen, different types of overall motions (tetrahedral jumps, methyl group rotation, 180° flips) give rise to quite different fast exchange line shapes, which at the same time – as outlined earlier – reflect the symmetry of the underlying molecular motions.

Model calculations showing the influence of motional processes in the intermediate time-scale are given in Figs. 5 to 7. The two series of ^{2}H NMR line shapes in Fig. 5 were obtained with the assumption of a 3-fold jump motion around a motional symmetry axis which is perpendicular to the C-^{2}H bond direction. As can be seen, the variation of the equilibrium populations of the jump sites p_1, p_2 and p_3 (with $p_1 = 1 - p_2 - p_3$; $p_2 = p_3$) has a significant influence on such ^{2}H NMR line shapes.

Likewise, Fig. 6 depicts theoretical ^{2}H NMR line shapes that are obtained by considering two superimposed overall rotations – modelled by degenerate 3-site jump motions (i.e., with equally populated jump sites) – about two motional symmetry axes that are oriented perpendicular on each other. It is quite obvious that the actual rate constants of both motions have a considerable

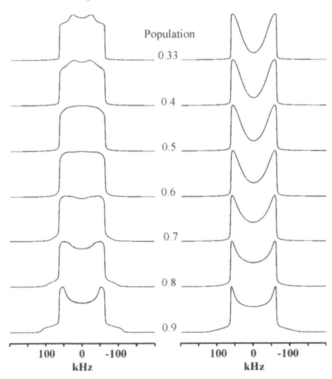

Fig. 5.   Theoretical ^{2}H NMR line shapes (quadrupole echo, $_e$ = 20 μs) for a non-degenerate 3-fold jump process with di erent populations p_1, as indicated, and at correlation times of 3   × 10^{-7} s (left) and 1 × 10^{-6} s (right). C-^{2}H bonds are oriented perpendicular with respect to motional symmetry axis

impact on the overall appearance of these ^{2}H NMR spectra. At the same time, it is found that the relative orientation of the two motional symmetry axes as well as the equilibrium populations of the jump sites also play a significant role on such NMR line shapes (spectra not shown). That is, both the types of motion and the motional correlation times can be determined with a high precision, since the change of these molecular quantities is directly reflected by the alterations in the ^{2}H NMR spectra.

In Fig. 7 three series of partially relaxed ^{2}H NMR spectra – calculated for the quadrupole echo sequence – are shown. These spectra were obtained on the basis of a degenerate 3-site jump motion, where three di erent angles between the motional symmetry axis and the C- ^{2}H bond direction have been chosen. Here, the observed changes in the partially relaxed ^{2}H NMR spectra, as a function of the pulse spacing $_e$, are a direct measure of the angular dependence of T_2 (or T_2 anisotropy) [11, 22], which strongly depends on the particular model assumptions. The present examples clearly demonstrate the influence of di erent opening angles between the motional symmetry axis

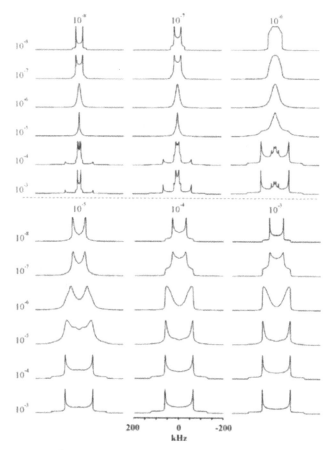

Fig. 6.   Theoretical ^{2}H NMR line shapes (quadrupole echo, τ_e = 20 μs) for two degenerate 3-fold jump processes (both motional symmetry axes are perpendicular on each other, C- ^{2}H bond oriented perpendicular with respect to first motional symmetry axis) for the correlation times given in the figure (rows: correlation times of outer rotation are varied; columns: correlation times of inner rotation are varied)

and the C- ^{2}H bond direction. In quite the same way, the T_2 anisotropy also is strongly affected by the actual equilibrium populations or the motional correlation time (data not shown).

In Figs. 8 and 9 model simulations are given that refer to partially relaxed ^{2}H NMR spectra from the modified inversion recovery sequence. The first example in Fig. 8 refers to a 3-fold jump motion in the fast exchange limit assuming a perpendicular orientation of the motional symmetry axis with re-spect to the C- ^{2}H bond direction. The three series of spectra demonstrate the influence of the actual equilibrium population p_1 on the spin-lattice relax-ation. As reflected by the characteristic changes of these spectra as a function of the relaxation period τ_r after the inversion pulse, the relaxation rate is not

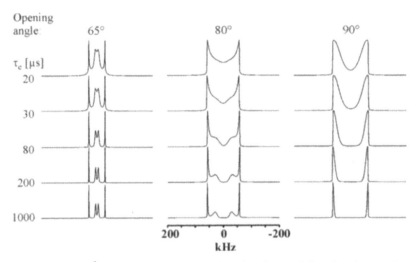

Fig. 7. Theoretical ^{2}H NMR line shapes (quadrupole echo, partially relaxed spectra) for a degenerate 3-fold jump process at di erent angles between the C- ^{2}H bond direction and the motional symmetry axis. The motional correlation time is 1 $\times 10^{-6}$ s

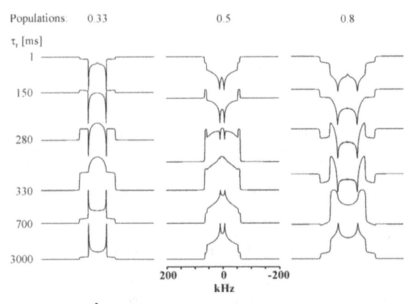

Fig. 8. Theoretical ^{2}H NMR line shapes (inversion recovery, partially relaxed spectra) for a 3-fold jump process and di erent equilibrium populations p_1. The angle between the C- ^{2}H bond direction and the motional symmetry axis is 90   . The motional correlation time is 1 $\times 10^{-11}$ s

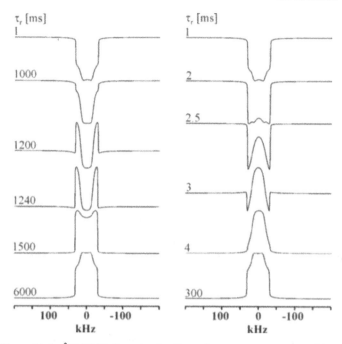

Fig. 9.   Theoretical ^{2}H NMR line shapes (inversion recovery,   = 20 μs) for two degenerate 3-fold jump processes (both motional symmetry axes are perpendicular on each other, C- ^{2}H bond oriented perpendicular with respect to first motional symmetry axis). The correlation times are left: 1   × 10^{-13} s (inner rotation), 8   × 10^{-7} s (outer rotation), and right: 5   × 10^{-9} s (inner rotation), 8   × 10^{-7} s (outer rotation)

identical across the ^{2}H NMR spectrum, which is a direct consequence of the angular dependence of T_1, i.e., T_1 anisotropy [11, 14, 36].

The second set of partially relaxed spectra, shown in Fig. 9, were calculated on the basis of two superimposed degenerate 3-fold jump motions with motional symmetry axes that are perpendicular on each other. The interesting point is that one motion occurs in the intermediate motional regime, giving rise to line shape (or T_2) e ects along with a characteristic line broadening. The second motion, however, takes place in the fast motional limit, and is responsible for the characteristic spin-lattice relaxation e ects [24].

Figure 10 shows a 2D exchange ^{2}H NMR spectrum. The simulation was performed with the assumption of a mutual exchange between two sites which are distinguished by their quadrupolar coupling constants. In fact, such a situation is encountered if six-membered ring hydrocarbons exhibit a fast overall rotation – leading to a di erent motional averaging of the quadrupolar coupling constants for the axial and equatorial deuterons – along with ultraslow ring inversion, the latter of which determines the 2D exchange pattern [40].

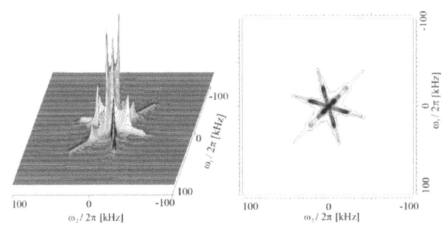

Fig. 10. Theoretical 2D exchange ^{2}H NMR spectrum assuming chemical exchange between axial and equatorial deuterons in six-membered ring hydrocarbons. Due to fast rotation around the molecular C$_3$-axis, the ratio of quadrupolar coupling constants is $_{ax}$: $_{eq}$ = 1 : (− 1/ 3). In addition, the condition of (1 / $_c$) $_m$ 1 ($_c$: correlation time for chemical exchange) holds

An analysis of the height of these exchange ridges provides the actual rate constants for the chemical exchange process. If the 2D exchange spectra are dominated by ultraslow reorientational motions, then the analysis of such exchange pattern provides valuable information about the underlying motional process (jump angle, distribution of correlation times, etc.) [13, 27, 37].

5 Applications for Guest-Host Systems

In the following we report on the application of dynamic ^{2}H NMR methods for the characterization of guest species in cyclophosphazene (CPZ) inclusion compounds [40, 44, 45]. The basic structure of the host matrix is given by parallel, hexagonal channels (see Fig. 11), in which various types of guest molecules can be incorporated. The examples discussed in the following refer to NMR studies on CPZ inclusion compounds with benzene-d$_6$ or pyridine-d$_5$ as guest molecules.

In general, it could be shown that these guest molecules exhibit a high mobility, which even holds for low temperatures (around 100 K). From a thorough data analysis it could be shown that the benzene guests undergo two motional processes, namely a fast rotation around the molecular C$_6$ symmetry axis (degenerate 6-fold jump process, C$_6$-axes perpendicular to channel axis), and a second rotation around the channel long axis (3-fold jump process with unequally populated jump sites).

The ^{2}H NMR line shapes and partially relaxed spectra (quadrupole echo experiment) in Figs. 12 and 13 thus are dominated by the latter motional

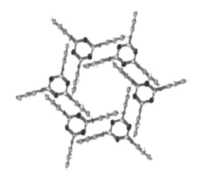

Fig. 11.   Host structure of CPZ inclusion compounds

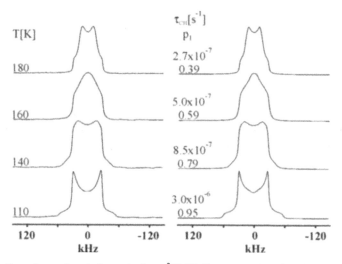

Fig. 12.   Experimental and theoretical ^{2}H NMR spectra (quadrupole echo, $_e$ = 20 μs) for benzene-d$_6$/CPZ at di erent temperatures. The relevant simulation para-meters are given in the Figure and in the text

contribution around the CPZ channel long axis, which occurs on a slower time-scale. The rotation around the molecular C$_6$ axis is significantly faster and dominates spin-lattice relaxation, as can be derived from the partially relaxed ^{2}H NMR spectra (inversion recovery experiment), depicted in Fig. 14. The general very good agreement between the experimental and theoretical data sets in Figs. 12 to 14 strongly supports the chosen model assumptions. The Arrhenius plots for the derived motional correlation times are given in Fig. 15, which yielded very low activation energies of 2.1 kJ mol^{-1} and 4.6 kJ mol^{-1} for the rotation around C$_6$-axis and the rotation around channel axis, respectively.

The pyridine guests again turned out to be highly mobile. However, due to their di erent chemical structure and lower symmetry, the motional behaviour is di erent from that discussed for the benzene guests. The experimental ^{2}H

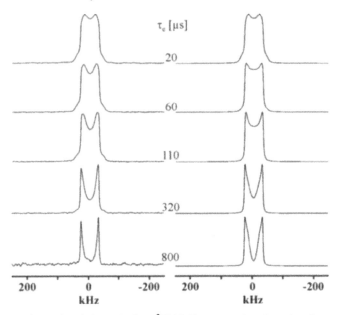

Fig. 13.   Experimental and theoretical ^{2}H NMR spectra (quadrupole echo, partially relaxed spectra,   T = 140 K) for benzene-d $_6$/CPZ. The correlation time for rotation around channel axis $_{CH}$ is 8.5×10^{-7} s. Other parameters are given in the text

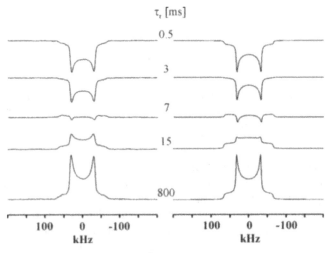

Fig. 14.   Experimental and theoretical ^{2}H NMR spectra (inversion recovery, partially relaxed spectra,   T = 60 K) for benzene-d $_6$/CPZ. The correlation time for rotation around channel axis $_{C6}$ is 2.2×10^{-10} s. Other parameters are given in the text

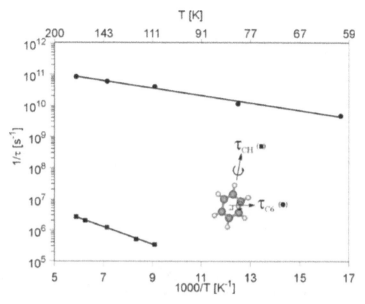

Fig. 15.   Arrhenius plot for the correlation times of C_6 rotation ($_{C6}$, circles) and rotation around the CPZ channel axis ($_{CH}$, squares) in benzene-d$_6$/CPZ

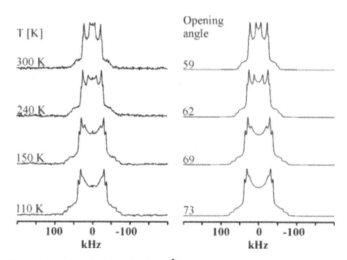

Fig. 16.   Experimental and theoretical ^2H NMR spectra (quadrupole echo, $_e =$ 20 µs) for pyridine-d$_5$/CPZ at di erent temperatures. The relevant simulation para-meters are given in the Figure and in the text

NMR spectra of the pyridine guests as well as the corresponding partially relaxed spectra (inversion recovery experiment), given in Figs. 16 and 17, can thus be reproduced by the assumption that the molecules undergo a fast rotation on a cone with on opening angle between 59   and 73 (see Fig. 18).

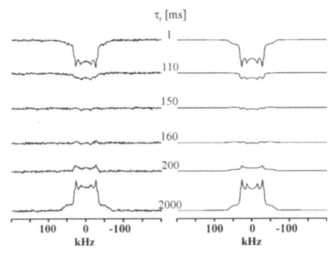

Fig. 17.   Experimental and theoretical ^{2}H NMR spectra (inversion recovery, partially relaxed spectra,   T = 210 K) for pyridine-d $_5$/CPZ. The correlation time for rotation on a cone $_{CH}$ is 1.1×10^{-12} s. Other parameters are given in the text

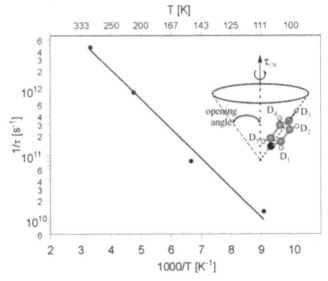

Fig. 18.   Arrhenius plot for the derived correlation times $_{CH}$ for rotation on a cone of pyridine-d $_5$ in the CPZ channels

As a result, the ^{2}H NMR spectra can be understood as a superposition of three subspectra due to magnetically non-equivalent deuterons (i.e., sub-spectra from deuterons 1, 5; deuterons 2, 4, and deuteron 3). The analysis of the spin-lattice relaxation data provided the correlation times for this motional process that are summarized in Fig. 18. As before, a relatively low activation energy of $8.7\,kJ\,mol^{-1}$ is found for the overall pyridine rotation, which, however, is consistent with the above results for the benzene guests and the published data on related compounds [39, 40, 46, 47].

6 Conclusions

In the present work the basics of dynamic NMR spectroscopy were briefly reviewed. For the case of dynamic ^{2}H NMR spectroscopy it has been shown that such techniques are very sensitive to motional processes that can occur on quite different time-scales. A comprehensive analysis of the experimental data on the basis of appropriate simulation programs can provide a very detailed picture about the motional characteristics and ordering features of quite different (motionally) disordered solids. As an example, results from the application of dynamic ^{2}H NMR spectroscopy during the characterization of the guest species in cyclophosphazene inclusion compounds were reported.

Acknowledgement

The authors would like to thank the Deutsche Forschungsgemeinschaft and the Fonds der Chemischen Industrie (FCI) for financial support.

References

1. J.I. Kaplan, G. Fraenkel: NMR of Chemically Exchanging Systems   (Academic Press, New York 1980)
2. J. Sandström: Dynamic NMR Spectroscopy   (Academic Press, London 1982)
3. R. Tycko (ed.):   Nuclear Magnetic Resonance Probes of Molecular Dynamics (Kluwer, Dordrecht 1994)
4. H.W. Spiess: Adv. Polym. Sci., 1985,   66 , 23
5. K. Müller, K.-H. Wassmer, G. Kothe: Adv. Polym. Sci.   95 , 1, (1990)
6. G.R. Luckhurst, C.A. Veracini: The Molecular Dynamics of Liquid Crystals (Kluwer, Dordrecht 1989)
7. R. Dong: Nuclear Magnetic Resonance of Liquid Crystals   (Springer, Berlin 1994)
8. R.G. Griffin: Methods Enzymol.   72 , 108, (1981)
9. J.H. Davis: Biochim. Biophys. Acta   737 , 117, (1983)
10. J. Ripmeester in Inclusion Compounds, Eds. J.L. Atwood, J.E.D. Davies, D.D. MacNicol: Oxford University Press, 1991; Vol.5, p 37
11. R.R. Vold: in NMR Probes of Molecular Dynamics, Ed. R. Tycko, Kluwer, Dordrecht, 1994, p 27
12. R.R. Ernst, G. Bodenhausen, A. Wokaun: Principles of Nuclear Magnetic Resonance in One and Two Dimensions   (Clarendon, Oxford 1987)
13. K. Schmidt-Rohr, H.W. Spiess: Multidimensional Solid-State NMR and Polymers (Academic Press, London 1994)

14. R.R. Vold, R.L. Vold: Adv. Magn. Opt. Res. 16 (1991) 85
15. C.A. Fyfe: Solid State NMR for Chemists   (CFC Press, Guelph 1983)
16. J.I. Kaplan: J. Chem. Phys.   28 , 278, (1958); 29 , 462, (1958)
17. S. Alexander: J. Chem. Phys.   37 , 967, (1962)
18. J. Jeener: Adv. Magn. Reson.   10 , 1, (1982)
19. M. Bak, J.T. Rasmussen, N.C. Nielsen: J. Magn. Reson.   147 , 296, (2000)
20. D.M. Brink, G.R. Satchler: Angular Momentum   (Clarendon, Oxford 1975)
21. H.W. Spiess in NMR, Basic, Principles and Progress, Eds. P. Diehl, E. Fluck,
    R. Kosfeld, Springer-Verlag, Berlin, 1978, Vol. 15, p 55
22. K. M¨uller, P. Meier, G. Kothe: Progr. Nucl. Magn. Reson. Spectrosc.   17 , 211,
    (1985)
23. H.W. Spiess, H. Sillescu: J. Magn. Reson.   42 , 381 (1981)
24. R.J. Wittebort, E.T. Olejniczak, R.G. Gri n: J. Chem. Phys.   86 , 5411, (1987)
25. A.J. Vega, Z. Luz: J. Chem. Phys.   86 , 1803, (1987)
26. M.S. Greenfield, A.D. Ronemus, R.L. Vold, R.R. Vold, P.D. Ellis, T.E. Raidy:
    J. Magn. Reson.   72 , 89, (1987)
27. C. Schmidt, B. Bl¨ umich, H.W. Spiess: J. Magn. Reson.   79 , 269, (1988)
28. Sample files are available from the authors
29. Mathcad ^{TM} , Mathsoft Engineering & Education, Inc., Cambridge, MA
30. MATLAB ^{TM} , The MathWorks, Inc., Natick, MA
31. M. Bloom, J.H. Davis, A.L. MacKay: Chem. Phys. Lett.   80 , 198, (1981)
32. R.K. Wangsness, F. Bloch: Phys. Rev.   89 , 728, (1953)
33. A.G. Redfield: Adv. Magn. Reson.   1 , 1, (1965)
34. A.G. Redfield: IBM J. Res. Develop.   1 , 19, (1953)
35. R.J. Wittebort, A. Szabo: J. Chem. Phys.   69 , 1722, (1978)
36. D.A. Torchia, A. Szabo: J. Magn. Reson.   42 , 107, (1982)
37. C. Boe el, Z. Luz, R. Poupko, A.J. Vega: Isr. J. Chem.   28 , 283, (1988)
38. B. Bl¨umich, H.W. Spiess: Angew. Chem.   100 , 1716, (1988)
39. J. Schmider, K. M¨ uller: J. Phys. Chem. A   102 , 1181, (1998)
40. A. Liebelt, A. Detken, K. M¨ uller: J. Phys. Chem. B   106 , 7781, (2002)
41. K. M¨uller: Phys. Chem. Chem. Phys.   4 , 5515, (2002)
42. B.T. Smith, J.M Boyle, B.S. Garbow, Y. Ikebe, V.C. Klema, C.B. Moler: Matrix
    Eigensystem Routines – EISPACK Guide   (Springer, Berlin 1976)
43. Further information about the simulation programs are available from the au-
    thors
44. H.R. Allcock, in Inclusion Compounds, (J.L. Atwood, J.E.D. Davies, D.D. Mac-
    Nicol, Eds.) Academic Press, New York (1984), Vol. 1, p 351
45. E. Meirovitch, S.B. Rananavare, J.H. Freed: J. Phys. Chem.   91 , 5014, (1987)
46. A. Liebelt, K. M¨ uller: Mol. Cryst. Liq. Cryst.   313 , 145, (1998)
47. J. Villanueva-Garibay, K. M¨ uller: J. Phys. Chem. B   108 , 15057, (2004)

En Route to Solid State Spin Quantum Computing

M. Mehring, J. Mende and W. Scherer

Physikalisches Institut, University Stuttgart, 70550 Stuttgart, Germany
m.mehring@physik.uni-stuttgart.de

Abstract.   We present routes to quantum information processing in solids. An introduction to electron and nuclear spins as quantum bits (qubits) is given and basic quantum algorithms are discussed. In particular we focus on the preparation of pseudo pure states and pseudo entangled states in solid systems of combined electron and nuclear spins. As an example we demonstrate the Deutsch algorithm of quantum computing in an S-bus system with one electron spin coupled to a many ^{19}F nuclear spins.

1 Brief Introduction to Quantum Algorithms

It was Feynman [1] who suggested more than twenty years ago to use quantum algorithms to simulate physical phenomena. A few years later Deutsch proposed the concept of a quantum computer [2, 3]. This initiated some exciting new ideas leading to quantum cryptography [4, 5] and quantum teleportation [6, 7]. This new area of science was stimulated enormously by the proposal of fast searching algorithms by Grover [8] and its NMR implementation by Chuang et al. [9] and more so by the quantum factoring algorithm proposed by Shor [10] which was implemented recently with liquid state NMR by Vandersypen et al. [11]. These quantum algorithms demonstrated the impressive parallelism of quantum computation which could speed up calculations tremendously well beyond classical computing.

A number of NMR experiments in the liquid state have demonstrated the concept of spin quantum computing [12, 13, 14, 15, 16, 17]. For an introduction to these concepts see [18]. After these initial experiments a number of other liquid NMR realizations have been published. It is beyond this overview to reference all these. A critical account on the quantum nature of these experiments can be found in [19].

In the following sections we want to summarize our initial steps for performing quantum computing with electron and nuclear spins in crystalline solids. The early proposal to perform solid state spin quantum computing by

M. Mehring et al.:   En Route to Solid State Spin Quantum Computing   , Lect. Notes Phys.   **684** ,
87–113 (2006)
www.springerlink.com   c Springer-Verlag Berlin Heidelberg 2006

Kane [20] is based on the nuclear spin of phosphorous in silicon. Here we utilize the combined states of the electron spin and nuclei in solids. We briefly introduce the concept of quantum gates and quantum algorithms. Next we consider the quantum states of an electron spin $S = 1/2$ coupled to a nuclear spin $I = 1/2$ and how basic quantum gates can be realized with such a spin system. As a speciality we treat the case of an electron spin $S = 3/2$ coupled to a nuclear spin $I = 1/2$. Finally we discuss the new S-Bus Concept, where an electron spin $S = 1/2$ couples to many nuclear spins $I_j = 1/2$ which allows to create multi spin correlated and entangled states. Moreover we present experiments with a qubyte+1 nuclear spin system in CaF_2:Ce in context of the S-Bus concept.

1.1 Basic Quantum Gates

Quantum gates are the building blocks for quantum computation. A quantum bit (qubit) is represented not only by the two binary states 0 and 1 like a classical bit, but by the whole two-dimensional Hilbert space representing the wavefunction

$$= c_1|0 + c_2|1 , \tag{1}$$

with complex numbers c_1 and c_2 obeying the condition $c_1 c_1 + c_2 c_2 = 1$. Here we have used the qubit basis states $|0$ and $|1$. Considering a spin 1/2 as the ideal qubit, we will often use the notation $| = |+ = |0$ and $| = |- = |1$ here.

The one bit gate which can easily be implemented as a quantum gate, is the NOT gate (see Fig. 1).

$$a \longrightarrow \times \longrightarrow \bar{a}$$

Fig. 1.   NOT gate

It simply inverts an arbitrary bit a. The corresponding unitary transformation can be represented in matrix form as

$$U_{NOT} = \begin{matrix} 0 & 1 \\ 1 & 0 \end{matrix} \quad \text{with basis states} \quad |0 = \begin{matrix} 1 \\ 0 \end{matrix} \quad \text{and} \quad |1 = \begin{matrix} 0 \\ 1 \end{matrix} . \tag{2}$$

The bit flip operation is readily implemented in a qubit system by applying the unitary transformation $P_y( ) = e^{-i \, y}$ which performs essentially the same operation as U_{NOT}. The bit flip can be considered as a classical operation.

An important aspect of qubits is the fact that the superposition of quantum states can be exploited. This is achieved by the Hadamard transformation

$$H = \frac{1}{\sqrt{2}} \begin{matrix} 1 & 1 \\ 1 & -1 \end{matrix} . \tag{3}$$

If applied to the state $|+\rangle = |0\rangle$ it transforms it to the superposition state $(|+\rangle + |-\rangle)/\sqrt{2} = (|0\rangle + |1\rangle)/\sqrt{2}$. This corresponds to the transverse components in the xy-plane in magnetic resonance. In magnetic resonance this is usually achieved by a $\pi/2$-pulse. However, the unitary transformation $P_y(\pi/2)$ does not correspond exactly to the Hadamard transform, but can be used instead if one obeys the fact that the Hadamard transform corresponds to its own inverse, whereas $P_y(-\pi/2)$ must be applied if the inverse operation is required. For completeness we mention that the Hadamard transform can be implemented by the composite pulse $P_y(-\pi/4)P_x(\pi)P_y(\pi/4)$.

A little more advanced is the two qubit CNOT (Controlled NOT) operation, sketched as a block diagram in Fig. 2.

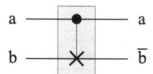

Fig. 2.   CNOT gate (see text)

It requires two qubits, namely qubit a which is called the control bit and qubit b which is the target bit. The target bit b is inverted only if the control bit a is in a particular state. The most common case is where qubit b is inverted if the control bit $a = 1$. The control bit a stays unchanged in this operation. The matrix representation of the CNOT operation is given by

$$U_{\text{CNOT}} = \begin{matrix} 1\,0\,0\,0 \\ 0\,1\,0\,0 \\ 0\,0\,0\,1 \\ 0\,0\,1\,0 \end{matrix} \quad . \tag{4}$$

These gates are reversible. In order to demonstrate the application of these quantum gates we discuss the preparation of the fundamental Einstein-Podolsky-Rosen (E-P-R) state $(|+-\rangle - |-+\rangle)/\sqrt{2} = (|01\rangle - |10\rangle)/\sqrt{2}$ which is at the heart of quantum mechanics, by starting from the two qubit product state $|--\rangle = |11\rangle$ [21]. The E-P-R state results by applying a Hadamard transformation to the first qubit followed by a CNOT gate (Fig. 3):

$$|11\rangle \xrightarrow{H} \frac{1}{\sqrt{2}}(|01\rangle - |11\rangle) \xrightarrow{\text{CNOT}} \frac{1}{\sqrt{2}}(|01\rangle - |10\rangle) . \tag{5}$$

It corresponds to a superposition state which intimately involves both qubits and cannot be expressed as a product state of qubits one and two. The state of each individual qubit is undetermined.

All four possible entangled states of a two qubit system are called Bell states:

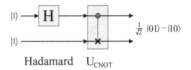

Fig. 3. Quantum gate for creating an entangled state

$$^\pm = \frac{1}{2}(|01 \pm| 10) \text{ and } ^\pm = \frac{1}{2}(|00 \pm| 11) . \tag{6}$$

Superposition and entanglement of qubits are the essential ingredients of quantum computing. They can be obtained from simple product states by the unitary transformations corresponding to the quantum gates NOT, H and CNOT and their combinations.

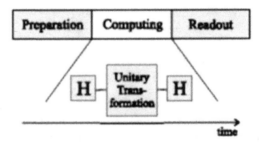

Fig. 4. General block diagram for quantum computing

The basic block diagram for quantum computing, as sketched in Fig. 4, comprises the preparation of a particular initial quantum state, a series of unitary transformations, representing the quantum algorithm and finally the detection of the outcome. We will exemplify this in the following for the Deutsch-Jozsa algorithm. For a summary on quantum gates and liquid state NMR applications see [18, 22].

1.2 The Deutsch-Jozsa Algorithm

The Deutsch-Jozsa (DJ) algorithm is a quantum algorithm which evaluates a binary function and decides if the function is   constant or balanced [2]. Let us consider a function f (ab) of two qubits a and b where the function returns only a single bit state. The following table lists some of the $2^4 = 16$ possible functions.

Note that the first two functions are constant. Their value is either 0 or 1 independent of the variable   ab. The other functions are called balanced, because their values represent an equal number of 0 and 1. There are six of those. Since the total number of possible functions is 16 there must be eight more functions which are neither constant nor balanced. Their values correspond to an odd number of 0 or 1.

In the context of the Deutsch-Jozsa algorithm one is only interested in distinguishing the constant and balanced functions. In order to implement the DJ algorithm one needs to represent the functions   f (ab) by unitary transformations. In Table 1 we have therefore included the label of the corresponding unitary transformations. They are constructed such that on the diagonal every 0 is represented by 1 and every 1 by -1. An example for U_{0101} is given in (7)

$$U_{0101} = \begin{matrix} 1 & 0 & 0 & 0 \\ 0 & -1 & 0 & 0 \\ 0 & 0 & 1 & 0 \\ 0 & 0 & 0 & -1 \end{matrix} . \tag{7}$$

Table 1.   Function f (ab)

f (ab)	00	01	10	11	
U_{0000}	0	0	0	0	constant
U_{1111}	1	1	1	1	constant
U_{0101}	0	1	0	1	balanced
U_{0011}	0	0	1	1	balanced
U_{1001}	1	0	0	1	balanced
U_{1010}	1	0	1	0	balanced
U_{1100}	1	1	0	0	balanced
U_{0110}	0	1	1	0	balanced

The block diagram of the DJ algorithm is presented in Fig. 5. Note that the output of qubit 1 can be 0 or 1. The DJ algorithm is constructed such that a constant function gives 0 and a balanced function gives 1. This quantum algorithm evaluates the functions in a single computational step, whereas the classical computer must evaluate each function separately in order to decide if the function is constant or balanced. Other quantum algorithms have been formulated and were demonstrated partially by NMR quantum computing. Space does not allow to go into details here.

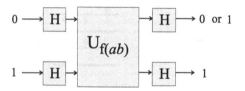

Fig. 5.   Block diagram for the DJ algorithm

In Sect. 5.6 we present the implementation of this algorithm within the S-Bus Concept in a single crystal of CaF$_2$:Ce.

2 Combined Electron Nuclear Spin States in Solids

Here we consider the situation, where a number of nuclei I_j are connected via hyperfine interaction to an electron spin S. In the context of quantum computing we have labelled this an S-Bus system to be discussed in more detail in Sect. 5.

2.1 Quantum States

The total Hamiltonian of the S-Bus system can be expressed as

$$H_{tot} = \omega_S S_z + \omega_I I_z + S_z \sum_{j=1}^{N} a_j I_{zj} + \sum_{j=k} D_{jk} I_{zj} I_{zk} , \qquad (8)$$

with the Larmor frequencies $\omega_S = g\mu_B B_0/\hbar$ and $\omega_I = -\gamma_I B_0$. In order to keep only the diagonal terms of the Hamiltonian we applied the approximation $\omega_S \omega_I > a_j D_{jk}$, where the absolute values of these parameters are considered in these inequalities. O-diagonal terms may exist, but will be rather small in most cases. The energy spectrum is schematically sketched in Fig. 6.

Representative electron spin resonance (ESR) transitions ($\Delta m_S = \pm 1$, $\Delta m_I = 0$) are indicated by solid lines and some nuclear spin transitions

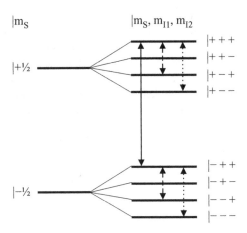

Fig. 6.   Schematic energy level scheme of two nuclear spins $I_{1,2} = 1/2$ coupled to an electron spin $S = 1/2$. Solid lines : ESR transitions.   Dashed lines : ENDOR transitions.   Dotted lines : Entangled states (see text))

($m_S = 0$, $m_I = \pm 1$) are drawn as dashed lines. Experiments are performed by applying microwave pulses at the electron spin trans   itions combined with pulsed irradiation in the radio frequency spectrum at the di erent nuclear transitions. Direct detection of the nuclear spin transitions is exceedingly difficult because of the small gyromagnetic ratio and most of all because of the low concentration and the low thermal polarisation. The nuclear transitions and coherences can, however, be observed indirectly by monitoring the electron spin signal while irradiating at the NMR transition. This type of double resonance is called Electron Nuclear Double Resonance (ENDOR).

2.2 Equilibrium Density Matrix

The initial state of a real physical spin system will correspond to an equilibrium state which is usually assumed to be the Boltzmann state. The Boltzmann equilibrium density matrix for the S-Bus system is defined as

$$
_B = \frac{e^{-\ H_{tot}}}{Tr(e^{-\ H_{tot}})} , \tag{9}
$$

where $= /k_B T$ and $Tr(\ A)$ implies taking the trace (sum o diagonal elements) of the corresponding matrix   A. This is the typical Boltzmann density matrix of an ensemble (large number) of spin clusters consisting of a single   S spin coupled to N nuclear spins I as expressed by the Hamiltonian according to (8).

Because of the dominance of $_S$ over all other interactions in the electron nuclear spin system discussed here, we assume for simplicity that the Boltzmann density matrix is represented by the electron Zeeman term $_S S_z$ which leads for an S spin 1/2 and arbitrary spins   I to

$$
_{BS} = \frac{1}{2(2I+1)^N} (I_0 - 2K_B S_z) \quad \text{where} \quad K_B = \tanh\ \frac{1}{2}\ _S , \tag{10}
$$

with $0 \quad K_B \quad 1$ and where I_0 is the $2(2I+1)^N \times 2(2I+1)^N$ identity matrix. We note that under this approximation there is no nuclear spin polarization or correlation whatsoever. Since we have no control over the identity matrix in the density matrix expression of (10) we ignore it usually in magnetic resonance and deal with the truncated equilibrium density matrix $_B = -K_B S_z$. In the high temperature low field approximation K_B is rather small. This is the usual case in magnetic resonance (MR) experiments.

In order to prepare for the concept of pseudopure states, first introduced by Cory and co-workers [12, 13], we use some freedom to rearrange the expression for the Boltzmann density matrix (10) in the following way.

$$
_{BS} = \frac{1}{2(2I+1)^N}\ 1 - \frac{K_B}{K}\ I_0 + \frac{K_B}{K}\ _{PB} , \tag{11}
$$

which defines the   pseudo Boltzmann density matrix

$$\rho_{PB} = \frac{1}{2(2I + 1)^N} (I_0 - 2KS_z) . \tag{12}$$

Note that the Boltzmann density matrix as expressed by (12) is still exact with arbitrary parameters $K > 0$. This allows us to express the pseudo Boltzmann density matrix ρ_{PB} at will. The idea behind this is to manipulate the operator part $2KS_z$ in ρ_{PB} in such a way that it is converted into a density matrix which has the same operator structure as a pure state. Let us consider the simplest possible case, namely spin $I = 0$ and $K = 1$. This would convert ρ_{PB} into

$$\rho_1 = \begin{array}{cc} 0 & 0 \\ 0 & 1 \end{array} , \tag{13}$$

which clearly would represent a density matrix of the pure state $|1\rangle$. Nevertheless, it is still nothing but the Boltzmann density matrix at temperature $T > 0$ and as such represents a mixed state. The usefulness of the pseudo pure density matrix becomes more obvious for the case, when we manage to convert, by some means of manipulations, $-KS_z$ into $S_z + I_z + 2S_zI_z$ for two spins S and I. This leads to the pseudopure density matrix

$$\rho_{PB,00} = |00\rangle\langle 00| = \frac{1}{4}I_0 + \frac{1}{2}(S_z + I_z) + S_zI_z = \begin{array}{cccc} 1 & 0 & 0 & 0 \\ 0 & 0 & 0 & 0 \\ 0 & 0 & 0 & 0 \\ 0 & 0 & 0 & 0 \end{array} . \tag{14}$$

This and related types of pseudopure density matrices we will use in the following as initial states. We note that such a pseudopure density matrix requires to introduce a correlation between the spins, represented by the operator product S_zI_z. The relevance of the spin correlation and pseudopure states for liquid state NMR quantum computing was discussed by Warren et al. [19] and Cory et al.[12, 13, 23].

3 Entanglement of an Electron and a Nuclear Spin $\frac{1}{2}$

Typically one discusses the entanglement between spins 1/2 of the same type, like either electrons or protons. Although quantum algorithms have been formulated independent of the type of qubit system, the entanglement of an electron and a nucleus is somewhat exotic because of their very different properties like coupling to external fields and their strong hyperfine interaction. In this section we will demonstrate the pseudo entanglement between and electron spin 1/2 and a nucleus with spin 1/2 namely a proton and in a separate section between an electron spin 3/2 and a ^{15}N nucleus with spin 1/2.

What we are aiming at are Bell states introduced in (6) and represented here by spin symbols, where the first spin represents the electron spin

$$\Phi_\pm = \frac{1}{\sqrt{2}}(|\uparrow\uparrow\rangle \pm |\downarrow\downarrow\rangle) \quad \text{and} \quad \Psi_\pm = \frac{1}{\sqrt{2}}(|\uparrow\downarrow\rangle \pm |\downarrow\uparrow\rangle) . \tag{15}$$

They correspond to a superposition of the states in Fig. 8 connected by dotted arrows.

3.1 An Electron Nuclear Spin Pair in a Single Crystal

This section is based on a recent publication observing pseudo entanglement between an electron spin 1/2 and a nuclear spin 1/2 in a crystalline solid [24]. Let us consider the intensively studied malonic acid radical in a single crystal. The basic molecular unit is sketched in Fig. 7. The electron spin density extends over whole molecular unit and beyond. The strongest hyperfine coupling is to the adjacent proton to the central carbon which carries most of the spin density. The hyperfine interaction is highly anisotropic. We consider here a special orientation of the magnetic field where we can to a good approximation describe the Hamiltonian and the corresponding energy levels of this spin pair by (8) with $a < 0$. The corresponding energy level scheme is sketched in Fig. 8.

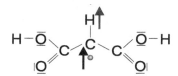

Fig. 7.   The malonic acid radical

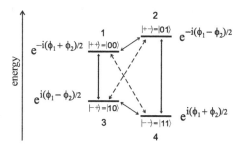

Fig. 8.   Four level scheme of a two qubit system

We have doubly labelled quantum states in Fig. 8 according to their spin orientation and the qubit terminology. We have also included the phase rotation property of each individual state under the operation $e^{-i\, _1S_z}\, e^{-i\, _2I_z}$. This will turn out to be useful when we discuss the tomography of the entangled states. The corresponding spectrum consists of four allowed transitions two at the ESR frequency ($m_S = \pm 1$, $m_I = 0$) and two at NMR frequencies

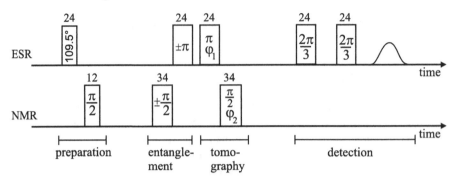

Fig. 9.   Complete pulse sequence for creating an entangled state of a two qubit
system by transition selective excitation

($m_S = 0$, $m_I = \pm 1$). All four transitions are well resolved in the spectra
of the malonic acid radical.

The complete pulse sequence for the preparation of the pseudopure initial
states, the creation of entanglement which applies a CNOT (controlled NOT)
operation and the density matrix tomography is shown in Fig. 9. The sequence
ends with the detection of the spin echo sequence for the electron spin.

The different building blocks of this sequence will be discussed in the
following sections.

3.2 Creating Pseudopure States by Selective Excitation

Due to the well resolved transitions we can apply   transition selective   excitations in contrast to most liquid state NMR quantum computing experiments,
where spin selective excitations have been performed. In order to prepare the
pseudopure state

$$\rho_{10} = |10\rangle\langle 10| = \frac{1}{4}I_0 - \frac{1}{2}S_z + \frac{1}{2}I_z - S_z I_z = \begin{matrix} 0\;0\;0\;0 \\ 0\;0\;0\;0 \\ 0\;0\;1\;0 \\ 0\;0\;0\;0 \end{matrix} \qquad (16)$$

we need to convert $-KS_z$ in (12) into $-S_z + I_z - 2S_z I_z$ which is readily
achieved by the pulse sequence

$$P_y^{(24)}(\arccos(-1/3)) - \tau_1 \; P_y^{(12)}(\pi/2) - \tau_2 \; \rho_{10}, \qquad (17)$$

with $K = 3/4$. This pulse sequence is sketched in Fig. 9 in the prepara-
tion segment. Here we use the notation $P_y^{(jk)}(\varphi)$ for a   -pulse in y-direction
at the transition   j   k   which corresponds to a unitary transformation
$U_y^{(jk)}(\varphi) = \exp(-i\,\varphi\, I_y^{(jk)})$. We will use this definition for   x- and y-pulses
throughout this article. The whole process of creating the pseudopure state

is, however, not a unitary process, because we eliminate after each pulse the o-diagonal components of the density matrix by allowing for   decoherence times $_1$ and $_2$ after pulses. In a similar way we prepare the pseudopure density matrix

$$_{11} = |11\rangle\langle 11| = \frac{1}{4}I_0 - \frac{1}{2}(S_z + I_z) + S_zI_z = \begin{array}{cccc} 0 & 0 & 0 & 0 \\ 0 & 0 & 0 & 0 \\ 0 & 0 & 0 & 0 \\ 0 & 0 & 0 & 1 \end{array} . \tag{18}$$

3.3 Tomography of the Pseudopure States

An elegant way of performing a spin density matrix tomography was introduced by Madi et al. [25] by utilizing the concept of two-dimensional NMR spectroscopy. A more recent account of density matrix tomography applied after a quantum algorithm in liquid state NMR quantum computing can be found in [26, 27]. Here we apply a di erent and simpler approach because ESR lines in solids are rather broad (inhomogeneous broadening) and moreover we can address every transition selectively.

In order to prove that we have prepared the wanted pseudopure state we need to perform a density matrix tomography. Because of the decoherence times we need to determine only the four diagonal elements of the density matrix. Due to the normalized trace of the density matrix we need to determine only three di erent parameters. This is readily obtained by measuring the population di erences $p_1 - p_2$, $p_1 - p_3$ and $p_3 - p_4$. This we could obtain by measuring the amplitudes of the Rabi precession of the corresponding transitions. This requires a proper calibration in particular since ESR and NMR transitions are to be compared. By virtue of relating all Rabi precessions to the change in the electron spin echo amplitude we were able to determine all parameters with reasonable precision. A matrix representation of the experimentally determined $_{10}$ gives

$$_{10} = \begin{array}{cccc} 0.01 & 0 & 0 & 0 \\ 0 & -0.06 & 0 & 0 \\ 0 & 0 & 1.02 & 0 \\ 0 & 0 & 0 & 0.03 \end{array} . \tag{19}$$

Similar results were obtained for $_{11}$.

3.4 Entangled States

For this spin system we have prepared all of the four Bell states according to (6, 15). The pulse sequence for creating the entangled states from the prepared pseudopure state comprises a   / 2-pulse at a selective NMR transition,

replacing the standard Hadamard transformation, followed immediately by   -
pulse at a selective ESR transition which corresponds to a CNOT operation
as is included in the pulse sequence Fig. 9.

This corresponds exactly to the quantum algorithm for creating entangled
states as presented in Sect. 1.1. The question arises how do we know that
we have indeed created an entangled state. In order to prove this we need
to perform a density matrix tomography. Before we discuss this procedure
and the results obtained we first take a look at the phase dependence of the
entangled states.

When applying the phase rotation operator exp(− i($_1 S_z$ + $_2 I_z$)) to the
E-P-R state, which corresponds to a phase rotation about the   z-axis, one
observes the relation

$$ -(_1, \ _2) = \frac{1}{2}(e^{-i\frac{1}{2}(_1 - \ _2)}| + - - \ e^{i\frac{1}{2}(_1 - \ _2)}| - +), \qquad (20)$$

$$ +(_1, \ _2) = \frac{1}{2}(e^{-i\frac{1}{2}(_1 + \ _2)}| + + \ + e^{i\frac{1}{2}(_1 + \ _2)}| - -), \qquad (21)$$

which identifies this entangled state through its phase di erence. By incre-
menting both phases $_1$ and $_2$ in steps phase interferograms like the one
shown in Fig. 10 are obtained. The phase increments can be related to a frequency which is defined as $_j$ = 2 $_j$ t . The phase interferograms shown
in Fig. 10 therefore correspond to the sum $_1$ + $_2$ (Fig. 10 (bottom)) and
di erence $_1$ − $_2$ (Fig. 10 (top)) of the applied phases. After Fourier transformation of the phase interferograms a spectrum is displayed as is shown
in Fig. 11, with lines appearing at particular frequencies. Here we used the
following individual frequencies: spin 1: $_1$ = 2 .0 MHz; spin 2: $_2$ = 1 .5 MHz.

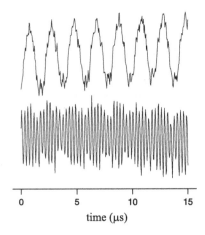

Fig. 10.   Phase interferograms,   top : $^-$ state, bottom : $^+$ state

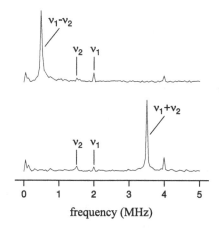

Fig. 11.   Fourier transform of the phase interferograms of Fig. 10

We expect to see a line at these frequencies for the superposition states of the individual spins. For the entangled states we expect to see lines at $\nu_1 \pm \nu_2$ depending on which of the Bell states were created. One clearly observes these features in Fig. 11. The upper spectrum represents the Ψ_- state, whereas the lower spectrum corresponds to the state Ψ_+. The characteristic frequencies $\nu_1 \pm \nu_2 = 2.0 \pm 1.5$ MHz are clearly identified as the dominant lines. The lines appearing at ν_1 and ν_2 are clearly contaminations of unwanted superposition states.

Density Matrix Tomography

The tomography of the entangled states requires to determine the diagonal part and the off-diagonal parts of the density matrix. The diagonal part is obtained in a similar way as already discussed in the case of the pseudop-ure density matrix by measuring the Rabi precession of the different allowed transitions and obtain from their amplitudes the diagonal elements. The off-diagonal elements are obtained from the phase rotation and the corresponding spectral amplitudes discussed in the previous section.
   The numerical values obtained by the tomography procedure are

$$
\Psi_+ = \begin{pmatrix}
0.49 & 0.00 & 0.00 & 0.49 \\
0.00 & -0.03 & 0.00 & 0 \\
0.00 & 0.00 & 0.02 & 0.00 \\
0.49 & 0 & 0.00 & 0.52
\end{pmatrix}, \tag{22}
$$

where the off-diagonal elements labelled 0.00 correspond to values 0.00 ± 0.05. The label 0 refers to values which could not be measured, but are expected to be small, because no excitation was performed at that transition. In a similar way we obtained Ψ_-. The following data were obtained for the density matrix

$$- = \begin{pmatrix} -0.02 & 0.00 & 0 & 0.00 \\ 0.00 & 0.55 & -0.47 & 0.00 \\ 0 & -0.47 & 0.50 & 0.00 \\ 0.00 & 0.00 & 0.00 & -0.03 \end{pmatrix} . \tag{23}$$

More details on this subject can be found in [28].

4 Entangling an Electron Spin $\frac{3}{2}$ with a Nuclear Spin $\frac{1}{2}$

ESR and ENDOR spectra of the molecule N@C$_{60}$ were first reported in [29]. The electron spin of the nitrogen atom is $S = 3/2$, whereas the ^{15}N nucleus has spin $I = 1/2$. There have been a number of proposals how one could use ^{15}N@C$_{60}$ in order to perform quantum computing [30, 31, 32]. However, these proposals neither addressed the problem of what is the relevant qubit in this system and how, realistically a quantum algorithm could be performed. In a recent publication we have defined the relevant qubit and performed a CNOT operation leading to entanglement [33]. We have also performed a density matrix tomography in order to evaluate the entangled state.

The corresponding Hamiltonian of ^{15}N@C$_{60}$ can be expressed in first order as

$$H = ({}_S S_z + {}_I I_z + a S_z I_z) , \tag{24}$$

with energy eigenvalues

$$E (m_S, m_I) = ({}_S m_S + {}_I m_I + a m_S m_I) . \tag{25}$$

According to the negative g-factor and the negative   of ^{15}N $_S$ $_I > 0$. Furthermore the approximation $|a|, $ $_I$ $_S$ has been applied in (24).

The eigenstates $|m_S m_I$ are labelled here

$$\{ |1 , |2 , \cdots |8 \} = \frac{3}{2}, \frac{1}{2} , \frac{3}{2}, -\frac{1}{2} , \cdots -\frac{3}{2}, -\frac{1}{2} . \tag{26}$$

The energy levels including the quantum states are shown in Fig. 12. The energy level structure can be considered as separated into two four level systems, each corresponding to an electron spin 3/2, where the levels $|1 , |3 , |5 , |7$ correspond to the nuclear spin quantum number $m_I = +1/2$, whereas the levels $|2 , |4 , |6 , |8$ correspond to the nuclear spin quantum number $m_I = -1/2$. We introduce two fictitious spins 3/2 by defining $S^{(m_I)}$ as $S^{(\pm)}$ for $m_I = \pm 1/2$.

In the following we will concentrate on the fictitious two qubit subsystem

$$\{ |ab \} = \frac{3}{2}, \frac{1}{2} , \frac{3}{2}, -\frac{1}{2} , -\frac{3}{2}, +\frac{1}{2} , -\frac{3}{2}, -\frac{1}{2} . \tag{27}$$

Although ESR pulses are applied at the two fictitious spin 3/2 systems, the pulse sequences are tailored such that the wanted state of the fictitious two qubit subsystem is reached.

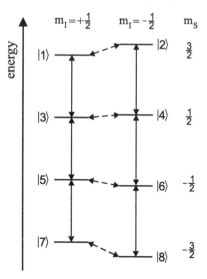

Fig. 12.   Schematic energy level scheme of ^{15}N@C$_{60}$

4.1 Pseudopure States

The preparation of the pseudopure density matrix $_{11}$ of the fictitious two qubit system as defined before is performed by a similar pulse sequence as in the electron spin 1/2 case as

$$ _{BP} \xrightarrow{P_y^{(+)}(_0)} t_1 \xrightarrow{P_y^{(12)}(/2)} t_2 = P_{11}. \tag{28}$$

First a selective $_0$-pulse, with $_0 = \arccos(-1/3)$, is applied at the fictitious spin 3/2 subsystem corresponding to $m_I = +1/2$ followed by the decay time t_1 which allows to decohere the o-diagonal states. After this a $/2$-pulse is applied to the 1 2 transition, followed again by the decay time t_2.

The resulting density matrix is given by

$$ P_{11} = \begin{matrix} 0\,0\,0\,0\,0\,0\,0\,0 \\ 0\,0\,0\,0\,0\,0\,0\,0 \\ 0\,0\,\frac{1}{3}\,0\,0\,0\,0\,0 \\ 0\,0\,0\,0\,0\,0\,0\,0 \\ 0\,0\,0\,0\,\frac{1}{6}\,0\,0\,0 \\ 0\,0\,0\,0\,0\,\frac{1}{2}\,0\,0 \\ 0\,0\,0\,0\,0\,0\,0\,0 \\ 0\,0\,0\,0\,0\,0\,0\,1 \end{matrix}, \tag{29}$$

where we have marked the fictitious two qubit sublevel system as bold face. Extracting from this the density matrix for the fictitious two qubit subsystem leads obviously to

$$\rho_{11} = \begin{array}{cccc} 0\ 0\ 0\ 0 \\ 0\ 0\ 0\ 0 \\ 0\ 0\ 0\ 0 \\ 0\ 0\ 0\ 1 \end{array} . \tag{30}$$

The preparation of the pseudopure density matrix ρ_{10} proceeds in a similar way as described for ρ_{11}. These are used as initial matrices for creating the entangled states of the fictitious two qubit subsystem.

4.2 Entangled States

We consider as entangled states of the fictitious two qubit subsystem the density matrices

$$\rho_{P}^{(\pm)} = \frac{1}{2}|2 \pm 7\rangle\langle 2 \pm 7| \tag{31}$$

and

$$\rho_{P}^{(\pm)} = \frac{1}{2}|1 \pm 8\rangle\langle 1 \pm 8| \tag{32}$$

in analogy to the two qubit Bell states (6, 15). Preparation of these entangled states is achieved by the unitary transformations $U_{\pm}^{(27)} = P_{Sy}^{(-)}(\pi) P_{Iy}^{(78)}(\pi/2)$ and $U_{\pm}^{(18)} = P_{Sy}^{(+)}(\pm\pi) P_{Iy}^{(78)}(\pi/2)$. The following results are obtained.

When extracting the corresponding density matrices of the fictitious two qubit subsystem, as discussed before, we obtain

$$\rho = \begin{array}{cccc} 0 & 0 & 0 & 0 \\ 0 & \frac{1}{2} & \pm\frac{1}{2} & 0 \\ 0 & \pm\frac{1}{2} & \frac{1}{2} & 0 \\ 0 & 0 & 0 & 0 \end{array} \tag{33}$$

and

$$\rho = \begin{array}{cccc} \frac{1}{2} & 0 & 0 & \pm\frac{1}{2} \\ 0 & 0 & 0 & 0 \\ 0 & 0 & 0 & 0 \\ \pm\frac{1}{2} & 0 & 0 & \frac{1}{2} \end{array} , \tag{34}$$

which correspond to the Bell states.

4.3 Density Matrix Tomography

The tomography of the diagonal part of the different density matrices is performed by measuring the amplitude of the Rabi precessions when particular transitions are excited as was discussed in the electron spin 1/2 case. These amplitudes are proportional to the difference of the corresponding diagonal elements.

A typical Rabi precession for two different transitions of the ρ_{10} state is shown in Fig. 13. As expected for this state, the population difference at the

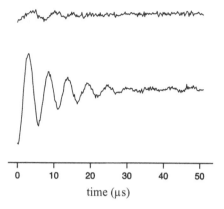

time (μs)

Fig. 13.   Rabi precession of the transitions $1 \rightarrow 2$ (top) and $7 \rightarrow 8$ (bottom) for the $_{10}$ state in ^{15}N@C$_{60}$

$1 \rightarrow 2$ transition is close to zero, whereas the large population difference at the $7 \rightarrow 8$ transition gives rise to a large amplitude as is expected for the $_{10}$ state.

The tomography of the entangled states proceeds via a detection sequence which basically converts the entangled state to an observable state. This will be in general a product state. In order to demonstrate the entanglement the characteristic phase dependence of the particular entangled state should appear in the detection signal.

As was already discussed in the case of the entangled spins $S = 1/2$ and $I = 1/2$ in Sect. 3.4 phase interferograms were obtained as is shown for the E-P-R state $-$ in Fig. 14.

We apply a similar pulse sequence as in the case of the electron spin 1/2, where first an ESR $-$pulse is applied at the $m_I = -1/2$ sublevel system represented by the unitary transformation $P_{Sx}^{(-)}(-, _1)$. Immediately after this a selective ENDOR $/2$-pulse at the $7 \rightarrow 8$ transition follows represented by unitary transformation $P_{Ix}^{(78)}(/2, _2)$. Phase rotation is applied here with frequencies $_1 = 2.0\,\text{MHz}$ and $_2 = 1.5\,\text{MHz}$. It is interesting to note that due to the electron spin $S = 3/2$ we now expect to see a tripled frequency for the ESR transition which is indeed the case as is seen in the interferogram (Fig. 14 middle) as well as in the correponding spectrum. The entangled state, however, of $_{p\pm}$ is expected to result in the detection signal

$$S_\pm(_1, _2) = \text{const}. \pm \frac{3}{20}\cos(3_1 - _2). \tag{35}$$

The phase difference leads to the difference frequency $3_1 - _2$ as seen in Fig. 15 (bottom). In a similar way all the other Bell states were analyzed. More details on this subject can be found in [28, 33].

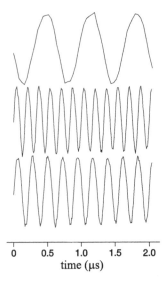

Fig. 14.   Phase interferogram obtained with frequencies $\nu_1 = 2.0$ MHz and $\nu_2 = 1.5$ MHz. Top: Variation of φ_2 ($\varphi_1 = 0$). Middle: Variation of φ_1 ($\varphi_2 = 0$). Bottom: $\varphi_1 = 0$ and $\varphi_2 = 0$

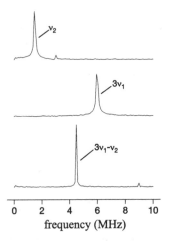

Fig. 15.   Spectrum corresponding to the phase interferogram of Fig. 14 with frequencies $\nu_1 = 2.0$ MHz and $\nu_2 = 1.5$ MHz

5 The S-Bus Concept

The S-Bus concept for spin quantum computing was first presented at the IS-MAR conference (Rhodos, Greece 2001) and is derived from multiple quantum ENDOR (MQE) [34] and was first published elsewhere [22, 35, 36]. It basically consists of a central spin, called   S spin, which acts as a sort of   server which is coupled to a network of nuclear spins, labelled   I spins. Only the S spin is observed similar to a Turing machine, where the   head moves along a tape and only state change of the   head is observed. Details will be laid out in the following sections. In this contribution the   S spin is always an electron spin whereas the I spins are nuclear spins. In this case the advantage of this concept is very pronounced, because the high spin polarization of the electron spin can be used to reach highly polarized and correlated states of the nuclear spins. The principle is, however, rather general and the   S and I spins could be any other spins.

5.1 S-Bus Structure

A typical topology of the S-Bus is presented in Fig. 16. The dominant coupling considered here is the hyperfine coupling a_j between the electron spin   S and the di erent nuclear spins I_j . The coupling constants a_j will in general be di erent for the di erent nuclei. The internuclear interaction will be dipole-dipole interaction in solid samples (considered here) or else scalar couplings in liquid samples. In any case the internuclear interactions are considered weak compared with the hyperfine interactions which are orders of magnitude larger, i.e. a_j, a_k D_{jk} with D_{jk} being the internuclear coupling. We will show in the following that one can prepare highly correlated nuclear spins, just through their interaction with the electron spins   S, even in the limit where $D_{jk} = 0$ and the initial correlation among the nuclear spins is zero.

The topology in Fig. 16 should not suggest that all coupling constants a_j are equal. In general the coupling constants will be di erent and the distance between the nuclei also varies considerably.

5.2 A Multi Qubit Solid State S-Bus System

In the preceding section we have investigated the entanglement between an electron spin and a single nuclear spin. The S-Bus system in its genuine sense

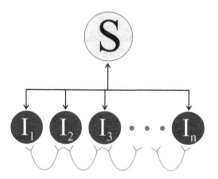

Fig. 16.   Basic S-Bus topology

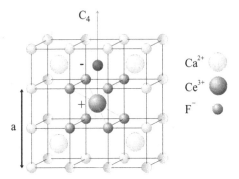

Fig. 17.   A qubyte, consisting of eight ^{19}F nuclei surrounding a Ce $^{3+}$ ion in CaF $_2$ with lattice constant   a = 0 .546 nm

implies a large number of nuclear spins coupled to a single electron spin. As an example we present here a Ce $^{3+}$ with e ective electron spin 1/2 replacing a Ca^{2+} ion in CaF $_2$ as displayed in Fig. 17.

The Ce$^+$ ions represents a fictitious electron spin   S = 1 / 2 with large g-anisotropy due to spin-orbit interaction. The combined orbital and spin states have been discussed in detail in the literature and will not be dwelled on here. This S spin together with the hyperfine coupled eight near neighbor ^{19}F nuclear spins I comprises our qubyte S-Bus system. There is also the charge compensating F $^-$ ion which also shows appreciable hyperfine interaction and can be considered as another qubit. This and the weaker hyperfine couplings to the further distant nuclei will not be considered here. We note that this center possesses C $_4$ symmetry and can be viewed as consisting of two layers of four fluorines, one near the F $^-$ ion (layer 1) and another layer 2 opposite to the F $^-$ ion. Their isotropic hyperfine interaction appears to be di erent, thus rendering the two layers inequivalent. Still all four fluorines of each layer would be magnetically equivalent without anisotropic hyperfine interaction. Due to this anisotropic interaction all eight fluorines become nonequivalent for certain orientations of the magnetic field. This allows to address any of the fluorines individually as is obvious from the related ENDOR spectrum displayed in Fig. 18.

The general pulse sequence for performing quantum computing in the S-Bus system is depicted in Fig. 19. It consists of a series of pulses applied to the electron spin S and individually to each nuclear spin   I$_j$ which takes part in the quantum algorithm. Detection of the final quantum state is through the electron spin echo.

The preparation segment is responsible for creating highly correlated nuclear spin states. The excitation at nuclear spin transitions creates pseudopure initial density matrices, performs quantum algorithms and applies density matrix tomography.

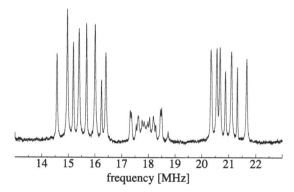

Fig. 18. ENDOR spectrum of the qubyte consisting of the eight near neighbours to the Ce $^{3+}$ S spin

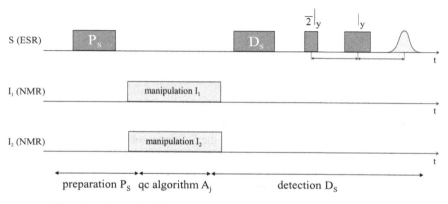

Fig. 19.   General Pulse sequence for S-Bus quantum computing

5.3 Creating Multi Nuclear Correlations

Multi nuclear correlations are readily correlated out of a totally uncorrelated Boltzmann state as represented just by the electron spin Zeeman interaction (10). Simply the application of two   / 2-pulses separated by a free evolution time   as is used in the Mims type pulsed ENDOR [22, 37] su ces to create highly correlated nuclear spin states. More elaborate sequences which filter out certain hyperfine interactions are possible.

In the two pulse sequence one usually applies the same phase (e.g. yy) for the / 2-pulse pair. If we let the residual coherences after the second pulse decay, the diagonal part of the density matrix can be expressed as $_0(\ ) = -S_z\ _I^{(ab)}(\ )$, where ab { yy, yx} and

$$_I^{(yy)}(\ ) = \text{Re} \sum_{j=1}^{N} P_j \qquad \text{with} \qquad P_j = e^{\ ia_j\ I\ _{zj}} \qquad (36)$$

and where a_j is the hyperfine interaction of nucleus j [22, 36]. Note that only even products of spin operators appear in this expression when one expands the factors. In order to get odd numbered operator products one could apply a yx phase for the initial pulse pair in the sequence. In this case the imaginary part of (36) applies

$$\rho_I^{(yx)}(\ ) = \mathrm{Im}\ \prod_{j=1}^{N} P_j \quad . \tag{37}$$

In the special case of $I = 1/2$ one arrives at

$$\rho_I^{(yy)}(\ ) = \mathrm{Re}\ \sum_{j=1}^{N} (c_j I_0 + i2\, I_{zj}\, s_j) \tag{38}$$

and

$$\rho_I^{(yx)}(\ ) = \mathrm{Im}\ \sum_{j=1}^{N} (c_j I_0 + i2\, I_{zj}\, s_j) \quad , \tag{39}$$

where I_0 is the one qubit identity matrix and we have used the abbreviations $c_j = \cos(\tfrac{1}{2} a_j)$ and $s_j = \sin(\tfrac{1}{2} a_j)$.

By this technique an arbitrary degree of nuclear spin correlations can be obtained depending on the hyperfine interactions and the delay time   [22, 36]. We further note that $\rho_I^{(ab)}(\ )$ represents the S-Bus or sublevel density matrix which refers either to the $m_S = +1/2$ or $m_S = -1/2$ electron spin state.

In general we want to extract a submatrix of a certain number n of spins in an N spin S-Bus system. As an example we extract the two qubit part for spin I_1 and I_2 which can be expressed as [35, 36]

$$\rho_{12}^{(yy)} = C_0 I_0^{(2)} - 2(C_1 I_{z1} + C_2 I_{z2}) - 4 C_{12} I_{z1} I_{z2} , \tag{40}$$

where $C_0 = K_R c_1 c_2$, $C_1 = K_I s_1 c_2$, $C_2 = K_I c_1 s_2$ and $C_{12} = K_R s_1 s_2$ and with

$$K_R^2 = \frac{1}{2}\left(1 + \prod_{j=3}^{N} \cos(a_j)\right) \quad \text{and} \quad K_I^2 = \frac{1}{2}\left(1 - \prod_{j=3}^{N} \cos(a_j)\right) \quad . \tag{41}$$

This procedure is readily extended to an arbitrary number of sub-spins as is shown elsewhere [35, 36].

5.4 Multiple Quantum ENDOR

In order to detect a nuclear spin correlated state we apply Multiple Quantum ENDOR (MQE) as first introduced in [34]. In its simplest version two $/ 2$-pulses are applied to every nuclear spin transition under consideration. For this discussion we set the delay time between the pulses to zero and introduce a phase shift ϕ + ψ to the second pulse. As an example we consider the case N = 2 (40).

After the first pulse all z-components of the nuclear spin operators are converted into x-operators. For ψ = 0 the second − y-pulse would just reconvert the x-operators back to z. In order to see how the density matrix σ_1 changes for ψ = 0 we perform the corresponding transformation and obtain

$$\sigma_{12}^{(yy)} = C_0 I_0 - 2(C_1 \cos\phi_1 I_{z1} + C_2 \cos\phi_2 I_{z2}) - 4C_{12} \cos\phi_1 \cos\phi_2 I_{z1} I_{z2}, \quad (42)$$

where we have assumed that all o -diagonal components have decayed after the second pulse. This state is reached after some delay time. The generalization to an arbitrary number N of nuclear spins can readily be written down [35, 36].

An early example of a multiple quantum ENDOR spectrum with non-selective excitation where all phase angles ϕ were varied in increments $\Delta\phi$ = $\Omega\Delta$t which defines the phase frequency Ω and the virtual time increment Δt was already published in [34]. After Fourier transform of the phase interferogram a spectrum is obtained with multiple quantum lines appearing at integers of the base frequency Ω. In Fig. 20 we demonstrate the e ect of applying individual phase rotation frequencies Ω_1 = 0.9 MHz and Ω_2 = 1.1 MHz to two di erent spins I_1 and I_2. Note the appearance of single quantum (1Q) lines at Ω_1 and Ω_2 due to the linear spin components in the density matrix and additional lines at $\Omega_1 \pm \Omega_2$ (zero (0Q) and double quantum (2Q) lines) due to the bilinear component. Extension to a larger number of correlated spins is straightforward [35, 36].

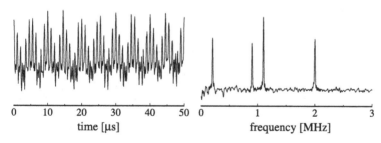

Fig. 20.   MQE-interferogram and MQE-spectrum of a two spin correlated state with single spin frequencies Ω_1 = 0.9 MHz and Ω_2 = 1.1 MHz

5.5 Creating Pseudopure Nuclear Spin States

The S-Bus density matrix represented by (36) or (37) depends on the hyperfine interactions a_j and the delay time τ. In general one will choose an optimum value of τ for a given distribution of hyperfine interactions. Except for fortuitous cases this will not correspond to a pure or pseudopure state. We therefore need to prepare a pseudopure state. This can be achieved simply by applying a $P_y(\ )$ pulse to each addressable nuclear spin after the S spin preparation sequence P_S. The prefactors in front of the spin operators can also be modified at will by the MQE sequence discussed in Sect. 5.4. After the decay of transient components the corresponding nuclear spin transition is scaled by $\cos(\ )$. For this we ignore the identity matrix which will be added appropriately later.

Suppose we want to prepare the following truncated pseudopure density matrix

$$\overset{(12)}{\underset{00}{}} = \text{const} \cdot I_0 + \frac{1}{2}I_{z1} + \frac{1}{2}I_{z2} + I_{z1}I_{z2} \,. \tag{43}$$

We simply need to modify the prefactors appropriately in order to prepare the pseudopure state. Examples of some pseudopure states are shown in Fig. 21. Note that the sign of the linear spin components of the density matrix is directly reflected in the sign of the 1Q signal, whereas the sign of the two spin correlated state depends on both spins. It can be read o from the MQE-spectrum directly from the 0Q and 2Q components. By evaluating the intensity of the MQE lines in reference to the as prepared intensities one can evaluate the density matrix components. Details of this procedure are published elsewhere [35, 36].

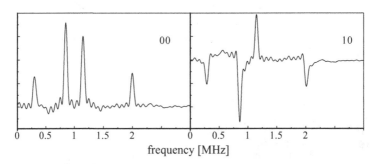

Fig. 21. MQE-spectra of the density matrices $_{00}$ and $_{10}$ with phase frequencies $_1 = 0.85\,\text{MHz}$ and $_2 = 1.15\,\text{MHz}$

5.6 Deutsch-Jozsa Algorithm in the S-Bus

The Deutsch-Jozsa algorithm was introduced in Sect. 1.2. It was already implemented in liquid state NMR [38, 39, 40, 41]. In order to implement it in the S-Bus system, we need to perform unitary transformations according to the block diagram Fig. 5. The first Hadamard transform was realized by $P_y(\ /\ 2)$ pulses at two nuclear spins and the second Hadamard transform by the inverse pulse. The unitary transformations representing the di erent functions f (ab) according to Table 1 were implemented by phase shifts [35, 36]. The two

balanced functions correspond to identity matrices and therefore represent a
NOP (no operation). As an example for the balanced functions we consider
the transformation

$$U_{0011} = e^{i I_{z2}} = \begin{matrix} 1 & 0 & 0 & 0 \\ 0 & 1 & 0 & 0 \\ 0 & 0 & -1 & 0 \\ 0 & 0 & 0 & -1 \end{matrix} . \tag{44}$$

In a similar way all other transformations can be implemented by applying
a   rotation to the other spin or both spins together. As an example we
display in Fig. 22 the result of the operations U_{0000} and U_{0011} on the initial
density matrix $_{00}$. More examples are presented elsewhere [35, 36]. Under the
U_{0000} transformation the initial density matrix $_{00}$ is unchanged as expected
since it represents a constant function. Note, however, that the DJ algorithm
changes the initial state if the transformation represents a balanced function.
This fulfills the requirement of the DJ algorithm, namely that a balanced
function leads to an output di erent from $_{00}$. The tomography of the final
state density matrices when applying the balanced transformations U_{0101},
U_{0011}, U_{1001} results in fact in $_{01}$, $_{10}$ and $_{11}$.
The other three balanced functions are equivalent in the sense that they
just involve a sign change of the unitary transformations which has no e ect
on the outcome. We note that no entangled states are involved in the two
qubit DJ algorithm as demonstrated here. In a three qubit DJ algorithm,
however, some of the balanced functions involve entangled states. We have
also implemented entangled nuclear spin states within the S-Bus system which
will be discussed elsewhere.

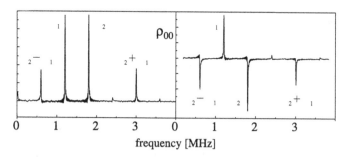

Fig. 22.   MQE-spectra of the density matrices after applying the DJ algorithm with
the constant function $U_{(0000)}$ (left) and the balanced function $U_{(0011)}$ (right). $_1$ and
$_2$ are the single spin frequencies

Acknowledgements

We gratefully acknowledge financial support by the BMBF, the Landesstiftung Baden W¨urttemberg and the Fond der Chemischen Industrie.

References

1. R.P. Feynman: Int. J. Theor. Phys.   21, 467 (1982)
2. D. Deutsch: Proc. R. Soc. Lond. A   400, 97 (1985)
3. D. Deutsch: Proc. R. Soc. Lond. A   439, 553 (1992)
4. C.H. Bennett, F. Besette, G. Brassard, L. Salvail, J. Smolin: J. Cryptology   5, 3 (1992)
5. C.H. Bennett, G. Brassard, C. Cr´ epeau, U.M. Maurer: Generalized privacy amplification. In   Proceedings of the IEEE Internatinal Conference on Computers, System and Signal Processing   (IEEE, New York 1994) p 350
6. C.H. Bennett, F. Besette, G. Brassard, L. Salvail, J. Smolin: Phys. Rev. Lett. 70, 1895 (1993)
7. D. Bouwmeester, J.-W. Pan, K. Mattle, M. Eibl, H. Weinfurter, A. Zeilinger: Nature (London)   390, 575 (1997)
8. L.K. Grover: Phys. Rev. Lett.   79, 325 (1997)
9. I.L. Chuang, N. Gershenfeld, M. Kubinec: Phys. Rev. Lett.   80, 3408 (1998)
10. P.W. Shor: SIAM J. Comput.   26, 1484 (1997)
11. L.M.K. Vandersypen, M. Ste en, G. Breyta, C.S. Yannoni, M.H. Sherwood, I.L. Chuang: Nature   414, 883 (2001)
12. D.G. Cory, A.F. Fahmy, T.F. Havel: Proc. Natl. Acad. Sci. U.S.A.   94, 1634 (1997)
13. D.G. Cory, M.D. Price, Timothy F. Havel: Physica D   120, 82 (1998)
14. N.A. Gershenfeld, I.L. Chuang: Science   275, 350 (1997)
15. I.L. Chuang, L.M.K. Vandersypen, D.W. Leung Xinlan Zhou, S. Lloyd: Nature 393, 143 (1998)
16. J.A. Jones, M. Mosca: J. Chem. Phys.   109, 1648 (1998)
17. E. Knill, I. Chuang, R. Laflamme: Phys. Rev. A   57, 3348 (1998)
18. M. Mehring: Appl. Mag. Reson.   17, 141 (1999)
19. W.S. Warren, N. Gershenfeld, I. Chuang: Science   277, 1688 (1997)
20. B.E. Kane: Nature   393, 133 (1998)
21. A. Einstein, B. Podolski, N. Rosen: Phys. Rev.   47, 777 (1935)
22. M. Mehring, V.A. Weberruß: Object Oriented Magnetic Resonance   (Academic Press, London 2001)
23. D.G. Cory, R. Laflamme, E. Knill, L. Viola, T.F. Havel, N. Boulant, G. Boutis, E. Fortunato, S. Lloyd, R. Martinez, C. Negrevergne, M. Pravia, Y. Sharf, G. Teklemariam, Y.S. Weinstein, W.H. Zurek: Fortschr. Phys.   48, 875 (2000)
24. M. Mehring, J. Mende, W. Scherer: Phys. Rev. Lett.   90, 153001 (2003)
25. Z.L. Madi, R. Br¨ uschweiler, R.R. Ernst: J. Chem. Phys.   109, 10603 (1998)
26. G. Teklemariam, E.M. Fortunato, M.A. Pravia, Y. Sharf, T.F. Havel, D.G. Cory: Phys. Rev. Lett.   86, 5845 (2001)
27. G. Teklemariam, E.M. Fortunato, M.A. Pravia, Y. Sharf, T.F. Havel, D.G. Cory, A. Bhattaharyya, J. Hou: Phys. Rev. A   66, 012309 (2002)

28. W. Scherer, M. Mehring: To be published.
29. T. Almeida Murphy, Th. Pawlik, A. Weidinger, M. Hoehne, R. Alcala, J.M. Spaeth: Phys. Rev. Lett. **77**, 1076 (1996)
30. W. Harneit: Phys. Rev. A **65**, 032322 (2002)
31. D. Suter, K. Lim: Phys. Rev. A **65**, 052309 (2002)
32. J. Twamley: Phys. Rev. A **67**, 052318-1, (2003)
33. M. Mehring, W. Scherer, A. Weidinger: Phys. Rev. Lett **93**, 206603 (2004)
34. M. Mehring, P. H¨ofer, H. K¨aß: Europhys. Lett. **6**, 463 (1988)
35. M. Mehring, J. Mende: To be published.
36. J. Mende, M. Mehring: To be published.
37. W.B. Mims: Proc. R. Soc. London **283**, 452 (1965)
38. Arvind, K. Dorai, A. Kumar: Pramana **56**, L705 (2001)
39. K. Dorai, Arvind, A. Kumar: Phys. Rev. A **61**, 042306/1-7 (2000)
40. O. Mangold: Implementierung des deutsch algorithmus mit drei ^{19}F-kernspins. Diploma Thesis, Universit¨at Stuttgart (2003)
41. O. Mangold, A. Heidebrecht, M. Mehring: Phys. Rev. A **70**, 042307 (2004)

Laser-Assisted Magnetic Resonance: Principles and Applications

D. Suter and J. Gutschank

Universit"at Dortmund, Fachbereich Physik, 44221 Dortmund, Germany
dieter.suter@physik.uni-dortmund.de

Abstract. Laser radiation can be used in various magnetic resonance experiments. This chapter discusses a number of cases, where laser light either improves the information content of conventional experiments or makes new types of experiments possible, which could not be performed with conventional means. Sensitivity is often the main reason for using light, but it also allows one to become more selective, e.g. by selecting signals only from small parts of the sample. Examples are given for NMR, NQR, and EPR spectra that use were taken with the help of coherent optical radiation.

1 Introduction

The interest in the field of magnetic resonance spectroscopy is based largely on the huge potential for applications: spins can serve as probes for their environment because they are weakly coupled to other degrees of freedom. In most magnetic resonance experiments, these couplings are used to monitor the environment of the nuclei, like spatial structures or molecular dynamics.

While the direct excitation of spin transitions requires radio frequency or microwave irrad iation, it is often possible to use light for polarizing the spin system or for observing its dynamics. This possibility arises from the coupling of spins with the electronic degrees of freedom: optical photons excite transitions between states that di er both in electronic excitation energy as well as in their angular momentum states.

1.1 Motivation

Some motivations for using light in magnetic resonance experiments include

- Sensitivity: In many cases, the possible sensitivity gains are the primary reason for using optical methods. Compared to conventional NMR, sensitivity gains of more than 10 orders of magnitude are possible. The ultimate

D. Suter and J. Gutschank: Laser-Assisted Magnetic Resonance: Principles and Applications
Lect. Notes Phys. 684 , 115–141 (2006)
www.springerlink.com

limit in terms of sensitivity was reached in 1993, when two groups showed that it is possible to observe EPR transitions in single molecules [1, 2]. The same technique was later used to observe also NMR transitions in a single molecule [3].

- Selectivity: Lasers can be used to selectively observe signals from specific parts of the sample, like surfaces, at certain times which may be defined by laser pulses with a resolution of 10^{-14} s, or from a particular chemical environment defined, e.g., by the chromophore of a molecule or the quantum confined electrons in a semiconductor.

- Speed: Magnetic resonance requires the presence of a population difference between spin states to excite transitions between them. In conventional magnetic resonance, this population difference is established by thermal relaxation through coupling with the lattice, i.e. the spatial degrees of freedom of the system. At low temperatures, this coupling process may be too slow for magnetic resonance experiments. In the case of optical excitation, the population differences are established by the polarizing laser light. Depending on the coupling mechanism, this polarization process can be orders of magnitude faster than the thermal polarization process, independent of temperature.

- Electronically excited states: If information about an electronically excited state is desired that is not populated in thermal equilibrium, it may be necessary to use light to populate this state. It is then advantageous to populate the different spin states unequally to obtain at the same time the polarization differences that are needed to excite and observe spin transitions.

1.2 What Can Lasers Do?

Light can support magnetic resonance experiments in different ways. They can, e.g., initiate a chemical reaction that one wishes to observe, like in photosynthetic processes. These light-induced modifications of the sample will not be considered here; instead we concentrate on the use of light for the magnetic resonance experiment, where light affects directly the spin degrees of freedom, rather than spatial coordinates. Typically, the laser is then used either to increase or to detect the spin polarization of nuclear or electronic spins.

These two approaches are largely independent of each other: It is, e.g., possible to use optical pumping to enhance the spin polarization and observe the transitions with a conventional NMR coil; conversely, optical detection can be used with or without increasing the population difference with laser light. In many cases, however, it is advantageous to combine both approaches. In some cases, a single laser beam may provide an increase of the spin polarization and an optical signal that can be related to a component of the magnetization. In others, a pump-probe setup separates the excitation and detection paths.

In addition to these applications of lasers, light can also be used to drive the dynamics of spin systems, e.g., through Raman transitions [4]. For this review, however, we will concentrate on the issues of increasing the spin polarization and on optical detection.

2 Optical Polarization of Spin Systems

Magnetic resonance spectroscopy requires a spin polarization inside the medium. In conventional magnetic resonance experiments, this polarization is established by thermal contact of the spins with the lattice. This process is relatively slow, especially at low temperatures, where relaxation times can be many hours, and it leads to polarizations that are limited by the Boltzmann factor. Photon angular momentum, in contrast, can be created in arbitrary quantities with a polarization that can be arbitrarily close to unity. If it is possible to transfer this polarization to nuclear or electronic spins, their polarization can increase by many orders of magnitude.

A number of di erent approaches have been used to achieve this goal. The oldest and best known approach is known as optical pumping [5]; it was originally demonstrated on atomic vapors [6] and later applied to condensed matter. While optical pumping allows one to create very high spin polarization in atomic vapors, it is less suitable for applications to anisotropic systems such as low symmetry solids. Other techniques were therefore developed, which can still be used in such an environment. While optical pumping was originally implemented with conventional light sources, most of the other approaches require the use of coherent optical radiation, i.e. laser light.

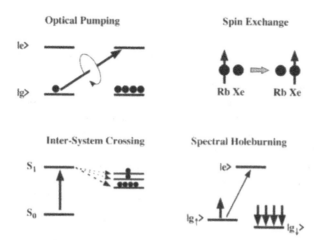

Fig. 1.   Four ways for optically increasing the spin polarization

2.1 Optical Pumping

The possibility to use optical radiation for exciting and detecting spin polarization can be traced back to the angular momentum of the photon. Photons as the carriers of the electromagnetic interaction carry one unit (   ) of angular momentum, which is oriented either parallel or antiparallel to the direction of propagation of the light. In an isotropic environment, angular momentum is a conserved quantity. When a photon is absorbed by an atom or molecule, its angular momentum must therefore be transferred to the atom. The resulting angular momentum of the atom is equal to the vector sum of its initial angular momentum plus the angular momentum of the absorbed photon.

The use of angular momentum conservation for increasing the population di erence between spin states was first suggested by Alfred Kastler [7, 8, 5]. If an atom is irradiated by circularly polarized light, the photons have a spin quantum number $m_s = +1$. Since the absorption of a photon is possible only if both, the energy and the angular momentum of the system are conserved, the atoms can only absorb light by simultaneously changing their angular momentum state by one unit.

After the atom has absorbed a photon it will reemit one, decaying back into the ground state. Spontaneous emission can occur in an arbitrary direction in space and is therefore not limited by the same selection rules as the excitation process with a laser beam of definite direction of propagation. The spontaneously emitted photons carry   away angular momentum with di erent orientations and the atom can therefore return to a ground state whose angular momentum state di ers by $m = 0, \pm 1$. The net e ect of the absorption and emission processes is therefore a transfer of population from one spin state to the other and thereby a polarization of the atomic system.

2.2 Spin Exchange

Spin polarization can be transferred between di erent reservoirs not only within one atomic species, but also between di erent particles. This was first demonstrated by Dehmelt who used transfer to free electrons to polarize them [9]. Another frequently used transfer process uses optical pumping of alkali atoms, in particular Rb and Cs and transfer of their spin polarization to noble gas atoms like Xe. These atoms cannot be optically pumped from their electronic ground state (although He can be pumped in the metastable state [10, 11]); spin exchange allows one to optically pump an alkali gas (typically rubidium) and transfer the spin polarization from there to the Xe nuclear spin. This method was pioneered by Happer [12], applied to the study of surfaces [13, 14], and used in a number of medical applications [15, 16, 11].

The transfer from alkali to noble gas atoms is relatively e cient when the two species form van der Waals complexes. During the lifetime of this quasi-molecule, the two spins couple, mainly by dipole-dipole interaction. This coupling allows simultaneous spin flips of the two species which transfer polarization from the Rb atoms to the Xe nuclear spin. Typical cross-polarization times are on the order of minutes, but the long lifetime of the Xe polarization permits to reach polarizations close to unity. The spin polarization survives freezing [17] and can be transferred to other spins by thermal mixing [18].

2.3 Excited Triplet States

In many classical optically detected magnetic resonance experiments, absorption of light excites the system into a singlet state that can, through non-radiative processes, decay into a triplet state, whose energy is below the excited singlet state. This intersystem conversion process as well as the decay of the triplet state can be spin-dependent, therefore creating a significant spin polarization of the triplet state. In many systems, these processes are quite e cient, even for unpolarized light, generating a high degree of spin polarization in the triplet state. Under certain conditions, this polarization of the electron spin can also lead to a polarization of the nuclear spin, which survives when the molecule returns to its ground state.

2.4 Spectral Holeburning

When the spin is located in a host material with low symmetry, the electronic angular momentum is quenched. Figure 2 shows the situation schematically: While angular momentum states with total angular momentum J are $2J + 1$ fold degenerate in free space, the Coulomb interaction of the atom or ion with neighboring charges (electrons and nuclei) lifts this degeneracy. The resulting states are usually no longer angular momentum eigenstates. While this argument applies directly only to orbital angular momentum, the spin-orbit interaction often is strong enough to also quench the electron spin.

If the angular momentum is quenched, optical pumping with circularly polarized light becomes ine cient for excitation of spin polarization. In these systems, other approaches may increase spin polarization. One possibility exists when the di erent spin states can be distinguished in frequency space,

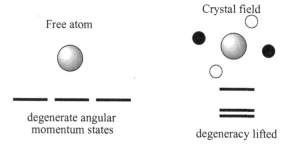

Fig. 2. Quenching of angular momentum by interaction with the crystal field

i.e. when the energy di erence between them is larger than the homogeneous width of a suitable optical transition. The situation is shown schematically in the lower right of Fig. 1. The laser only excites those ground state atoms whose spin is in the   state. Since the excited state can decay to both ground states, the population accumulates in the |g   state. This allows one to use a laser to selectively depopulate one of the spin states, while increasing the population of the other states.

Since the inhomogeneous width of the optical transitions is usually large compared to the energy of magnetic resonance transitions, it is rarely possible to address only a single spin state. The laser frequency selects then a subset of all the spins, for which the resonance condition is fulfilled; only for those systems, the spin polarization will be increased. This situation is known as spectral holeburning, since the depopulation of specific spin states reduces the absorption of light at the frequency of the pump laser beam. Additional details are discussed in the context of optical detection.

3 Optical Detection

Any magnetic resonance experiment includes a scheme for detection of time-dependent components of the spin polarization, usually as a macroscopic magnetization. In NMR, the precessing transverse magnetization changes the magnetic flux through the radio frequency (rf) coil. According to Faraday's law, the time derivative of the flux induces a voltage over the coil, which is detected as the free induction decay (in pulsed experiments) or as a change in the impedance of the coil (in continuous wave experiments).

The optical detection schemes that we discuss here can sometimes replace this inductive detection. They can be used together with optical polarization or they can be combined with conventional excitation schemes.

In suitable systems, optical detection provides a number of advantages over the conventional method: First, optical radiation introduces an additional resonance condition, which can be used to distinguish di erent signal components and thereby separate the target signal from backgrounds such as impurities. Second, optical radiation can be detected with single photon sensitivity (in contrast to microwave or radio freq   uency radiation). This has made detection of single spins possible in suitable systems. A third possible use of the optical radiation is that the laser beam breaks the symmetry of isotropic samples, such as powders or frozen solutions. As we discuss in Sect. 5.3, this allows one to derive the orientation of tensorial interactions, such as electron g-tensors or optical anisotropy tensors from non-oriented samples.

3.1 Circular Dichroism

An early suggestion that magnetic resonance transitions should be observable in optical experiments is due to Bitter [19]. The physical process used in

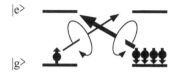

$|e\rangle$

$|g\rangle$

Fig. 3.   Optical detection through circular dichroism

such experiments may be considered as the complement of optical pumping: the spin angular momentum is transferred to the photons and a polarization selective detection measures the photon angular momentum.

Figure 3 illustrates this for the same model system that we considered for optical pumping. Light with a given circular polarization interacts only with one of the ground state sublevels. Since the absorption of the medium is directly proportional to the number of atoms that interact with the light, a comparison of the absorption of the medium for the two opposite circular polarizations yields directly the population di erence between the two spin states. This population di erence is directly proportional to the component of the magnetisation parallel to the laser beam.

Early experimental implementations of these techniques were demonstrated in atomic vapors [20, 21, 5], where angular momentum conservation is exact and the principle is directly applicable. Similar considerations hold also for solid materials [22], although, as we discussed above, angular momentum is not always a conserved quantity in such systems. It depends therefore on the symmetry of the material if absorptive detection is possible [23]. Nevertheless, even small optical anisotropies can be measured; changes in these parameters upon saturation of the spins provide a clear signature of magnetic resonance transitions [24].

While most implementations measure the longitudinal spin component by propagating a laser beam parallel to the static magnetic field, it is also possible to observe precessing magnetization with a laser beam perpendicular to the static field [25]. The two approaches provide complementary information [26] and a combination of longitudinal and transverse measurements is therefore often helpful for the interpretation of the spectra.

3.2 Photoluminescence

Photoluminescence is another important tool for measuring spin polarization. Depending on the system, the intensity or the polarization of spontaneously emitted photons can be a measure of the spin polarization in the ground – or in an electronically excited state. In free atoms, angular momentum conservation imposes correlations between the direction and polarization of the spontaneously emitted photons which depend on the angular momentum state of the excited atom. Photoluminescence has therefore long been used to measure spin polarization in electronically excited states [27, 28].

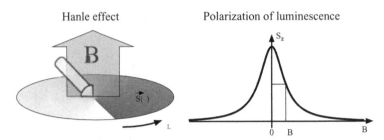

Fig. 4. Hanle e ect: A magnetic field perpendicular to a circularly polarized excitation laser forces Larmor precession of the electron spins. The spin polarization of
the excited electrons decreases therefore with increasing magnetic field strength

Spin polarization in the electronic ground state also a ects the photoluminescence, since the absorption of polarized light depends on the spin state.
If the spin orientation prevents absorption of light, the intensity of the photoluminescence decreases correspondingly. The intensity and polarization of the
photoluminescence can therefore serve for detecting ground state spin polarization and, e.g., by saturation with a resonant rf field, for detecting magnetic
resonance transitions [29, 30].

The e ect of Larmor precession on the spin polarization of excited states
has been observed as early as 1924 by Hanle [31]. He noticed that the
polarization of the photoluminescence decreases if a magnetic field is applied
perpendicular to the direction of the spin polarization (Fig. 4). The observed
polarization of the photoluminescence changes with the field B_0 as

$$S_z = \frac{\Delta B^2}{\Delta B^2 + B_0^2},$$ (1)

where the width $\Delta B = (\gamma_r + \gamma_s)/\gamma$ is determined by the gyromagnetic ratio
γ and the relaxation rates γ_s and γ_r of the spin and excited state population.

The Hanle e ect can also be observed in four-wave mixing experiments [32]
in atomic vapors as well as in crystals [33]; in this case, significant polarization of the photoluminescence is only obtained if the crystal has high enough
symmetry and mechanical strain is small enough to avoid depolarization. It is
particularly suitable for measuring spin polarization in semiconductors with
a direct band gap, such as GaAs [34].

3.3 Coherent Raman Scattering

Raman processes are optical scattering processes in which the frequency (and
therefore the wavelength) of the scattered light di ers from that of the incident
light [35]. The energy di erence between the incident and the scattered photon
is absorbed (or emitted) by excitations of the material in which the scattering
occurs. While this excitation of the material is often a vibration, it can also be

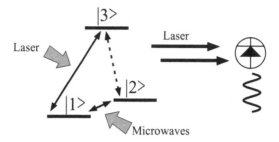

Fig. 5.   Coherent Raman scattering from a three-level system. Laser excitation creates a coherence between levels |1⟩ and |3⟩ and microwaves between |1⟩ and |2⟩. The resulting non-linear polarisation in the third transition creates a Raman wave

associated with spin degrees of freedom, in which case the scattering process can be used to detect magnetic resonance transitions.

Figure 5 shows the relevant process for the simplest possible case: The two states |1⟩ and |2⟩ represent two spin states of the electronic ground state, while |3⟩ is an electronically excited state. If a micr owave field (rf in the case of nuclear spin transitions) resonantly excites the transition between states |1⟩ and |2⟩, it creates a coherence between the two spin states. Similarly, the laser excites an optical coherence in the electronic transition |1⟩ ↔ |3⟩. Since the two transitions share state |1⟩, the two fields create a superposition of all three states, which contains coherences not only in the two transitions that are driven by the external fields, but also in the third transition |2⟩ ↔ |3⟩. If this transition has a non-vanishing electric dipole moment, this coherence is the source of a secondary optical wave, the Raman field. As the figure shows, the frequency of this wave di ers from that of the incident wave by the frequency of the microwave field. It has the same sp atial dependence as the incident laser field and therefore propagates in the same direction. If the two optical fields are detected on a usual photodetector (photodiode or photomultiplier), they interfere to create a beat signal at the microwave freq uency.

The type of scattering process used for magnetic resonance detection is referred to as "coherent" Raman scattering [36] since the Raman field is phase-coherent with the micr owave as well as with the incident laser field. This is an important prerequisite for the detection process: If the laser frequency drifts, the frequency of the incident field as well as that of the Raman field are shifted by the same amount. As a result, the di erence frequency is not a ected and the resolution of the measurement is not a ected by laser frequency jitter or broad optical resonance lines [37]. Coherent Raman processes provide therefore a combination of high resolution with high sensitivity.

Like in conventional magnetic resonance experiments, the excitation of the magnetic resonance transition indicated in Fig. 5 can be performed either in a continuous (cw) [38] or pulsed [39, 40] mode. Furthermore, the microwave or rf

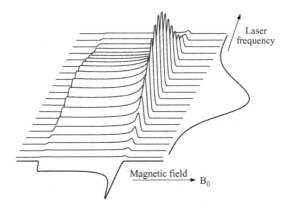

Fig. 6. Two-dimensional optically detected EPR (ODEPR) spectrum as a function of the laser frequency and the magnetic field strength. The result is a complete microwave resonance spectrum for each laser wavelength. The projections on the axes represent the conventional EPR and absorption spectra

field can be replaced by optical fields, applied to the two electronic transitions, that can excite the spin coherence by another Raman process [41, 42, 43].

Since the coherence that generates the signal is excited by two resonant fields, it depends on the frequencies of both fields. As shown in Fig. 6, the resulting signal is doubly resonant and contains therefore information about the optical as well as the magnetic resonance transition. As with other two-dimensional experiments, it allows one to correlate information from the two frequency dimensions. Examples that demonstrate this feature will be discussed in Sect. 5. While we have discussed the process here as involving magnetic resonance transitions in the ground state, equivalent processes are also possible that relate to spins of electronically excited states.

3.4 Spectral Holeburning

In Sect. 2.4, we discussed how narrowband lasers that cause spectral holeburning can increase the polarization of spins, in analogy to optical pumping. In most such experiments, a second laser beam, whose frequency can be swept around the frequency of the pump beam, is used to monitor the changes in the populations. The resulting spectra are known as holeburning spectra [44].

As shown in Fig. 7, holeburning requires a pump and a probe laser beam. The pump laser modifies the population of those atoms for which the laser frequency matches an electronic transition frequency. When the probe laser hits the same transition, the absorption is reduced in line with the smaller population of the relevant ground state. The population that has been removed from this state is accumulated in the other spin state. When the probe laser frequency is tuned to the transition from this ground state to an electronically excited state, it finds increased absorption, which is referred to

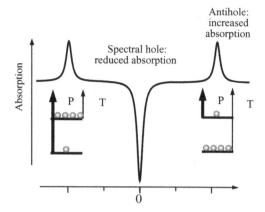

Fig. 7.   Schematic representation of a holeburning spectrum. The hole represents reduced absorption, the "antiholes" increased absorption. The separation between hole and antihole is equal to the transition frequency of the two spin states

as an "antihole". The separation between the hole and antihole matches the energy di erence between the two spin states and the hole burning spectrum can therefore measure magnetic resonance transition frequencies [45]. While this discussion has centered on spin transitions between electronic ground states, the procedure also allows one to measure energy di erences between spin states of electronically excited states [46].

   The optical detection techniques discussed here were chosen to represent the most frequently used approaches. There are several additional techniques which cannot be discussed, which include purely optical techniques like photon echo modulation [47, 48].

4 Applications to NMR and NQR

4.1 Rare Earth Ions

Ions of rare earth elements have been studied extensively with high resolution optical spectroscopy [49, 50]. The relevant optical transitions are between f electron states and have relatively small homogeneous and inhomogeneous broadening.

   Figure 8 shows the relevant energy levels for the ^{141}Pr ion doped into the host material YAlO$_3$. The electronic ground state as well as the electronically excited states are split into substates that di er with respect to their nuclear spin coordinates. The separation between these spin states is due to nuclear quadrupole coupling and second order hyperfine coupling, which combine into an e ective quadrupole interaction [51, 52].

   One approach to measure the NQR transition frequencies is by holeburning spectroscopy, as shown in Fig. 9. Since pump and probe laser beam can each

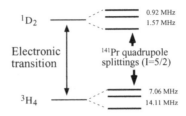

Fig. 8.   Relevant level scheme for ^{141}Pr doped into YAlO$_3$

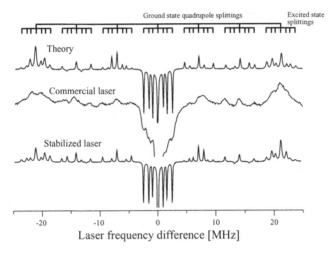

Fig. 9.   Holeburning spectrum of Pr: YAlO$_3$

be resonant with nine diﬀerent transitions (from three ground- to three excited states), the holeburning spectrum, which depends on the diﬀerence between the pump and probe laser frequencies, has a total of 81 resonances. A number of these resonances have identical frequencies (e.g., $_P - _T = 0$), resulting in a total of 49 distinguishable frequencies.

The width of each resonance line increases with the laser frequency jitter. As the comparison of the middle and lower traces shows, it is therefore important to use a narrowband laser for measuring these spectra. In this example, the laser linewidth was   30 kHz; the width of the observed resonance lines was therefore close to the homogeneous width of the optical transition [46].

The amplitudes of the individual resonance lines depend on the optical transition matrix elements and are proportional to the overlap integral $_g|_e$ of the ground and excited state nuclear spin states. While the nuclear spin is not involved in the electronic transition, the electronically excited state can have diﬀerent quantization axes than the ground state if, as in this example, the eﬀective quadrupole interaction changes with the electronic excitation. The precise measurement of the holeburning spectrum allows one then to determine the relative orientation of the principal axis system between the ground- and excited states [46].

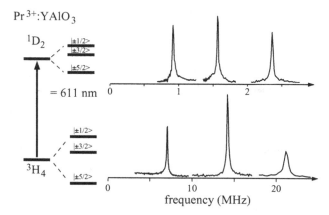

Fig. 10. Raman-heterodyne detected NQR spectra of Pr:YAlO$_3$. All six resonances can be detected in a single wide frequency scan, but the low frequency part, which is associated with the NQR transitions in the electronically excited state has been expanded in the upper part of the figure

The spectrum can be simplified considerably if the Raman scattering experiment is used instead. In this case, every nuclear spin transition gives rise to a single resonance at the transition frequency. Figure 10 shows an example of a coherent Raman spectrum of Pr:YAlO$_3$ in zero magnetic field, which was recorded by irradiating the optical transition at 611 nm with a laser beam and measuring the Raman-heterodyne signal while sweeping the rf frequency. In the example shown here, the three NQR transitions at 7, 14, and 21 MHz occur within the electronic ground state, while the three low-frequency transitions (< 3 MHz) belong to the electronically excited state.

4.2 Sign Information

An interesting case of additional information that is unavailable with conventional techniques is the sign of the nuclear quadrupole interaction. As is well known [53], conventional magnetic resonance experiments cannot provide the sign of the quadrupole coupling. In the simplest case of axial symmetry, the Hamiltonian H_Q of the nuclear quadrupole interaction is given by a coupling constant D times the square of the nuclear spin operator I_z, $H_Q = DI_z^2$. The coupling constant D is determined by the size of the nuclear quadrupole moment and the electric field gradient. It can be measured either in the absence of a magnetic field, which corresponds to the case of pure quadrupole coupling, or in a high magnetic field, which corresponds to the case of high-field NMR.

Figure 11 shows schematically the NQR (zero magnetic field) and NMR spectra of a spin $I = 5/2$ with axial quadrupole interaction. The spectra for the positive and negative coupling constant D are identical, unless the spin

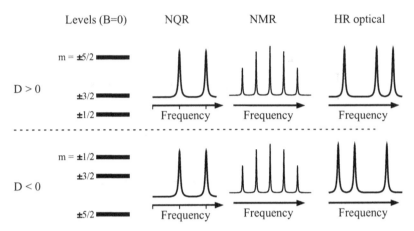

Fig. 11.   Measurement of the nuclear quadrupole coupling by NQR, NMR and optical spectroscopy. Only the optical spectra distinguish the sign of the interaction

temperature becomes very low or dipolar couplings are resolved. The optical spectrum, however, shown on the right, clearly distinguishes the two cases.

In practice, the inhomogeneous broadening of the optical transitions prevents one from measuring such spectra directly. It is nevertheless possible to use optical and optical-rf double resonance experiments that produce spectra that clearly distinguish the two cases. In the case of Pr^{3+} in the host material YAlO$_3$, the coupling constant turned out to be negative [54].

4.3 Semiconductors

Semiconductors have become a very active area for applications of optically enhanced magnetic resonance. Detection usually relies on changes in the polarization of the photoluminescence. Depending on the sample, the photoluminescence may be dominated by light from trapping sites such as point defects or by recombination of conduction electrons. Point defects usually dominate in indirect semiconductors and amorphous materials [55], while very pure, MBE-grown III/V materials show predominantly interband recombination. Under these conditions, the magnetic resonance signal may originate from the whole sample, while it provides information on localized parts of the material if the recombination is due to defects or heterostructures [56].

Optical pumping occurs by first polarizing the electron spin system [57]. Using a photon energy close to the bandgap, optical excitation creates spin-polarized electron-hole pairs. In bulk materials, this spin polarization is not complete, because optical pumping excites different degenerate transitions that lead to different spin states. In quantum well materials, however, the degeneracy is lifted and electron spin polarization can reach values close to unity [58, 59]. The hyperfine coupling transfers part of this polarization to the nuclear spins [60], which can reach polarizations of more than 50%.

Detection often relies on the Hanle e ect: the hyperfine interaction of polarized nuclear spins creates an e ective magnetic field, which tends to depolarize the photoluminescence [61, 62]. A number of specific rf excitation schemes have been developed to optimize the optical detection process. The conventional procedure relies on saturation of the nuclear spins, which may lead to power broadening. Other techniques include two-dimensional procedures [63] or beat signals between di erent isotopes [64]. It is also possible, however, to apply an rf pulse and observe the free induction decay as a modulation of the polarization of the photoluminescence [65].

Materials like GaAs crystallize in a cubic lattice; the symmetry at the site of the nuclei is high enough that the electric field gradient (EFG) tensor vanishes and the three allowed dipole transitions of the   I = 3 / 2 nuclei are degenerate in an ideal crystal. Optically detected NMR spectra of quantum wells show quadrupole splittings of several tens of kHz, indicating that the EFG tensor does not vanish in this case. One cause for the nonvanishing quadrupole coupling is a distortion under the influence of mechanical strain [66]. Other possibilities include electrical fields [67]: Since the site symmetry of the nuclei does not include an inversion center, electric fields also can induce a nonvanishing EFG [68, 69].

Figure 12 demonstrates how laser-assisted NMR is capable of measuring small variations of the crystal structure in a multiple quantum well sample with a spatial resolution of some tens of nanometers. The spectra from the individual quantum wells all show distinct quadrupole splittings, which become smaller as the quantum well thickness decreases. At the same time, the width of the satellite lines increases, indicating an increase in the variation of structural distortion and/or electric field distribution.

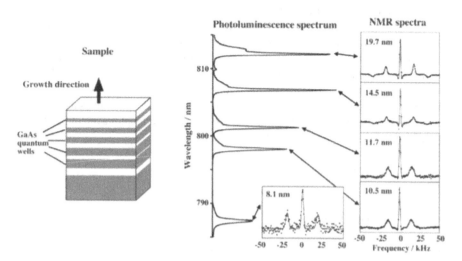

Fig. 12.   Variation of quadrupole splittings in a multiple quantum well sample

4.4 Surfaces and Interfaces

Quasi-twodimensional systems have always proved di cult to investigate by conventional NMR, since the number of spins in these systems is quite small [70, 71]. Most of the NMR work on surfaces has therefore concentrated on systems with large surface to volume ratio like Zeolites [72], where most of the atoms are close to the interface. Increasing the spin polarization by optical pumping has significantly improved the sensitivity of this type of experiments. In particular the transfer of spin polarization from optically pumped alkali atoms to Xe nuclear spins [12] has allowed to study the e ect of surfaces on the magnetic resonance spectrum [73, 74]. For experiments with oriented surfaces, the use of light for optical pumping as well as for detection brings, apart from the sensitivity advantages, also the possibility to select signal contributions that originate from atoms that are close to the surface. For this purpose, changes in the penetration depth of light with wavelength [75] have been used, but more frequently, the selection is achieved by reflecting a laser beam from the interface being investigated. The reflection coe cient for the laser beam depends on the refractive indices on both sides of the interface and is therefore a ected by atoms close to the interface that are resonant with the laser light. The changes in the reflection coe cient, which can be measured through changes of the amplitude and polarization of the reflected beam contain therefore information on the atoms close to the interface. The combination of optical pumping with this type of optical detection provides su cient sensitivity, so that it is no longer necessary to use samples with high surface to volume ratios and allows therefore studies on oriented surfaces. One method that relies on such a technique was used to study nuclear quadrupole resonances of Pr $^{3+}$ in LaF $_3$ [76]. In this case, the beam was reflected from an optically dense material. The reverse is also possible: if the laser beam undergoes total internal reflection at an interface to an optically less dense medium, an evanescent wave penetrates into the thinner medium by a distance of the order of the optical wavelength. Atoms in this evanescent wave can thus modify the reflected laser beam by absorbing light from it. Similarly, the presence of resonant atoms changes the refractive index of the medium and thereby the reflection coe cient. Both e ects can be used for measuring magnetic resonance spectra of atoms that are within an optical wavelength of the reflecting surface [77, 78, 79, 80, 81].

5 EPR

5.1 Experimental Approach

Optically detected magnetic resonance has long been used to study electron spin resonance in various environments. The classical technique measures the photoluminescence while irradiating an EPR transition with a micr   owave field and sweeping either the frequency or a magnetic field. A number of reviews has appeared that discuss such experiments [82, 28, 83]. We therefore concentrate here on a more recent development, where coherent Raman scattering is used to probe EPR transitions. The examples that we discuss will be mostly metalloproteins, where the information gained with optical-microwave double resonance experiments has proved very useful for identifying the electronic structure of the active centers.

As discussed in Sect. 3.3, coherent Raman scattering is driven by two electromagnetic fields: microwave radiation and laser light, which are tuned to an optical and a magnetic dipole transition in the sample. The sample is placed in a magnetic field to lift the degeneracy of the Zeeman levels. If the laser and microwave fields are both resonant with a transition in the sample, the transmitted laser beam is modulated at the microwave frequency. This modulation is picked up by a fast photodiode The signal can be phase-sensitively down-converted (lock-in detected) with microwaves to yield the optically detected EPR (ODEPR) signal.

While the underlying process can be understood as a coherent Raman process [84], the experiment may also be discussed in terms of modulated circular dichroism [26]. In this model, the resonant microwave irradiation excites transverse magnetisation precessing at the microwave frequency around the static field. As shown in Fig. 13, the precessing magnetisation modulates the circular dichroism and thereby the absorptivity of the sample for circularly polarized light. The modulated signal component is thus proportional to the EPR signal as well as to the magnetic circular dichroism (MCD) of the sample.

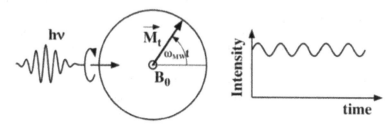

Fig. 13. Continuous wave excitation of EPR creates transverse magnetization M_t that rotates at the angular frequency ω_{MW} around the static magnetic field B_0. The absorption of circularly polarized light is therefore modulated sinusoidally

From this rotating MCD, we expect a proportionality of the ODEPR signal to the MCD, as well as to the classical EPR signal. The proportionality to MCD was experimentally demonstrated on cytochrome c551 by comparing MCD and ODEPR data of the same sample [26]. The proportionality factor between longitudinal and transverse MCD is determined by the ratio of the transverse vs. longitudinal magnetization components, $\frac{M_{xy}}{M_z} = \omega_{Rabi} T_2$, where ω_{Rabi} is the Rabi frequency and T_2 is the phase memory time.

5.2 Experimental

The experimental setup required for ODEPR experiments is based on a conventional EPR spectrometer, extended by a laser and some optical components for controlling the laser beam. Detection is based on the modulation signal from a fast photodiode rather than the microwave signal r   eflected from the resonator [85]. For many samples of interest, the relevant wavelength range cannot be covered by a single laser system. It is then necessary to use di erent continuous wave lasers including dye, semiconductor, solid state and gas lasers.

While the microwave modu lation of the transmitted laser beam can be measured with a single circularly polarized laser beam, it is in practice advantageous to alternate the polarization of the light between left and right circular and use the di erence as the actual signal. The modulation can be generated by a photoelastic modulator (PEM) that is placed in the laser beam before it passes through the sample.

The experimental examples that we will discuss are from frozen solutions of metalloproteins. These solutions are placed in a cylindrical cuvette of 0.5 mm inner length and 3 mm inner diameter. The cuvette is mounted inside a rectangular TE $_{102}$ microwave cavity with its micr owave magnetic field B_1 parallel to the direction of propagation of the laser beam. The cavity is located inside a helium bath cryostat and has two openings for transmitting the laser beam. The static magnetic field B_0 of the superconducting split coil magnet is perpendicular to the propagation of the light and thus also to the micr   owave field B_1. The modulated light (or local oscillator and Raman side-band) is detected with a fast photodiode which is connected to the micr   owave r eceiver setup.

ODEPR spectra are measured in field-sweep mode, in close analogy to conventional EPR spectra, except that field modulation is not required. The resulting spectra are therefore directly absorption and dispersion mode, rather than their derivatives. After a proper calibration of the instrument [85], the signal amplitude can be represented as the di erence in absorbance   A or extinction coe cient   between opposite circular polarisations [26].

5.3 Information Content of ODEPR Spectra

The signal generated in this experiment depends on the optical as well as on the magnetic resonance condition. The additional resonance condition, compared to conventional EPR spectroscopy, provides a possible mechanism for distinguishing di erent paramagnetic centres. This is particularly important when pure samples are di cult or impossible to obtain.

Figure 14 illustrates how the ODEPR spectra provide orientational information. The lineshape of the ODEPR spectrum (Fig. 14c,d) di ers significantly from that of the conventional EPR spectrum (Fig. 14e). The di erence arises from an orientational selectivity: The contribution of every molecule to the total signal is weighted with its MCD sensitivity for the direction of

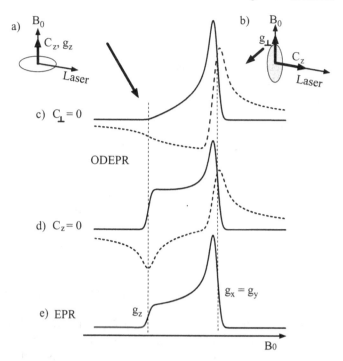

Fig. 14.   ODEPR lineshapes contain orientational information. C_z represents the MCD sensitivity in the molecular z-direction and the shaded disk represents the plane of the axially symmetric molecule. (a) Molecules with MCD sensitivity only along their z-axis do not contribute when the z-axis is perpendicular to the laser beam. (b) They contribute strongly when the z-axis is parallel to the laser beam. (c) Calculated ODEPR absorption (solid line) and dispersion (dashed line) of an axially symmetric molecule with MCD sensitivity along its z-axis only. (d) Calculated ODEPR when the MCD sensitivity is only perpendicular to the z-axis. (e) Calculated conventional EPR absorption

propagation of the laser beam. If a given molecule has the highest MCD sensitivity along the molecular z-axis (Fig. 14a, b), the main signal contribution (Fig. 14c) arises from molecules whose z-axis is oriented perpendicular to the static magnetic field. For these molecules, the resonant magnetic field is determined by their g_x and g_y values. The signal is reduced around the g_z position of the spectrum, since molecules with the z-axis along the direction of the magnetic field contribute little to the modulated MCD.

The spectra can be calculated quantitatively by a theory that combines EPR (to calculate spectral positions) and MCD (to calculate the amplitude). We discuss here only the case of an axially symmetric system. We calculate the difference of the extinction coefficients $_x$ (parallel to the laser beam) for circularly polarised light [86, 87, 88] by averaging over the contributions from all possible molecular orientations:

$$\int_0^{\pi/2} \sin\theta \, d\theta \, T(\theta) f(\theta) \left[C_z g_z \frac{g_z^2}{g^2} \sin^2\theta + C_{\perp} g_{\perp} \frac{g_z^2}{g^2} \cos^2\theta + 1 \right] . \quad (2)$$

Here θ is the angle between the molecular z-axis and the static magnetic field, $T(\theta) = \tanh(g(\theta)\mu_B B_0 / 2kT)$ is the Boltzmann factor, and $f(\theta)$ describes the transverse magnetization as a function of molecular orientation, amplitude and frequency of the microwave field. $g_{\perp}$ and g_z are the principal values of the g-matrix perpendicular to and along the molecular z-axis and $g^2 = g_z^2 \cos^2\theta + g_{\perp}^2 \sin^2\theta$. $C_{\perp}$ and C_z are the principal values of the optical anisotropy tensor, which describes the MCD sensitivity.

To obtain the g and C values and the orientation from the experimental spectrum, we fit the conventional as well as the optically detected EPR spectrum with the same parameter set. For the additional analysis, it is convenient to calculate the ratio

$$\frac{C_{\perp}}{C_z} = \tan\psi , \quad (3)$$

which parametrises the direction of the optical anisotropy with respect to the g-tensor axis.

A comparison of ODEPR spectra measured at different optical wavelengths shows strong variations of the amplitude and lineshapes of the spectra. This variation arises because the optical anisotropy tensor C is a characteristic property of each optical transition. As the laser interacts with different transitions, the optical anisotropy changes and, according to (2), also the ODEPR spectrum.

To obtain the anisotropy parameters for each optical transition, we first evaluate the orientation ψ as a function of the optical frequency ν. We then fit this angle together with the longitudinal MCD Δ_z (z indicating parallel to B_0) to a sum of contributions i from each optical transition at position p_i with width w_i,

$$\Delta_z(\nu) = \sum_i \Delta_{zi} \, e^{-(\nu - p_i)^2 / 2w_i^2} \quad (4)$$

and

$$\psi(\nu) = \arctan \frac{\sum_i C_{\perp i} \, e^{-(\nu - p_i)^2 / 2w_i^2}}{\sum_i C_{zi} \, e^{-(\nu - p_i)^2 / 2w_i^2}} . \quad (5)$$

5.4 Example 1: Azurin

EPR is used extensively to probe the active centres of metalloproteins [89]. Here we use metalloproteins to illustrate the procedure of extracting orientational information from ODEPR spectra of frozen solutions. Our first example is Pseudomonas aeruginosa azurin [88]. The conventional EPR spectra

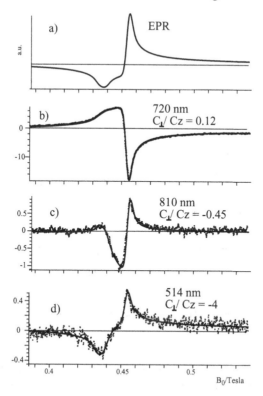

Fig. 15. (a) Simulation of conventional EPR absorption (solid line) and dispersion (dashed line). (b) to (d) Dispersion type ODEPR spectra of azurin at di erent optical wavelengths in units of · 10^{-3} M^{-1} cm^{-1} (dashed line) and fit curves (solid line) [88]

of azurin have the typical shape of an axially symmetric system, as shown in (Fig. 15a). The spectrum can be fitted with the following g-values g_z = 2 .26, g = 2 .045, the hyperfine coupling constants A_z = 172 MHz, A = 27 MHz, and the EPR linewidth EPR = 55 MHz.

The corresponding ODEPR spectra (see Fig. 15) are dispersion phase spectra, since the absorption (i.e. in phase) component of the ODEPR signal is strongly saturated under the experimental parameters typically used in these experiments (T = 1 .8 K, microwave power = 100 mW). This behaviour is exactly analogous to conventional EPR, where the absorption phase of inhomogeneously broadened lines saturates much faster than the dispersion component [90].

A comparison of the ODEPR spectra in Fig. 15 shows that the lineshape varies significantly with the laser wavelength. This variation indicates that di erent optical transitions are involved, and that the optical anisotropy co-e cients are di erent for these transitions. Fitting the ODEPR spectra with

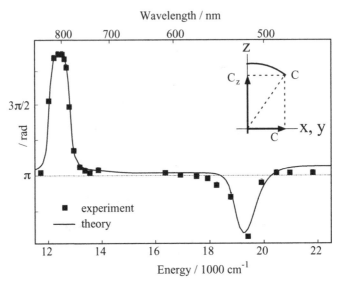

Fig. 16.   Variation of the orientational angle   with the laser wavelength. Filled squares represent experimental data, the   solid line the theory. The inset shows the definition of the orientational angle   in the molecular coordinate system

(2), the orientation angle   = arctan $\frac{C}{C_z}$ can be determined for each wavelength (Fig. 16). Over a large wavelength range, the angle is close to   = , indicating that the optical anisotropy reaches a maximum for light propagating parallel to the z-axis of the g-tensor (i.e. molecular symmetry axis) and that the MCD is negative. Close to 800 nm, the MCD becomes positive (2), and in the region close to 520 nm, the angle reaches   = / 2.

This variation is a strong indication of an underlying band structure of the optical transitions. To determine this band structure, the theoretical wavelength-dependence of the orientational angle (5) was fitted to the experimental data simultaneously with a fit of the MCD spectrum. A convincing agreement between the theory and all available experimental data is obtained if six optical transitions are considered (see Fig. 17).

As discussed in detail elsewhere [88], these optical transitions can be assigned to transitions between electronic states of the metalloprotein. The three higher-energy resonances are charge transfer transitions, the three resonances at lower energy correspond to ligand field transitions. This assignment allows one to calculate the orientation of the optical anisotropy tensor   C with respect to the molecular axis system. Since the orientation of the   C with respect to the g tensor was determined experimentally, one thereby obtains the orientation of the g-tensor within the molecule. The result obtained for azurin agrees well with results from the related compound plastocyanin [91], which had been determined by single crystal EPR.

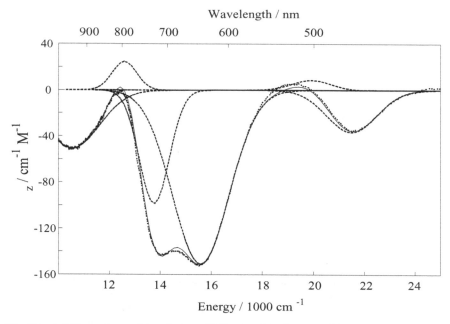

Fig. 17.   MCD spectrum of Azurin (solid line) and its decomposition in six Gaussian bands (dotted lines)

5.5 Example 2: Rubredoxin

The first high spin system, to which ODEPR was applied, is oxidised rubre-doxin from Clostridium pasteurianum   [92]. This electron transfer protein contains a single high-spin iron sulfur cluster. The optical spectrum has 6 charge transfer bands in the visible and near UV region. To cover the most interesting part of this spectral range, dierent lasers with wavelengths between 459 nm and 560 nm were used.

The EPR spectrum of rubredoxin (conventional and ODEPR) can be ex-plained with a zero field splitting of   D = +46 .3 GHz and a strong rhombic distortion of E/D = 0 .25, where E is the axial and D the rhombic coe cient. The spectra showed significant deviations from the ideal spectrum expected for these parameters, which can be explained as   E/D   strain, i.e. a statisti-cal distribution around the mean value of 0.25. This result indicates that the protein conformation is quite variable even in the frozen solution.

The strong variation of the ODEPR lineshape with the optical excitation wavelength allowed us to identify four optical transitions in the wavelength range covered by our measurements. As in the low-spin cases, the ODEPR and MCD spectra were fitted with a single parameter set. Figure 18 shows the MCD of rubredoxin and its decomposition into the four relevant optical transitions.

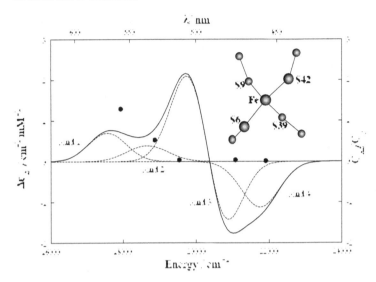

Fig. 18. MCD spectrum (solid line, left hand scale) of Rubredoxin with the four bands (dotted lines). The ratio of C_{xy}/C_y from ODEPR is indicated with dots (right hand scale). The inset shows the active site of rubredoxin, with the iron centre and the four adjacent sulfur atoms. The pseudo-S_4-axis is perpendicular to the plane of drawing

Even though optical and EPR experiments had already provided most of the relevant parameters in rubredoxin, details like the orientation of the optical and magnetic tensors had remained elusive. With the combination of ODEPR and conventional EPR it was possible to find the positive sign of the zero field Hamiltonian. Further, the orientation of the g-tensor could be identified and the orientation of the optical symmetry axis was found to be along the direction of largest g-value (perpendicular to the plane of the drawing, inset of Fig. 18).

6 Conclusions

This article summarizes some of the applications of laser radiation to magnetic resonance spectroscopy. The discussion concentrates on applications where the laser light does not modify the sample (apart from the spin system) and therefore excludes, e.g., EPR spectroscopy of photogenerated states. We have distinguished between optical processes that increase the spin polarization and others that are used for the detection. Applications from NMR and EPR were chosen to illustrate the potential of optical techniques. In particular the EPR examples show that the laser is not merely an aid for magnetic resonance, but that the combination of optical and magnetic resonance techniques provides information that is not accessible from separate single resonance optical and magnetic resonance spectra.

Acknowledgments

We gratefully acknowledge contributions from many members of the workgroup. Experimental examples were provided by T. Blasberg, B. Enkisch, M. Eickho , R. Klieber, and A. Michalowski. Financial support was provided by the Deutsche Forschungsgemeinschaft .

References

1. J. K"ohler, J. Disselhorst, M. Donckers, E. Groenen, J. Schmidt, W. Moerner: Nature 363 , 242–243 (1993)
2. J. Wrachtrup, C.V. Borczyskowski, J. Bernard, M. Orrit, R. Brown: Nature 363 , 244–245 (1993)
3. J. Wrachtrup, A. Gruber, L. Fleury, C.V. Borczyskowski: Chem. Phys. Lett. 267 , 179–185 (1997)
4. T. Blasberg, D. Suter: Phys. Rev. B   51 , 6309–6318 (1995)
5. A. Kastler: Science   158 , 214–221 (1967)
6. J. Brossel, A. Kastler, J. Winter: J. Phys. Radium   13 , 668–668 (1952)
7. A. Kastler: J. Phys. Rad.   11 , 255–265 (1950)
8. A. Kastler: J. Opt. Soc. Am. B   47 , 460–465 (1957)
9. H. Dehmelt: Phys.Rev.   109 , 381–385 (1958)
10. N. Bigelow, P. Nacher, M. Leduc: J. Phys. II France   2, 2159–2179 (1992)
11. J.C. Leawoods, D.A. Yablonskiy, B. Saam, D.S. Gierada, M.S. Conradi: Concepts Magn Reson. 13 , 277–293 (2001)
12. W. Happer, E. Miron, S. Schaefer, D. Schreiber, W.V. Wijngaarden, X. Zeng: Phys. Rev. A 29 , 3092–3110 (1984)
13. D. Raftery, H. Long, T. Meersmann, P. Grandinetti, L. Reven, A. Pines: Phys. Rev. Lett. 66 , 584–587 (1991)
14. H. J"ansch, T. Hof, U. Ruth, J. Schmidt, D. Stahl, D. Fick: Chem. Phys. Lett. 296 , 146–150 (1998)
15. H.-U. Kauczor, M. Ebert, K.-F. Kreitner, H. Nilgens, R. Surkau, W. Heil, D. Hofmann, E.W. Otten, M. Thelen: JMRI   7, 538–543 (1997)
16. D. Levron, D.K. Walter, S. Appelt, R.J. Fitzgerald, D. Kahn, S.E. Korbly, K.L. Sauer, W. Happer, T.L. Earles, L.J. Mawst, D. Botez, M. Harvey, L. DiMarco, J.C. Connolly, H.E. M" oller, X.J. Chen, G.P. Cofer, G.A. Johnson: Appl. Phys. Lett. 73 , 2666–2668 (1998)
17. G. Cates, D. Benton, M. Gatzke, W. Happer, K. Hasson, N. Newbury: Phys. Rev. Lett. 65 , 2591–2594 (1990)
18. C. Bowers, H. Long, T. Pietrass, H. Gaede, A. Pines: Chem. Phys. Lett.   205 , 168–170 (1993)
19. F. Bitter: Phys. Rev.   76 , 833–835 (1949)
20. H.G. Dehmelt: Phys. Rev.   105 , 1924–1925 (1957)
21. E.W. Bell, A.L. Bloom: Phys. Rev.   107 , 1559–1565 (1957)

22. N. Bloembergen, P.S. Pershan, L.R. Wilcox: Phys. Rev.   120 , 2014–2023 (1960)
23. I. Wieder: Phys. Rev. Lett.   3, 468–470 (1959)
24. C.P. Barrett, J. Peterson, C. Greenwood, A.J. Thomson: J. Am. Chem. Soc. 108 , 3170–3177 (1986)
25. S.J. Bingham, D. Suter, A. Schweiger, A.J. Thomson: Chem. Phys. Letters   266 , 543–547 (1997)
26. B. B̈orger, S.J. Bingham, J. Gutschank, M.O. Schweika, D. Suter: J. Chem. Phys. 111 , (18), 8565–8568 (1999)
27. A.B. Dennison: Magnet. Resonance Rev.   2, (1), 1–33 (1973)
28. R.H. Clarke, editor: Triplet State ODMR Spectroscopy   (John Wiley, New York 1982)
29. J. Brossel, F. Bitter: Phys. Rev.   86, 308–316 (1952)
30. F. Jelezko, I. Popa, A. Gruber, C. Tietz, J. Wrachtrup, A. Nizovtsev, S. Kilin: Appl. Phys. Lett.   81 , 2160–2162 (2002)
31. W. Hanle: Z. Physik   30 , 93–105 (1924)
32. N. Bloembergen, Y. Zou, L. Rothberg: Phys. Rev. Lett.   54, 186–188 (1985)
33. J. Brossel, S. Geschwind, A. Schawlow: Phys. Rev. Lett.   3, 548 (1959)
34. D. Paget, G. Lampel, B. Sapoval, V. Safarov: Phys. Rev. B   15 , 5780–5796 (1977)
35. C. Raman, K. Krishnan: Nature   121 , 501–502 (1928)
36. J.A. Giordmaine, W. Kaiser: Phys. Rev.   144 , (2) (1966)
37. Y. Bai, R. Kachru: Phys. Rev. Lett.   67, 1859–1862 (1991)
38. N. Wong, E. Kintzer, J. Mlynek, R. DeVoe, R. Brewer: Phys. Rev. B   28 , 4993–5010 (1983)
39. R. Shelby, C. Yannoni, R. Macfarlane: Phys. Rev. Lett.   41, 1739–1742 (1978)
40. L. Erickson: Phys. Rev. B   43, 12723–12728 (1991)
41. R. Shelby, A. Tropper, R. Harley, R. Macfarlane: Opt. Lett.   8, 304–306 (1983)
42. R. Shelby, R. Macfarlane: J. Luminesc.   31, 839–844 (1984)
43. T. Blasberg, D. Suter: Optics Commun.   109 , 133–138 (1994)
44. S. V̈olker: Ann. Rev. Phys. Chem.   40, 499–530 (1989)
45. N.B. Manson, N. Rigby, B. Lou, J.P. Martin: J. Luminesc.   53, 251–254 (1992)
46. R. Klieber, A. Michalowski, R. Neuhaus, D. Suter: Phys. Rev. B   67, 184103 (2003)
47. Y. Chen, K. Chiang, S. Hartmann: Phys. Rev. B   21, 40–47 (1980)
48. A. Szabo: J. Opt. Soc. Am. B   3, 514–522 (1986)
49. R.M. Macfarlane: Journal of Luminescence   100 , 1–20 (2002)
50. A. Kaplyanskii, R. Macfarlane: Spectroscopy of solids containing rare earth ions (Elsevier 1987)
51. B. Bleaney: Physica   69, 317–329 (1973)
52. M. Teplov: Sov. Phys. JETP   26, 872 (1968)
53. A. Abragam: The Principles of Nuclear Magnetism   (Oxford University Press, Oxford 1961)
54. T. Blasberg, D. Suter: Phys. Rev. B   48, 9524–9527 (1993)
55. K. Morigaki: Jap. Journal of Applied Physics   22, 375–388 (1983)
56. E. Glaser, J. Trombetta, T. Kennedy, S. Prokes, O. Glembocki, K. Wang, C. Chern: Phys. Rev. Lett.   65, 1247–1250 (1990)
57. M. D'Yakonov, V. Perel: Sov. Phys. JETP   33, 1053–1059 (1971)
58. G. Flinn, R. Harley, M. Snelling, A. Tropper, T. Kerr: J. Luminesc.   45, 218–220 (1990)
59. T. Uenoyama, L. Sham: Phys. Rev. Lett.   64, 3070–3073 (1990)

60. M. D'Yakonov, V. Perel: Sov. Phys. JETP   36, 995–1000 (1973)
61. D. Paget: Phys. Rev. B   24, 3776–3793 (1981)
62. M. Krapf, G. Denninger, H. Pascher, G. Weimann, W. Schlapp: Solid State Comm. 78, 459–464 (1991)
63. S.K. Buratto, D.N. Shykind, D.P. Weitekamp: Phys. Rev. B   44, 9035–9038 (1991)
64. J. Marohn, P. Carson, J. Hwang, M. Miller, D. Shykind, D. Weitekamp: Phys. Rev. Lett. 75, 1364–1367 (1995)
65. M. Eickho, D. Suter: J. Mag. Res.   166, 69–75 (2004)
66. D. Guerrier, R.T. Harley: Appl. Phys. Lett.   70, 1739–1741 (1997)
67. M. Eickho, B. Lenzmann, D. Suter, S.E. Hayes, A.D. Wieck: Phys. Rev. B   67, 085308 (2003)
68. D. Gill. N. Bloembergen: Phys. Rev.   129, 2398–2403 (1963)
69. K. Dumas, F. Soest, A. Sher, E. Swiggard: Phys. Rev. B   20, 4406–4415 (1979)
70. C. Slichter: Ann. Rev. Phys. Chem.   37, 25 (1986)
71. T. Duncan, C. Dybowski: Surf. Science Reports   1, 157 (1981)
72. B. Chmelka, D. Raftery, A. McCormick, L. DeMenorval, R. Levine, A. Pines: Phys. Rev. Lett. 66, 580 (1991)
73. Z. Wu, W. Happer, M. Kitano, J. Daniels: Phys. Rev. A   42, 2774 (1990)
74. R. Butscher, G. W" ackerle, M. Mehring: J. Chem. Phys.   100, 6923–6933 (1994)
75. D.J. Lepine: Phys. Rev. B   6, 436–441 (1972)
76. M. Lukac, E. Hahn: J. Luminesc.   42, 257–265 (1988)
77. D. Suter, J. Aebersold, J. Mlynek: Opt. Commun.   84, 269–274 (1991)
78. S. Grafstr" om, T. Blasberg, D. Suter: J. Opt. Soc. Am. B   13, 3–10 (1994)
79. S. Grafstr" om, D. Suter: Optics Letters   20, 2134–2136 (1995)
80. S. Grafstr" om, D. Suter: Phys. Rev. A   54, 2169–2179 (1996)
81. S. Grafstr" om, D. Suter: Zeitschrift f" ur Physik D   38, 119–132 (1996)
82. B. Cavenett: Adv.Phys.   30, 475–538 (1981)
83. J. K"ohler: Physics Reports   310, 261–339 (1999)
84. M.O. Schweika-Kresimon, J. Gutschank, D. Suter: Phys. Rev. A   66, (4), 043816 (2002)
85. S.J. Bingham, B. B" orger, D. Suter, A.J. Thomson: Rev. Sci. Instrum.   69, (9), 3403–3409 (1998)
86. S.J. Bingham, B. B" orger, J. Gutschank, D. Suter, A.J. Thomson: JBIC   5, 30–35 (2000)
87. S.J. Bingham, J. Gutschank, B. B"   orger, D. Suter, A.J. Thomson: J. Chem. Phys. 113, 4331–4339 (2000)
88. B. B"orger, J. Gutschank, D. Suter, A.J. Thomson, S.J. Bingham: J. Am. Chem. Soc. 123, 2334–2339 (2001)
89. G. Palmer: Methods for Determining Metal Ion Environments in Proteins   In Vol. 2 of Advances in Inorganic Biochemistry   , Chapter in 6   Electron Paramagnetic Resonance   (Elsevier, Amsterdam 1980)
90. A.M. Portis: Phys. Rev.   91, (5), 1071–1078 (1953)
91. K.W. Penfield, R.R. Gay, R.S. Himmelwright, N.C. Eickman, V.A. Norris, H.C. Freeman, E.I. Solomon: J. Am. Chem. Soc.   103, 4382–4388 (1981)
92. B. B"orger, D. Suter: J. Chem. Phys.   115, (21), 9821–9826 (2001)

Multiple-Photon Transitions
in EPR Spectroscopy

Moritz K¨alin, Matvey Fedin, Igor Gromov and Arthur Schweiger

Laboratory for Physical Chemistry, ETH Zurich, 8093 Zurich, Switzerland
schweiger@esr.phys.chem.ethz.ch

Abstract. An overview of the various multiple-photon processes in electron paramagnetic resonance (EPR) spectroscopy is given. First, we describe di erent types of multiple-photon transitions that can be observed with monochromatic and bichromatic microwave fields in spin systems with unequally spaced energy levels and in two-level systems. Then we discuss multiple-photon processes that are based on a bichromatic radiation field consisting of a transverse microwave field and a longitudinal radio frequency field. Two semiclassical methods, namely the tilted frame and the toggling frame approach, as well as the quantized radiation field formalism (second quantization), and Floquet theory are used for the theoretical description of the e ects. We discuss how these processes manifest in the conventional field modulated continuous wave EPR experiment and demonstrate how multiple-photon transition can be observed in pulse EPR. Finally, new features of multiple-photon transitions induced by bichromatic radiation fields are introduced. We particularly stress on the phenomenon of   -photon-induced transparency and describe the characteristics of these transitions and their potential applications.

1 Introduction

Many phenomena in infrared spectroscopy and coherent optics are based on the interaction of electromagnetic radiation with matter, where several photons are simultaneously absorbed or emitted [1, 2, 3, 4, 5, 6, 7]. Such multiple-photon transitions [1], have already been described in the first half of the last

---

[1] In literature the term "multiple-quantum transition" is often used instead of "multiple-photon transition". However, since in magnetic resonance this term is also used for transitions between states which di er in the magnetic quantum number by   m >   1, we use the term multiple-photon transition throughout this work. During a multiple-quantum transition the radiation frequency matches the energy di erence between the two levels involved in the transition so that, like in a single-quantum transition, only one photon is absorbed or emitted. On the other hand, during a multiple-photon transition several photons are absorbed and/or emitted.

M. K¨alin et al.: Multiple-Photon Transitions in EPR Spectroscopy   , Lect. Notes Phys.   684 , 143–183 (2006)
www.springerlink.com

century [8]. Also in magnetic resonance multiple-photon phenomena have been known for a long time, but up to now extensive use of them as a spectroscopic tool has not yet been made.

In any transition between two energy levels energy and angular momentum are conserved. As a consequence, during a multiple-photon process in magnetic resonance the energy and the polarization of the photons involved in the process must correspond to the total change in energy and angular momentum in the spin system.

Electrons have an intrinsic angular momentum characterized by the spin quantum number $S = 1/2$. When an electron spin is subjected to a static magnetic field B_0 oriented along the laboratory z-axis, the projection of the angular momentum along this axis is either $/2$ or $-/2$, and the energy levels corresponding to the magnetic quantum numbers $m_S = 1/2$ and $m_S = -1/2$, are split owing to the Zeeman interaction.

Photons have an intrinsic angular momentum, characterized by the spin quantum number $J = 1$. Photons in state $m_J = -1$ have angular momentum $-$ relative to the direction of propagation and are called $^-$ photons (negative helicity), whereas photons in state $m_J = 1$ have angular momentum and are called $^+$ photons (positive helicity) [3]. Left and right circular fields perpendicular to B_0 can be associated with $^-$ and $^+$ photons propagating along B_0.

The state with projection $m_J = 0$ is forbidden, since a photon has a vanishing mass. However, a linear field oriented parallel to B_0 can be associated with so-called photons propagating perpendicular to B_0 and described by the wave function $= (\overline{2}/2)^+ + (\overline{2}/2)^-$. The eigenstate of this function thus corresponds to $m_J = 0$, and absorption or emission of photons does not change the total angular momentum of the spin system.

In magnetic resonance radio frequency (rf) or microwave (mw) radiation fields are applied. A radiation field with only one frequency is called a monochromatic field. Correspondingly, we call a radiation field with two frequencies a bichromatic field, independent on whether the frequencies are very close to each other or differ by orders of magnitudes.

During a multiple-photon process, several photons are simultaneously absorbed and/or emitted from the radiation field. The sum of the angular momenta of all absorbed photons must equal the change in angular momentum of the electron spin system expressed by the change in m_S. There is no restriction to the number of absorbed or emitted photons, since they do not add to the angular momentum.

Several theoretical approaches have been developed to describe multiple-photon processes [9]. In magnetic resonance the Bloch equations [10, 11], or semiclassical theories, with the radiation field treated as a classical observable, are frequently used. However, to understand the physics behind a multiple-photon process second quantization, where not only the spins but also the radiation fields are quantized [12, 13, 14], and Floquet theory [15, 16, 17] has to be applied. A brief introduction to second quantization and Floquet theory is given in the Appendix.

In this Lecture Notes we discuss different types of multiple-photon processes that can be observed in electron paramagnetic resonance (EPR) spectroscopy. Completeness in the cited literature is not aimed at. In pulse EPR spectroscopy multiple-photon resonances are also attractive from the point of view of spin dynamics and potential applications. In Sect. 2 we give a brief overview of the different types of multiple-photon transitions that can be observed in EPR with monochromatic or bichromatic mw fields. Section 3 is devoted to multiple-photon transitions induced by a bichromatic radiation field consisting of a transverse mw field and a longitudinal rf field. In Sect. 3.1 we describe how these processes manifest in continuous wave (cw) EPR. Although this type of multiple-photon transitions is omnipresent in cw EPR, most spectroscopists are not aware of the fact that they are dealing with them in their daily routine work. In Sect. 3.2 we demonstrate how multiple-photon transitions can be observed in pulse EPR. Finally, in Sect. 3.3 a new type of multiple-photon transitions induced by a bichromatic field is introduced. We particularly stress on the phenomenon of -photon-induced transparency and describe the characteristics of these transitions and their potential applications.

2 Different Types of Multiple-Photon Transitions in EPR

The energies of states of a paramagnetic species with an electron spin S and nuclei with spins I subject to a static magnetic field B_0 are described by the static spin Hamiltonian H_0. The Hamiltonian consists of terms which describe the interactions between the electron spin and B_0, as well as the interaction between electron spins and between electron and nuclear spins.

The coupling between an electron spin and a linear radiation field $B_{mw}(t) = 2B_1 \cos(\omega_{mw} t)$ with amplitude $2B_1$ and mw frequency ω_{mw} is expressed by the perturbation Hamiltonian $H_1(t)$ [18]. In this Lecture Notes energies and amplitudes of radiation fields are usually given in angular frequency units, i.e., $\omega = E/\hbar$ and $\omega_1 = -\gamma_e B_1$.

In the great majority of EPR studies, the linear response of a spin system to a weak radiation field is recorded. The transition amplitudes are then determined by the matrix elements of the perturbation operator $H_1(t)$. The transition amplitude for a single-photon transition between the two eigenstates with energies ω_1 (labeled by 1) and ω_2 (labeled by 2) of the static Hamiltonian having wave functions ψ_1 and ψ_2 is given by

$$b_{12} = \langle 1|H_1|2 \rangle , \tag{1}$$

where H_1 is the time-independent perturbation Hamiltonian in a frame that rotates with the mw frequency ω_{mw} (rotating frame). The computation of b_{12}

is most convenient in the eigenbasis of H_0, where b_{12} is the matrix element of H_1 connecting the states 1 and 2.

Due to state mixing, the magnetic quantum numbers of the electron and nuclear spins may no longer be good quantum numbers, and forbidden EPR transitions of the type ($m_S = \pm 1$; $m_I = \pm 1, \pm 2, \dots$), or ($m_S = \pm 2, \pm 3, \dots$) become weakly allowed. Since for these forbidden transitions, $|m_S| > 1$ or $|m_I| > 0$, they are often called multiple-quantum transitions. Note, however, that such a transition, where only one photon is absorbed or emitted, is still a single-photon process.

During a single-photon transition, the spin system absorbs or emits one photon of a radiation field perpendicular to the quantization direction of the spin, so that the angular momentum is conserved. During an electron spin transition with a positive g value or a nuclear spin transition with a negative g_n value, the absorbed or emitted photon is a $^+$ photon (Fig. 1a). In the rare case of a negative g value [19], or for a nucleus with positive g_n, the absorbed or emitted photons are $^-$ photons (Fig. 1b).

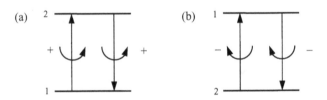

Fig. 1. Energy level diagram for a single-photon transition in a two-level system, fulfilling energy and angular momentum conservation (curved arrows symbolize the polarization of the absorbed or emitted photon). (a) $(m_S = 1/2) > (m_S = -1/2)$: One $^+$ photon is absorbed or emitted. (b) $(m_S = -1/2) > (m_S = 1/2)$: Order of the energy levels is reversed, one $^-$ photon is absorbed or emitted

In magnetic resonance linear radiation fields are used with a very few exceptions [19, 20, 21]. Such a linear field consists of both $^-$ and $^+$ photons, since it can be considered as a superposition of a left and a right circular field. The generation of a circular field is experimentally more demanding, but has the advantage that the power needed to generate the same circular field amplitude $_1$ is reduced to one quarter compared to a linear field, and that the Bloch-Siegert shift (see below) is absent. Note, however, that for many of the multiple-photon processes described in these Lecture Notes, both $^-$ and $^+$ photons are required, and transitions can not be induced by a circular radiation field.

2.1 Multiple-Photon Transitions in Spin Systems with Unequally Spaced Energy Levels

In EPR spectroscopy multiple-photon transitions, where the magnetic quantum number changes by $\Delta m_s > 1$, have been observed in various spin systems with unequally spaced energy levels (see for example [22, 23, 24, 25, 26]). In this kind of multiple-photon transitions, which requires real intermediate energy levels, the final state is reached without directly altering the populations of the intermediate states.

For the description of these multiple-photon transitions, higher-order time-dependent perturbation theory is usually applied. It has been found in cw EPR (and cw NMR), that (a) the unsaturated linewidth of an n-photon transition is $1/n$ of the linewidth of the single-photon transition, (b) in the absence of saturation effects the intensity of the signal for an n-photon transition is proportional to B_1^{2n-1}, (c) the intensity of an n-photon transition between levels with energy ε_a and ε_b depends on the position of the intermediate energy levels with respect to the $(n-1)$ equally spaced virtual levels between ε_a and ε_b [27].

One Radiation Field Perpendicular to B_0

In this type of multiple-photon transitions, two or more σ^+ photons of the same frequency are absorbed when the energy levels are almost equally spaced. In EPR these transitions can be observed either with a linear or a right circular radiation field perpendicular to B_0 (Fig. 2 a).

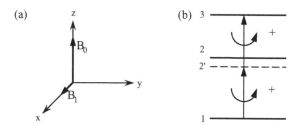

Fig. 2. (a) Field configuration in the laboratory frame and, (b) energy level diagram for a two-photon transition induced in a three-level system with unequally spaced energy levels by one right circular radiation field perpendicular to B_0 (σ^+ photons)

As an example, we consider the simple situation of a right circular radiation field applied to a spin system consisting of three energy levels labeled by 1, 2, and 3 (Fig. 2b). In addition to the two single-photon transitions with $\Delta m_s = \pm 1$ between level 1 and 2, and level 2 and 3, a two-photon transition can be induced between level 1 and 3. If the states are not mixed the transition amplitude of the single-photon transition with $m_S = \pm 2$ is zero. For the two-photon transition, where two $^+$ photons of the same frequency $_{mw} = {}^{(13)}/2$ are absorbed, the transition amplitude is given by [28]

$$b_{13} \quad {}_{1}^{2}\frac{{}_1|S^+|_2 \quad {}_2|S^+|_3}{{}^{(32)} - {}^{(31)}/2}, \qquad (2)$$

with ${}^{(32)} = {}_3 - {}_2$, ${}^{(31)} = {}_3 - {}_1$, and the raising operator $S^+ = S_x + iS_y$. Such a two-photon transition can be considered as a forbidden transition between level 1 and 3 with $m_S = \pm 2$, carried out in two allowed steps, first from level 1 to an intermediate (virtual) level 2', and then from level 2' to level 3. The transition amplitude b_{13} in (2) is proportional to the square of the mw field amplitude $_1$ and inversely proportional to $= {}^{(32)} - {}^{(31)}/2$, where

is the energy di erence between level 2 and the virtual level 2'. Thus, to observe a two-photon transition with su cient intensity, the amplitude of the pumping field has to be large and   has to be small. Corresponding formulas hold for transitions where three or more photons are absorbed. For multiple-photon transitions in spin systems with unequally spaced energy levels, the number of absorbed $^+$ photons is equal to the total change in m_S of the electron spin, so that angular momentum is conserved.

Two examples of multiple-photon transitions observed in electron spin systems with unequally spaced energy levels are given in Fig. 3. The spectrum of a Ni^{2+} -doped MgO single crystal with $S = 1$ is shown in Fig. 3a. When the spectrum is recorded with high mw power a broad single-photon transition

(a)   (b)

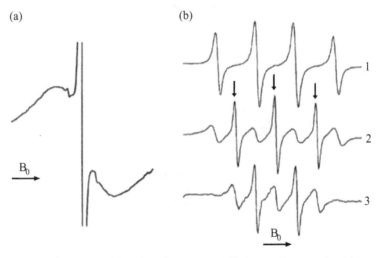

Fig. 3.   Multiple-photon transitions in spin systems with unequally spaced energy levels. (a) Two-photon transition (narrow center line   ) of Ni^{+2} in a MgO single crystal (adapted from [29]). (b) Multiple-photon transitions of atomic oxygen. 1: mw power 0.6 mw; 2: 4.5 mW (arrows mark two-photon transitions); 3: 42 mW (adapted from [30])

superimposed by a narrow double-photon line is observed [29]. The broadening of the single-photon line is caused by lattice strains, which shift the state $|0\rangle$ by different amounts relative to the states $|-1\rangle$ and $|1\rangle$. Since for the two-photon transition state $|0\rangle$ plays only the role of an intermediate level, the line remains narrow.

The second example (Fig. 3b) shows the spectral changes observed in atomic oxygen in the gas phase as a function of the mw power [30]. At low power (0.6 mW) only single-photon transitions are observed. But already at a power of 4.5 mW the spectrum is dominated by three two-photon transitions (marked by arrows). At a power of 42 mW the two three-photon transitions are the most intense lines, while the single-photon transitions are buried in the noise.

In all these EPR experiments where the multiple-photon transitions are directly observed, the line intensity is proportional to B_1^{2n-1}. The situation is different when the multiple-photon transitions are indirectly detected via the change in polarization of another transition. This is the case, for example, in cw electron-nuclear double resonance (ENDOR) experiments where nuclear frequencies are detected via desaturation of an EPR line. For a non-saturating rf field the intensity of an n-photon transition is then proportional to B_1^{2n}.

Figure 4 shows two examples for the observation of two-photon ENDOR transitions (marked by arrows) in a copper complex with two magnetically equivalent nitrogen ligands [28]. The intensity of the two-photon transitions depends on the parameter which can be varied via the orientation-dependent nuclear quadrupole interaction by rotating the crystal.

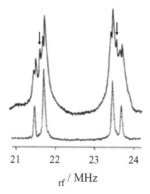
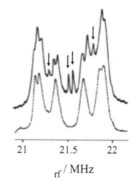

Fig. 4. Two-photon cw ENDOR transitions observed in bis(salicylaldoximato) copper(II), Cu(sal)$_2$, diluted into a Ni(sal)$_2$ single crystal. Two different orientations of the crystal are shown in the left and right panel. Bottom: low rf power, no two-photon transitions are observed; top: high rf power, arrows denote two-photon transitions (adapted from [28])

Two Radiation Fields Perpendicular to B_0

The restriction $^{(32)}$ $(\omega_3 - \omega_1)/2$ for the intermediate energy level may be dropped by using a bichromatic field perpendicular to B_0 with frequencies $\omega_{mw,1}$ and $\omega_{mw,2}$ and amplitudes $\omega_1 = \gamma_e B_1$ and $\omega_2 = -\gamma_e B_2$, as is shown in Fig. 5a.

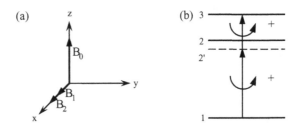

Fig. 5. (a) Field configuration in the laboratory frame and, (b) energy level diagram for a two-photon transition induced in a three-level system with unequally spaced energy levels by two right circular radiation fields perpendicular to B_0 (σ^+ photons)

In this scheme, $\omega_{mw,1}$ and $\omega_{mw,2}$ fulfilling the resonance condition $\omega_{mw,1} +$ $\omega_{mw,2} = \omega^{(31)}$, may be chosen such that $\Delta = \omega^{(21)} - \omega_{mw,1}$ is small for any value of ω_2. The transition amplitude for this type of multiple-photon transitions is then given by

$$b_{13} \simeq \omega_1 \omega_2 \frac{{}_1\langle S^+\rangle_2 \; {}_2\langle S^+\rangle_3}{\omega^{(21)} - \omega_{mw,1}}. \qquad (3)$$

Again corresponding formulas hold for transitions where three or more photons are absorbed.

An example for a two-photon transition induced by two mw frequencies is shown in Fig. 6 [31]. The system is again a Ni^{2+}-doped MgO single crystal. Line (a) or (c) is observed when the absorption of mw with frequency $\omega_{mw,1}$ or $\omega_{mw,2}$ is recorded. Line (b) in the center of the spectrum is attributed to the two-photon transition, where one photon of each radiation field is absorbed.

2.2 Multiple-Photon Transitions in Two-Level Systems

Multiple-photon transitions can also be induced between two energy levels when no real intermediate level exist. In a two-level system, or a multi-level system where only two levels are involved in the experiment, one has to distinguish between multiple-photon processes where either an odd or an even number of photons is absorbed.

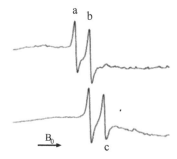

Fig. 6.   Two-photon transitions of Ni $^{+2}$ in a MgO single crystal induced by two mw fields of different frequencies (adapted from [31]).   Top: $\omega_{mw,1}$ is swept, bottom: $\omega_{mw,2}$ is swept. (a), (c) indicate single-photon transitions, (b) indicates two-photon transition

One Linear Radiation Field

Consider a two-level system with electron spin $S = 1/2$ interacting with a linear radiation field with frequency ω_{mw}. For an angle θ between this field and the static field B_0, the laboratory frame Hamiltonian is given by

$$H(t) = \omega_s S_z + 2 \omega_1 \cos(\omega_{mw} t)[\cos(\theta)S_z + \sin(\theta)S_x] , \qquad (4)$$

with the electron Zeeman frequency $\omega_s = -\gamma_e B_0$. The two eigenstates of the unperturbed spin system are denoted by α and β, with energies $\varepsilon_\alpha = \omega_s/2$ and $\varepsilon_\beta = -\omega_s/2$.

To follow the evolution of the spin system under the Hamiltonian in (4) the Schrödinger equation for the corresponding evolution operator has to be solved. To eliminate the time dependence of the Hamiltonian, Shirley implemented the Floquet theorem [15] as is outlined in the Appendix. Following this approach the transition probability between the two eigenstates of the time-dependent Hamiltonian $H(t)$ can be found by using the corresponding Floquet Hamiltonian with matrix elements

$$\langle \alpha n |H_F| \beta m \rangle = H^{n-m}_{\alpha\beta} + n \omega_{mw} \delta_{\alpha\beta} \delta_{nm} , \qquad (5)$$

where $|\alpha n\rangle$ represent the Floquet states, α describes the spin state and n denotes the component in the Fourier expansion $H(t) = \sum_n H^n e^{in\omega_{mw} t}$. The Hamiltonian in (5) has non-zero off-diagonal elements, which couple states that differ in the number of photons by $\Delta n = 1$ and in the spin projections by $\Delta m_s = 0, \pm 1$. The possible transitions can be revealed by checking the behavior of the energy levels in the vicinity of the ($n-m$)-resonance conditions $\varepsilon_\alpha - \varepsilon_\beta = (n - m)\omega_{mw}$. The corresponding energy levels $\varepsilon_{\alpha n} = \varepsilon_\alpha + n\omega_{mw}$ and $\varepsilon_{\beta m} = \varepsilon_\beta + m\omega_{mw}$ are degenerated when the oscillating field is zero. For $\omega_1 = 0$, the level crossings $\varepsilon_{\alpha n} = \varepsilon_{\beta m}$ are lifted. The existence of multiple-photon transitions thus implies a level anticrossing, and   vice versa.

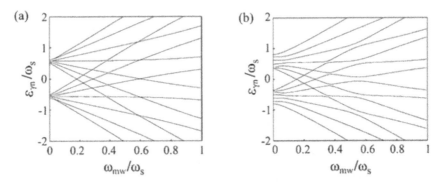

Fig. 7.   Eigenvalues of the Floquet Hamiltonian calculated as a function of the normalized photon energy ω_{mw}/ω_s for $\omega_1/\omega_s = 0.2$ and two angles   between B_0 and B_1. (a) Eigenvalues for $= 90$. (b) Eigenvalues for $= 45$

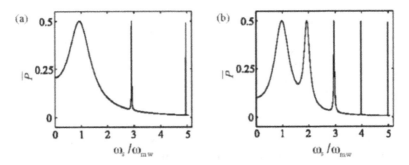

Fig. 8.   Time-averaged transition probabilities as a function of the normalized Larmor frequency ω_s/ω_{mw} for $\omega_1/\omega_s = 0.25$. (a) Radiation field perpendicular to B_0. (b) Direction of the radiation field tilted by $= 45$ (adapted from [32])

To demonstrate the existence of anticrossings, the eigenvalues of H_F have been calculated as a function of ω_{mw}/ω_s with $= 90$ and 45, and $\omega_1/\omega_{mw} = 0.25$ (Figs. 7a and b). Anticrossings are found in the vicinity of $\omega_{mw}/\omega_s = 1/n$. For an oscillating field perpendicular to B_0 an anticrossing is observed at $\omega_{mw}/\omega_s = 0.38$, corresponding to the position of a three-photon resonance (Fig. 7a). For $= 45$, anticrossings occur at $\omega_{mw}/\omega_s = 0.55, 0.37$ and 0.26; i.e., at two-, three- and four-photon resonances (Fig. 7b). A radiation field at an angle of 45 to B_0 may thus induce both   odd and even resonances.

Numerical computations of the time-averaged transition probability performed for H_F as a function of the normalized Larmor frequency ω_s/ω_{mw} are plotted in Figs. 8a and b for $= 0$ and 45. For a radiation field with two frequencies, the analysis is similar.

- One Linear Radiation Field Perpendicular to B_0

In a two-level system with electron spin $S = 1/2$ and energy levels 1 and 2, multiple-photon transitions with an   odd number of photons can be induced by a linear radiation field   perpendicular   to B_0 (Fig. 9a). In a $(2n+1)$-photon transition, $(n+1)^+$ photons and n^- photons are absorbed, so that the angular momentum is conserved.

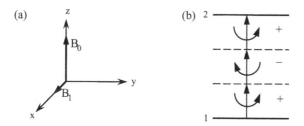

Fig. 9. (a) Field configuration in the laboratory frame and, (b) energy level diagram for a three-photon transition induced between the two states of an $S = 1/2$ spin system by a linear radiation field perpendicular to B_0 ($^+$ and $^-$ photons)

This type of multiple-photon transitions can only be induced by a   linear field which consists of both $^+$ and $^-$ photons. The energy level scheme for a three-photon transition $^+ + {}^- + {}^+$ with $\omega_{mw} = \omega_s/3$, where two $^+$ photons and one $^-$ photon are absorbed, is given in Fig. 9b. For this process the transition amplitude is found to be

$$b_{12}^{(3)} \quad {}_e^{(3)} = \quad {}_1^3/2 \, \omega_{mw}{}^2 . \tag{6}$$

The observation of such multiple-photon resonances has been reported for Cr^+-doped gallium phosphide [33]. Figure 10 shows the X-band cw EPR spectrum induced by red light.

In addition to the single-photon transition at 328 mT, also the two-photon transition at 656 mT and the three-photon transition at 984 mT (arrow) are induced. At high mw power multiple-photon transitions with   n up to 7 could be observed. It is assumed that the very long relaxation time of this material allows the observation of the multiple-photon transitions. In this experiment with the mw field oriented perpendicular to B_0, not only odd resonances as predicted by theory, but also even resonances could be observed (see below).

- One Linear Radiation Field at Arbitrary Orientation

A linear radiation field, which includes an angle < 90 with the static field B_0, can be decomposed into a linear field component   perpendicular to B_0 consisting of $^-$ and $^+$ photons, and a field component   parallel to B_0

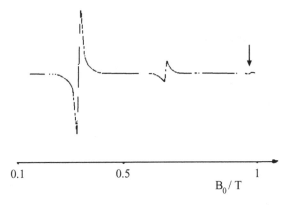

Fig. 10.   Light-induced EPR spectrum of chromium-doped gallium phosphide showing a single-photon, a two-photon, and a three-photon transition (marked by arrow)
(adapted from [33])

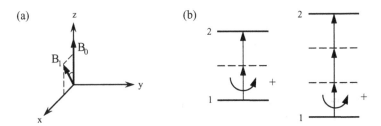

Fig. 11.   (a) Field configuration in the laboratory frame and, (b) energy level diagram for a two-photon and a three-photon transition induced between the two states
of an $S = 1/2$ spin system by a linear radiation field which includes an angle $\vartheta < 90$
with B_0 ($+$, $-$, and photons)

consisting of photons (Fig. 11a) [34]. Consequently, multiple-photon transitions with both an even and an odd number of absorbed photons can be
observed, as is shown in Fig. 11b for the transitions $+ + -$ and $+ + 2 \times$.

For a two-photon transition induced by a radiation field with frequency
$\omega_{mw} = \omega_s/2$, the transition amplitude given by

$$b_{12} \propto \frac{\omega_1^2 \sin(2\vartheta)}{2\omega_s} \tag{7}$$

is maximum for $\vartheta = 45$.

First observations of multiple-photon transitions of the type $+ + k \times$
have been reported by Winter in an optical pumping experiment [7]. The observation of even resonances in the spectrum of Cr^+-doped gallium phosphide
mentioned above (Fig. 10) is explained by a distortion of the mw field caused
by the high dielectric constant of the sample, which results in a component
of the radiation field along B_0.

The so-called second-harmonic-detected EPR experiments described by Bosca-ino and coworkers [35, 36, 37, 38, 39] are also based on two-photon transitions of the type $^+ +$. In this approach a bimodal cavity with two resonant modes in the frequency range 2– 6 GHz is used. In the first mode, a linear radiation field with frequency $_{mw}$ $_s / 2$, which includes with B_0 an angle of $= 45$, induces two-photon transitions in an $S = 1/2$ spin system. The resulting second-harmonic signal at the Larmor frequency $_s$ is detected via the second mode perpendicular to B_0. This scheme has the advantage that the signal at frequency $_s$ can be recorded during excitation, since the radiation field at half this frequency is far o -resonant from the observation frequency. The experimental scheme has been applied in various investigations of two-level systems (see for example [40]). Very recently, second-harmonic detection has been used to study the decay properties of stimulated nutation echoes [41].

The two-pulse sequence used to generate this type of echoes is shown in Fig. 12a. Both pulses with frequency $_{mw}$ $_s / 2$ drive the two-photon transition $^+ +$. The first pulse of length $_1$ burns a polarization pattern into an inhomogeneous line [18], which is recalled by the second pulse after a waiting time $_1$. The stimulated nutation echo is formed at time $t = _1$ after the onset of the second pulse (Fig. 12b [41]).

Two or More Radiation Fields Perpendicular to B_0

Multiple-photon processes become more complex when transitions are induced in a two-level system by linear bichromatic or tetrachromatic radiation fields perpendicular to B_0. Multiple-photon transitions with an odd number n of photons can be induced by two linear radiation fields with frequencies $_{mw,1}$ and $_{mw,2}$ and amplitudes $_1 = _2$ (Fig. 13a), when the resonance condition

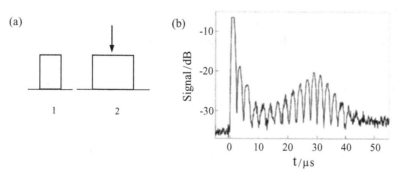

(a)

(b)

1

2

Signal / dB

-10

-20

-30

0   10   20   30   40   50

t / μs

Fig. 12.   Stimulated nutation echo induced by a linear radiation field ($= 45$ and recorded via the second-harmonic signal. (a) Pulse sequence, arrow marks echo maximum. (b) Stimulated nutation echo in -irradiated quartz glass (adapted from [41])

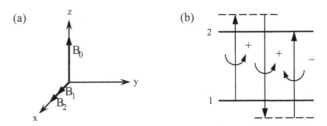

Fig. 13. (a) Field configuration in the laboratory frame and, (b) energy level diagram for a three-photon transition induced between the two states of an $S = 1/2$ spin system by two linear radiation fields perpendicular to B_0 ($^+$ and $^-$ photons)

$$\omega_s = \omega_{mw,1} \pm [n^2(\omega_{mw,2} - \omega_{mw,1})^2 - \omega_1^2]^{1/2} \tag{8}$$

is fulfilled [42]. A three-photon transition of this type is shown in Fig. 13b.

Recently, Hyde and co-workers introduced a multiple-photon EPR experiment, where the spin system is excited by two (or four) cw radiation fields with closely spaced mw frequencies [43, 44, 45, 46, 47]. In the case of two radiation fields with frequencies $\omega_{mw,1} = \omega_{mw} + \frac{\omega}{2}$ and $\omega_{mw,2} = \omega_{mw} - \frac{\omega}{2}$ and amplitudes ω_1, ω_2, the multiple-photon process creates coherences that manifest as sidebands at frequencies $\omega_{mw} \pm (k + \frac{1}{2})\omega$. Corresponding experiments have also been reported in pulse NMR [48, 49] and cw NMR [12, 50, 51], and in the optic regime [52].

Each individual sideband can be down-converted with respect to the reference frequency ω_{mw} and detected with a phase-sensitive detector. The same detection scheme has also been used in cw electron-electron double resonance [53] and cw ENDOR [54] experiments.

This approach can be used to measure pure absorption EPR spectra [55]. As an example, in Fig. 14 the conventional cw EPR spectrum of nitrous oxide reductase is compared with the three-photon EPR spectrum recorded without field modulation, and the corresponding first-harmonic spectrum obtained by pseudo-modulation [56]. The difference between the two spectra (marked by an arrow) is traced back to the enhanced suppression of the forbidden transitions in the multiple-photon spectrum.

This type of multiple-photon process can also be studied with pulse excitation. Recently, an inversion recovery experiment with the pulse sequence shown in Fig. 15a and a tetrachromatic radiation field has been carried out on γ-irradiated quartz glass.

The tetrachromatic pulses were created by mixing the mw frequency ω_{mw} with two radio frequencies $\omega_{rf,1}$ and $\omega_{rf,2}$. An odd number of photons is absorbed from the tetrachromatic radiation field, which induces up-down transitions and creates polarization at frequency ω_s, corresponding to the center frequency ω_{mw} of the inversion pulse. The resonance condition, $\omega_s = \omega_{mw} + n_1 \omega_{rf,1} + n_2 \omega_{rf,2}$, is fulfilled for example for the three-photon

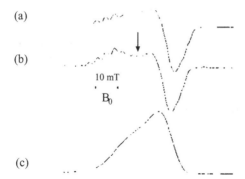

Fig. 14.   EPR on nitrous oxide reductase substituted with ^{63}Cu, temperature 20 K.
(a) Cw EPR spectrum. (b) First harmonic of the spectrum in (c), using pseudo-modulation. (c) Three-photon EPR spectrum (adapted from [55])

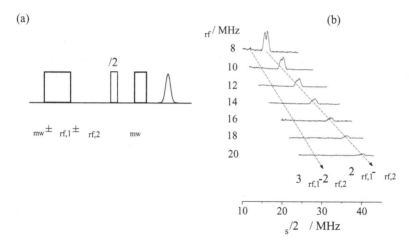

Fig. 15.   Three- and five-photon resonances in   -irradiated quartz glass at X-band frequency. (a) Inversion recovery pulse sequence with a tetrachromatic inversion pulse. (b) Multiple-photon resonances as a function of   rf,2 at di erent fixed frequencies   rf,1 of the tetrachromatic pulse. Three-photon resonances are observed for 2 rf,1 − rf,2 = 0 (dashed line). Very weak five-photon resonances with 3 rf,1 − 2 rf,2 = 0 (dashed line) can be observed at the two top traces. For convenience, the spectra are inverted

transition with $n_1 = 2$ and $n_2 = -1$, and the five-photon transition with $n_1 = 3$ and $n_2 = -2$. The integrated echo intensities as a function of   rf,2 at di erent fixed frequencies   rf,1 are plotted in Fig. 15b. In addition to the three-photon resonances, weak five-photon resonances become visible with radio frequencies of 8 and 10 MHz.

3 Effects of Oscillating Longitudinal Field

In this section we describe multiple-photon transitions induced by a bichromatic radiation field consisting of a linear or right circular mw field in the gigahertz range, perpendicular to B_0, and an rf field in the megahertz range, parallel to B_0 (Fig. 16a). Such a bichromatic field induces multiple-photon transitions of the type $\omega_{mw} + k \times \omega_{rf}$. In this section we understand by a bichromatic field always this kind of radiation field, and for the multiple-photon transitions we use the abbreviated form $\omega + k \times \omega$.

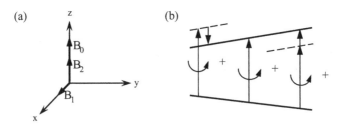

Fig. 16. (a) Field configuration in the laboratory frame and, (b) energy level diagram for the experiments with one mw field perpendicular to B_0 and one rf field parallel to B_0, showing two-photon resonances with $\omega_s = \omega_{mw} \pm \omega_{rf}$

Two-photon transitions $\omega + \omega$ of this type (Fig. 16b) have first been reported by Winter in 1958 [34] on the free radical DPPH, and three years later by Burget and coworkers and Hashi on DPPH and other two-level systems [57, 58]. Recently, such two-photon transitions has also been observed in electron-nuclear multi-level systems, where they appear in cw ENDOR spectra at the corresponding hyperfine frequencies [59].

3.1 The Field Modulation Used in cw EPR as a Multiple-Photon Process

In this section we demonstrate that field modulation commonly used in cw EPR spectroscopy, together with the mw field, represents a bichromatic radiation field which gives reason to complex multiple-photon processes.

Since the dawn of cw EPR spectroscopy, modulation of the static magnetic field B_0 with subsequent phase-sensitive detection is used to improve signal-to-noise. As a consequence of this field modulation the derivative of the absorption signal is observed. When the modulation amplitude or the modulation frequency are larger than the width of a particular line, the line shape is distorted. The former situation is known as modulation broadening due to overmodulation, and in the latter case sidebands appear in the spectrum [60].

A first theoretical description of modulation effects was given by Karplus [61] for mw rotational spectroscopy, where either the excitation frequency or the energy levels of the molecule are modulated. In magnetic resonance this theory was used by Smaller [62], and was further developed by several authors to obtain quasi steady-state solutions of the modified Bloch equations [63, 64, 65, 66, 67]. These equations [67] give a correct description for any harmonics of the cw EPR signal, including distortions due to overmodulation.

In the classical approach the modulation of B_0 with frequency ω_{rf}, leads to the time-dependent field $B(t) = B_0 + 2 B_2 \cos(\omega_{rf} t)$. The corresponding signal can be expanded to a Taylor series

$$S(t) = A(B_0) + \sum_{n=1} \frac{1}{n!} \frac{d^n A(B_0)}{dB_0^n} [2B_2 \cos(\omega_{rf} t)]^n , \qquad (9)$$

where $A(B_0)$ is the absorption spectrum (Fig. 17a). For modulation amplitudes small compared to the linewidth, only the first-order term with $n = 1$ has to be considered. The first-harmonic absorption spectrum, the part of the absorption signal oscillating with frequency ω_{rf}, then consists of lines which are the first derivative $d A(B_0)/ dB_0$ of the absorption lines (Fig. 17b).

(a)   (b)   (c)

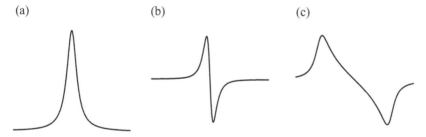

Fig. 17.   (a) Lorentzian absorption line with width Γ. (b) First-harmonic line, modulation amplitude $\Gamma / 20$ (first derivative of the absorption line). (c) First-harmonic line with strong overmodulation, modulation amplitude 6Γ

With increasing modulation amplitude higher-order terms of the Taylor series in (9) become relevant (Fig. 17c) [68, 69, 70]. For modulation amplitudes large compared to the linewidth, the spacing between the positive and negative peak maximum of the distorted line corresponds to the modulation amplitude $2B_2$. This effect is routinely used to calibrate the modulation coils of EPR spectrometers.

We now discuss two semi-classical approaches, which in contrast to the modified Bloch equations can also be applied to pulse experiments.

Multiply Tilted Rotating Frames

The influence of an rf field parallel to B_0 on a spin system can be described by a series of transformations to tilted rotating frames [71, 11, 32]. This approach is restricted to cases where higher-order effects caused by strong radiation fields can be neglected.

An $S = 1/2$ electron spin system subject to an external field B_0 and a bichromatic radiation field can be described by the Hamiltonian

$$H_{lab}(t) = w_S S_z + 2\omega_1 \cos(\omega_{mw} t)S_x + 2\omega_2 \cos(\omega_{rf} t)S_z , \qquad (10)$$

where $\omega_2 = -\gamma_e B_2$ is the amplitude of the radio frequency in angular frequency units. In a first step the Hamiltonian in (10) is transformed to a frame rotating with frequency ω_{mw}. Omitting the counter-rotating component of the mw field we find

$$H_{SRF} = \Omega_S S_z + \omega_1 S_x + 2\omega_2 \cos(\omega_{rf} t)S_z , \qquad (11)$$

with the resonance offset $\Omega_S = \omega_S - \omega_{mw}$. The Hamiltonian in this singly rotating frame is shown graphically in Fig. 18a.

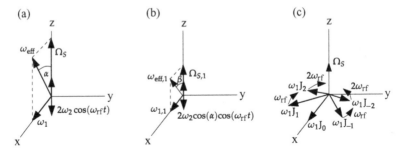

Fig. 18.   Pictorial representation of the cw EPR Hamiltonian. (a) Singly rotating frame. (b) First tilted rotating frame, tilted from the z-axis by angle α and rotating with ω_{rf} (β is the tilt angle for the next transformation). (c) Toggling frame with $k = 0$ and $\omega_z = 2\omega_2/\omega_{rf}$

The effective field vector ω_e, and the z-axis include an angle $\alpha = \arctan(\omega_1/\Omega_S)$. The frame is then tilted around the y-axis by this angle, resulting in a new z-axis parallel to ω_e. Finally, this frame is transformed to a frame rotating with frequency ω_{rf} (Fig. 18b). In this doubly rotating frame the equilibrium magnetization lies on a cone with apex angle 2β. For small angles β, corresponding to $\omega_1 \ll \Omega_S$, the magnetization is approximatively oriented along to the z-axis. After omitting the counter-rotating term and the remaining linear component, the Hamiltonian is given by

$$H_{TRF,1} = \Omega_{S,1} S_z + \omega_{1,1} S_x , \qquad (12)$$

with the new resonance o set $\Omega_{S,1} = \sqrt{\Omega_1^2 + \Omega_S^2} - \omega_{rf} \approx \omega_S - \omega_{rf}$. For the e ective field amplitude $\omega_{1,1}$ in the first tilted rotating frame we find $\omega_{1,1} = -\omega_2 \sin(\ )$. Near resonance ($\Omega_{S,1} \approx \omega_{rf}$), $\omega_{1,1}$ can be simplified to

$$b_{12} \approx \omega_{1,1} \approx \frac{-\omega_1 \omega_2}{\omega_{rf}}, \qquad (13)$$

leading to the Hamiltonian

$$H_{TRF,1} = (\omega_S - \omega_{rf})S_z - \omega_{1,1}S_x$$
$$= (\omega_S - \omega_{rf})S_z + \frac{\omega_1 \omega_2}{\omega_{rf}}S_x . \qquad (14)$$

Equation (14) describes a two-photon transition, resonant with $\omega_S = \omega_{mw} + \omega_{rf}$.

By repeating this procedure, a Hamiltonian for any multiple-photon transition with frequency $\omega_S = \omega_{mw} + k\omega_{rf}$ can be derived. For the three-photon transition, with frequency $\omega_S = \omega_{mw} + 2\omega_{rf}$, for example, the Hamiltonian in the triply rotating frame is found to be

$$H_{TRF,2} = (\omega_S - 2\omega_{rf})S_z + \frac{\omega_1 \omega_2^2}{2\omega_{rf}^2}S_x . \qquad (15)$$

The tilted frame approach delivers only correct results for rf amplitudes much weaker than the fields usually used in EPR experiments.

Toggling Frames

The toggling frame approach does not su er from this drawback [72, 73]. Starting with the Hamiltonian in the singly rotating frame (11), the time-dependent longitudinal component of the Hamiltonian is replaced by time-dependent transverse components using the rotation operator

$$R(t) = e^{i[k\omega_{rf}t + \frac{\omega_2^2}{\omega_{rf}}\sin(\omega_{rf}t)]S_z} . \qquad (16)$$

We then arrive at the toggling frame Hamiltonian

$$H_{TF,k} = R(t)H_{SRF}R^{-1}(t) - k\omega_{rf}S_z - 2\omega_2 \cos(\omega_{rf}t)S_z$$
$$= (\omega_S - k\omega_{rf})S_z + \omega_1 J_{-k}(z) S_x$$
$$+ \sum_{n=-\infty}^{n=-k} \omega_1 J_n(z) e^{i(k+n)\omega_{rf}tS_z} S_x e^{-i(k+n)\omega_{rf}tS_z}, \qquad (17)$$

with the normalized rf amplitude $z = \omega_2/\omega_{rf}$, and the Bessel function of the first kind $J_n(z)$. $H_{TF,k}$ contains the resonance o set ($\omega_S - k\omega_{rf}$), a time-independent e ective field amplitude $\omega_1 J_{-k}(z)$, and the sum of the remaining time-dependent transverse perturbations. For a better understanding H_{TF} is visualized in Fig. 18c for $k = 0$. The component $_1 J_0(z)$ points along the x-axis, while all the other transverse components rotate around the z-axis with frequencies that are multiples of $_{rf}$. For a properly chosen value of k, the remaining resonance o set is minimum and the time-dependent terms of the Hamiltonian can be neglected in first-order. A multiple-photon transition of the type $^+ + k \times$ has thus the transition amplitude

$$b_{12} \quad _{1,k} = \ _1 J_{-k}(z) \ . \tag{18}$$

Although the toggling frame is a convenient tool for the understanding of multiple-photon processes, it remains a semiclassical description.

Fully Quantum Mechanical Description

For a fully quantized description of the cw EPR experiment with field modulation, only two modes of the radiation field, corresponding to the two frequencies, are of importance (see Appendix). All other frequencies are either strongly damped or far o -resonant. The fully quantized Hamiltonian is then given by

$$H = \ _S S_z + \ _{mw} \ a_{mw}^T a_{mw} + \frac{1}{2} 1 \ + \ _{rf} \ a_{rf}^T a_{rf} + \frac{1}{2} 1$$
$$+ \ \frac{_1}{N_{mw}} \ a_{mw}^T + a_{mw} \ S_x + \frac{_2}{N_{rf}} \ a_{rf}^T + a_{rf} \ S_z. \tag{19}$$

The function space is spanned by $|m_S, n, m = |m_S \ | \ n \ | \ m$, where n and m are the numbers of mw and rf photons.

Due to the low frequency of the mw and rf fields, H can be replaced by a Hamiltonian in a two-mode Floquet space, as is shown in Appendix A.1–A.3. This approach provides a simpler mathematical treatment of the problem and is used for further evaluation. However, it is important to mention that Fourier indices of the Floquet states correspond to the photon state occupation numbers of the fully quantized Hamiltonian in (19). This correspondence proves that the modulation sidebands discussed later are caused by multiple-photon transitions .

The resonance condition for a multiple-photon transition of the type $^+ + k \times$ between the two levels $, 0, 0|$ and $, 1, k|$ is fulfilled when $, 0, 0|H_F|, 0, 0 \quad , 1, k|H_F|, 1, k$, or $_S \quad _{mw} + k \ _{rf}$. Coupling between the degenerate levels will then lead to a coherent transition in the form of Rabi oscillations.

Figure 19 a shows a segment of the matrix representation of the two-mode Floquet Hamiltonian. The two levels $|, 0, 0$ and $|, 1, 1$ are degenerate, corresponding to the two-photon transition $^+ +$. Since there is no direct coupling between the elements, only second- and higher-order e ects are relevant. The two pathways for the two-photon transition are indicated by arrows and

(a)

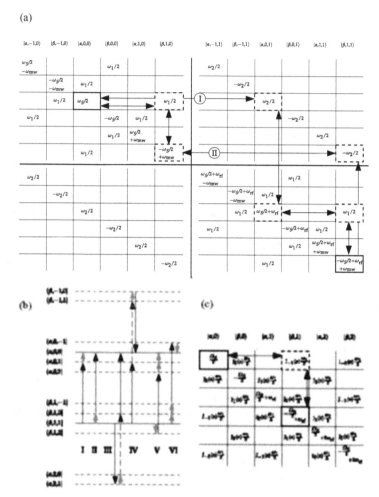

Fig. 19. (a) Segment of the two-mode Floquet space Hamiltonian H_F. Full rectangles: Degenerate levels $|,0,0\rangle$ and $|,1,1\rangle$. Dashed rectangles : Coupling elements and intermediate levels of the two-photon transition. The two pathways I and II are indicated by arrows. (b) Energy level scheme with transitions of the two two-photon pathways I and II and the four four-photon pathways III to VI. Black arrows : ω_{mw}^{+} photons, dashed arrows : ω_{mw}^{-} photons, grey arrows : ω_{rf} photons. (c) Segment of the single-mode Floquet Hamiltonian $H_{F,TF,0}$ (adapted from [72])

correspond to the transitions I and II in Fig. 19b. The next possible transition pathways are those of the four-photon transitions, $\omega^{-} \omega^{+} + \omega^{+} + 2\omega^{+}$, $2\omega^{+} + \omega^{-} \omega^{+}$, $\omega^{-} \omega^{+} + \omega^{+} + 2\omega$, and $2\omega + \omega^{+} \omega^{-}$ (transitions III to VI in Fig. 19b).

Using a perturbation approach [74], the effective Hamiltonian for any two coupled Floquet state can be derived. For two-photon transitions, I and II in

Fig. 19, the Hamiltonian is given by

$$H_e = \left(\omega_s - \omega_{rf} + \frac{\omega_1^2}{2\omega_{rf}} + \frac{\omega_1^2}{4\omega_{mw}} \right) S_z + \omega_{1,1} S_x , \tag{20}$$

where $\omega_{BS} = \omega_1^2/(4\omega_{mw})$ is the Bloch-Siegert shift [75]. The effective field amplitude $\omega_{1,1}$ is approximately ($\omega_{-1}^2/\omega_{rf}$), a result which is also obtained with the tilted frame approach (13).

The perturbation approach or a numerical evaluation of the two-mode Floquet Hamiltonian provide solutions for all possible transitions and transition amplitudes. However, it is better to reduce the number of modes in the Floquet space to get a more compact but still complete presentation of related states and transitions. This can be done by using a toggling frame approach, which as we have seen is a convenient choice for a semiclassical description of $^+ + k \times$ transitions. The parameter k of the toggling frame may be chosen arbitrarily and does not have to fulfill any kind of resonance condition, since the resonant transition is selected later. We simply set $k = 0$ and get, according to (17), the toggling frame Hamiltonian

$$H_{TF,0} = (\omega_s + \omega_{BS}) S_z + \sum_{n=-\infty}^{\infty} \omega_1 J_n(z) e^{in \omega_{rf} t S_z} S_x e^{-in \omega_{rf} t S_z} . \tag{21}$$

In contrast to the semiclassical approach in (17), the time-independent single-mode Floquet Hamiltonian $H_{F,TF}$ is derived from (21). A small segment of the matrix representation is shown in Fig. 19c. The Hamiltonian $H_{F,TF}$ has a direct coupling element between two degenerate levels, taking into account all pathways where one mw $^+$ photon is absorbed. The coherence is described in the function space of $H_{F,TF}$. The back-transformation to the singly rotating frame splits this single coherence into a set of coherences corresponding to all involved virtual levels. For the effective Hamiltonian of a resonant $^+ +$ $k \times$ multiple-photon transition calculated from $H_{F,TF}$ (see Fig. 19c), we find [72, 73]

$$H_e = ([\omega_s + \omega_{BS} + \omega_k] - [\omega_{mw} + k \omega_{rf}]) S_z + \omega_{1,k} S_x , \tag{22}$$

with a Bloch-Siegert-like resonance shift (in second order)

$$\omega_k = \frac{2\omega_1^2}{\omega_{rf}} \sum_{l=k} \frac{J_l^2(z)}{(k-l)} , \tag{23}$$

and the effective field amplitude[2]

$$\omega_{1,k} = \omega_1 \left(c_k^{(0)} + c_k^{(3)} \right)$$

$$= \omega_1 \left(J_{-k}(z) + \frac{\omega_1^2}{2\omega_{rf}} \sum_{l=k} \sum_{m=0} \frac{J_{-l}(z) J_{m-l}(z) J_{m-k}(z)}{(l-k)m} \right) . \tag{24}$$

---

[2] Note that the expression for the term $c_k^{(3)}$ in [72] contains a sign error.

For $\omega_1 \ll \omega_{rf}, \omega_k$, the higher-order terms in (24) and the Bloch-Siegert shift ω_{BS} can be neglected, resulting in

$$H_e = (\omega_s - k\omega_{rf}) S_z + \omega_1 J_{-k}(z) S_x . \tag{25}$$

This expression is equivalent to the Hamiltonian derived from the semiclassical toggling frame approach in (17) after omitting the time-dependent perturbations.

Description of Sidebands

The cw EPR absorption signal is proportional to the transverse component of the steady-state magnetization out-of-phase with the mw field. The steady-state solution is most easily obtained from the effective Hamiltonian given in (25) with the assumption that each $|+\rangle + k \times$ transition can be treated separately. The density operator at thermal equilibrium, $\sigma_{SRF} = -S_z$, is transformed to the kth toggling frame by $\sigma_{TF,k}(t) = R(t)\,\sigma_{SRF}(t)R^{-1}(t)$.

Since the detection operator $D = S_y$ is expressed in the singly rotating frame, the density operator has to be transformed back to this frame,

$$\sigma_{SRF}(t) = R^{-1}(t)\,\sigma_{TF,k}(t)R(t)$$

$$= \sum_{n=-\infty}^{\infty} J_n(z)\,e^{-i(k+n)\omega_{rf} t S_z}\,\sigma_{TF,k}(t)\,e^{i(k+n)\omega_{rf} t S_z} . \tag{26}$$

The time-independent density operator in the toggling frame is then transformed to a series of terms in the singly rotating frame, oscillating with multiples of the modulation frequency. Summation over all possible multiple-photon resonances results in the general solution for the cw EPR absorption signal

$$S_y = \mathrm{tr}[\sigma_{SRF}\,S_y] = \omega_1 \sum_{n=-\infty}^{\infty} \sum_{k=-\infty}^{\infty} J_n(z) J_{-k}(z)$$

$$\times \frac{T_2 \cos([k+n]\omega_{rf} t) - (\omega_s - k\omega_{rf}) T_2^2 \sin([k+n]\omega_{rf} t)}{1 + \omega_1^2 J_k^2(z) T_1 T_2 + (\omega_s - k\omega_{rf})^2 T_2^2} . \tag{27}$$

Apart from the saturation term $\omega_1^2 J_k^2(z) T_1 T_2$ in the denominator, (27) is identical to the formula derived from the Bloch equations [67]. Using (27), the absorption spectrum for any harmonic of the modulation frequency can be calculated. For the zeroth-harmonic spectrum, only the constant terms are relevant

$$S_{y,0} = \sum_{k=-\infty}^{\infty} \frac{T_2\,\omega_1 J_k^2(z)}{1 + \omega_1^2 J_k^2(z) T_1 T_2 + (\omega_s - k\omega_{rf})^2 T_2^2} . \tag{28}$$

In cw EPR the first-harmonic absorption spectrum, given by

$$S_{y\ 1} = \sum_{k=-\infty}^{\infty} \frac{T_2\ {}_1J_{-k}(z)\ J_{1-k}(z) + J_{-1-k}(z)}{1 + {}_i^2 J_k^2(z)\,T_1 T_2 + (\omega_s - k\,\omega_{rf})^2\,T_2^2}\,\cos(\omega_{rf}t)$$

$$-\frac{(\omega_s - k\,\omega_{rf})\,T_2^2\ {}_1J_{-k}(z)\ J_{1-k}(z) - J_{-1-k}(z)}{1 + {}_i^2 J_k^2(z)\,T_1 T_2 + (\omega_s - k\,\omega_{rf})^2\,T_2^2}\,\sin(\omega_{rf}t)\ ,\quad (29)$$

is usually measured. These equations becomes more clear from the plots shown in Fig. 20, which represent the field-swept EPR absorption spectra of the zeroth-harmonic and the first- and second-harmonic signal with the components in-phase and out-of-phase with respect to the rf field.

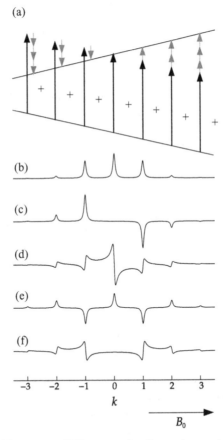

Fig. 20.   Simulated field-swept cw EPR spectra for different harmonics of the modulation frequency. k indicates the number of absorbed rf photons. (a) Multiple-photon transitions. (b) Zeroth-harmonic spectrum. (c) First-harmonic spectrum, in-phase with the radio frequency. (d) Corresponding out-of-phase spectrum. (e) Second-harmonic spectrum, in-phase with the radio frequency. (f) Corresponding out-of-phase spectrum (adapted from [72])

The zeroth-harmonic spectrum is symmetric (Fig. 20b) and consists of a center line (single-photon transition) and sidebands (multiple-photon transitions) with spacings $\Delta B = \omega_{rf}/\gamma_e$. All lines have an absorptive shape, with the intensity of the kth sideband proportional to $J_k^2(z)$. The first-harmonic spectrum consists of in-phase components with absorptive lineshapes (Fig. 20c), and out-of-phase components with dispersive lineshapes (Fig. 20d). The spectra are anti-symmetric with respect to the center line. The second-harmonic spectra shown in Figs. 20e,f are again symmetric.

In cw EPR spectroscopy the first-harmonic absorption signal in phase with the modulation frequency is usually measured (Fig. 20c). The derivative-like lineshape is then a superposition of a large number of overlapping sidebands separated by ω_{rf}. The two multiple-photon transitions with $|k|$ of approximately $2\omega_2/\omega_{rf}$ give the maximum contributions to the signal.

For a homogeneous line and $\omega_2 \gg 1/T_2$, (29) reduces to the commonly observed derivative lineshape

$$S_{y,1} \approx 2\omega_2 \frac{d}{d\omega_s} \frac{\omega_1 T_2}{1 + \omega_s^2 T_2^2} . \tag{30}$$

The signal amplitude is proportional to the modulation amplitude, and does not depend on the modulation frequency.

Equation (30) can be extended to the more general case of an inhomogeneous broadened line, which consists of a distribution of unresolved homogeneous lines. As long as the modulation amplitude does not exceed the inhomogeneous linewidth, again the derivative of the absorption line is observed.

It is noteworthy that in the first-harmonic in-phase cw EPR spectrum the observed derivative lines consist of a large number of multiple-photon transitions, and that the signal intensity is zero at the position of the single-photon transition. For example, in a standard cw EPR experiment with 100 kHz field modulation and a modulation amplitude of 0.1 mT ($\omega_2/2\pi \approx 3$ MHz), the two multiple-photon transitions constituting the peak maximum and the peak minimum are $\omega^+ \pm 28 \times \omega$ transitions, where 28 rf photons are absorbed and emitted.

The finding that the lineshapes observed in cw EPR experiments are caused by multiple-photon transitions was verified by experiments on a single crystal of lithium phthalocyanine with a linewidth of 4–8 μT. Two series of first-harmonic in-phase EPR spectra with different modulation amplitudes are shown in Fig. 21. The spectra recorded with a 1 MHz field modulation (Fig. 21a) show the patterns predicted by (29).

When a 100 kHz field modulation is used (Fig. 21b), the individual multiple-photon transitions are no longer resolved. The out-of-phase signal (not shown) is averaged to zero, since the individual multiple-photon transitions with derivative lineshape (see Fig. 20d) compensate each other.

Although it was not realized during many years since the beginning of EPR, in the traditional cw experiment we actually observe multiple-photon

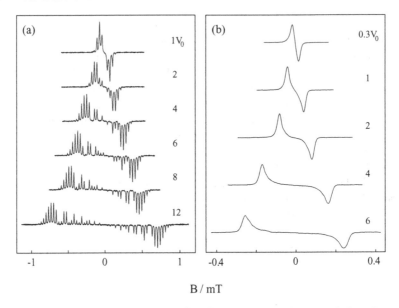

Fig. 21.   First-harmonic EPR spectra of lithium phthalocyanine recorded at Q-band with different modulation amplitudes in phase with the modulation frequency. Numbers are proportional to the signal voltage. (a) Modulation frequency 1 MHz. (b) Modulation frequency 100 kHz (adapted from [72])

transitions . This recent finding gives a deeper insight into the origin of the observed phenomena and the interpretation of the results.

3.2 Pulse EPR Experiments

We now demonstrate that multiple-photon transitions created by a bichromatic radiation field can also be observed in pulse EPR experiments. In particular, the creation of multiple-photon echoes is discussed.

The pulse sequence for the generation of multiple-photon echoes is shown in Fig. 22a. The mw part consists of the conventional two-pulse sequence [18]. The sequence of rf pulses with the field along B_0 is applied synchronously with the mw pulses. The first bichromatic pulse creates coherence at ω_{mw} and $\omega_{mw} + k\omega_{rf}$. For simplicity we consider lowest (two-photon) transitions only $\omega_{mw} \pm \omega_{rf}$. During the free evolution period of time between the first and the second pulse the coherence dephases. The second bichromatic pulse inverts the directions of spin precession, resulting in the formation an echo at time after the second pulse. The echo appears again at ω_{mw} and at $\omega_{mw} \pm \omega_{rf}$. By varying the mw and rf amplitude, as well as by choosing a proper B_0 value, the echoes observed at different frequencies can be optimized or separated from each other, as is demonstrated in next paragraph [32, 73].

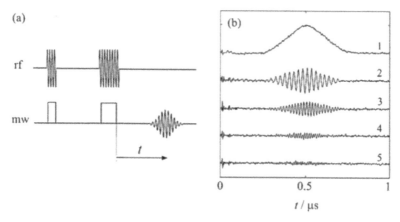

Fig. 22.   Multiple-photon pulse EPR with bichromatic pulses. (a) Pulse sequence
for a multiple-photon echo. (b) Observed time-domain echo signals. 1: Single-photon
echo, $_s$ = $_{mw}$, $_1/2$ = 2 .5 MHz. 2-5: Two-photon echoes measured at $_s -$ $_{mw}$
$_{rf}$, with $_s/2$ (in MHz) 40 (2), 60 (3), 80 (4), and 100 (5) (adapted from [32])

Two-photon echoes on   -irradiated quartz glass are recorded at Q-band
frequencies [32]. The lengths of the bichromatic pulses are 100 ns and 200 ns
and the mw and rf amplitudes are both approximately 10 MHz. With these
pulse sequences the signal of the conventional echo is maximum for an mw
amplitude of about 2.5 MHz (Fig. 22b, trace 1). The echo traces 2–5 in Fig. 22b
are obtained at B_0 fields corresponding to resonance o sets of 37.8, 58.2,
78.4 and 97.4 MHz, which correspond to the radio frequencies used in these
experiments.

The echo-detected two-photon EPR spectra obtained by Fourier transformation of the time-domain traces and recorded with radio frequencies of 40
and 80 MHz are shown in Figs. 23b,c. As a reference, the single-photon EPR
spectrum is measured with the same sequence. The two-photon EPR spectra
are symmetrically placed to the position of the single-photon EPR spectrum,
with shapes that are exact replicates of the latter. The observed peak-to-peak
splittings of the two-photon EPR spectra are approximately twice the radio
frequency.

3.3   -Photon-Induced Transparency

In this section we describe a new transparency phenomenon which can be
observed in spin systems. For a two-level system that is exposed to a bichromatic radiation field consisting of an mw and an rf field, where the two-level
system is resonant with the mw frequency ($_{mw}$ = $_s$), energy and angular
momentum are not only conserved for the single-photon transition but also
for all multiple-photon transitions of the type $- m \times$ $^+$ $^+ + m \times$, with
$1 < m <$ (Fig. 24a).

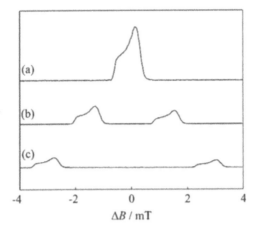

Fig. 23.   Echo-detected single- and two-photon EPR spectra of   -irradiated quartz glass. (a) Single-photon spectrum, $_1/2$ = 2 .5 MHz. (b) Two-photon spectrum with $_{rf}/2$ = 40 MHz, $_1/2$ = $_2/2$ = 10 MHz. (c) Two-photon spectrum with $_{rf}/2$ = 80 MHz, $_1/2$ = $_2/2$ = 15 MHz (adapted from [32])

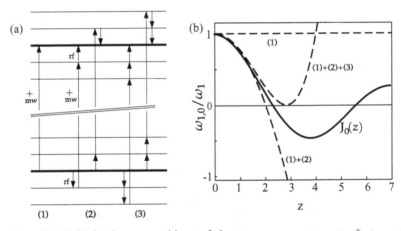

Fig. 24.   (a) Multiple-photon transitions of the type $- m \times$ $+$ $^+$ $+ m \times$. (1) Single-photon process $^+$; (2) three-photon processes, $+$ $^+ -$ and $- +$ $^+ +$; (3) five-photon processes 2 $+$ $^+ - 2$ and $- 2 +$ $^+ + 2$. (b) Normalized e ective field $_{1,0} =$ $_1 J_0(z)$ for the centre band at $_{mw} =$ $_s$, as a function of $z = 2$ $_2/$ $_{rf}$ (bold solid line). The contributions of the processes (1), (1)+(2) and (1)+(2)+(3) are also given (dashed lines) Adapted from [73]

    In the single-photon process (1), one mw $^+$ photon is absorbed, whereas in the two three-photon processes (2) with   m = 1, one mw $^+$ photon is absorbed, and one rf   photon is absorbed and one is emitted. Also five-photon transitions (3) with   m = 2 and transitions of higher order are induced.

For properly chosen experimental parameters of the rf field, the spin system becomes transparent owing to destructive interference of the single-photon and the multiple-photon processes [76]. Since this phenomenon is caused by rf photons, we call it -photon-induced transparency .

For the description of this type of multiple-photon transitions the phases ϕ_{mw} and ϕ_{rf} of the mw and rf field have to be taken into account. Starting from the laboratory frame Hamiltonian

$$H_{lab}(t) = \Omega_S S_z + 2\omega_1 \cos(\omega_{mw}t + \phi_{mw})S_x + 2\omega_2 \cos(\omega_{rf}t + \phi_{rf})S_z, \quad (31)$$

one arrives at the effective spin Hamiltonian in the toggling frame

$$H_{TF,k} = ([\Omega_S + k] - [\omega_{mw} + k\omega_{rf}])S_z + \omega_{1,k} e^{-i\phi_k S_z} S_x e^{i\phi_k S_z}, \quad (32)$$

where $\phi_k = \phi_{mw} + k\phi_{rf}$ is the phase of the effective field. The Hamiltonian in (32) essentially describes the spin system exposed to an effective radiation field with frequency $\omega_{mw} + k\omega_{rf}$, phase ϕ_k, and amplitude $\omega_{1,k}$.

The effective transition amplitudes for the single-photon transition and the contributions of the processes (1) , (1)+(2) and (1)+(2)+(3) are plotted in Fig. 24b as a function of the normalized rf amplitude $z = 2\omega_2/\omega_{rf}$. The effective transition amplitude of the sum of all multiple-photon processes is then given by

$$\omega_{1,0} = \omega_1 \sum_{m=0} \frac{(-1)^m \left(\frac{z}{2}\right)^{2m}}{(m!)^2} = \omega_1 J_0(z). \quad (33)$$

For the flip angle of a bichromatic pulse we find

$$\beta_e = \omega_{1,0} t_p = \omega_1 J_0(z) t_p. \quad (34)$$

Corresponding expressions $\omega_{1,k}$ can be derived for each of the sidebands.

Of special interest is the finding that the effective transition amplitudes $\omega_{1,k}$ can be zero; the two-level system then becomes transparent. For the center band with $\Omega_S = \omega_{mw}$, the effective transition amplitude $\omega_{1,0}$ is zero at the zero-crossings of the Bessel function $J_0(z)$. The first zero-crossing is at $z = j_{0,1} \approx 2.4048$, the second one at $z = j_{0,2} \approx 5.5201$, etc.

Figure 25 describes the motion of the tip of the magnetization vector $\mathbf{M}$ in the rotating frame during bichromatic radiation with an mw field of amplitude ω_1 along the x-axis and an rf field $2\omega_2 \sin(\omega_{rf}t)$, with $\omega_{rf}/2\pi$ MHz, along the z-axis.

The situation for a weak mw field with amplitude $\omega_1/2\pi = 0.01$ MHz, typically used in cw EPR experiments and $z = 2.4$ (first transparency condition) is shown in Fig. 25a. When the radiation field is turned on ($t = 0$), the effective field vector is given by $\omega_e = \omega_1$ and $\mathbf{M}$, which is oriented along the z-axis, starts to nutate around ω_x. Since $\omega_2 \gg \omega_1$, the angle between the effective field ω_e and the z-axis gets rapidly very small, so that $\omega_e \approx 2\omega_2$. After one quarter of the rf period the magnetization is in the xz-plane, with a

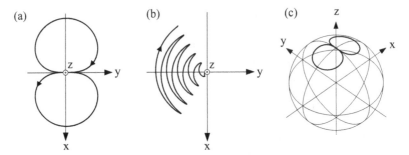

Fig. 25.   Trajectories of the magnetization vector   M under a bichromatic field,
starting from thermal equilibrium (M along the   z-axis). (a) Weak mw field,
$_1/2 = 0.01$ MHz, resonant with the single-photon transition. Transparency condition fulfilled,   z = 2.4048. Projection onto the   xy-plane. (b) Transparency condition
not fulfilled,   z = 1. The trajectory is shown for the first five rf periods. The scaling
of the axes is reduced by the factor 20 compared to (a). (c) Strong on-resonant mw
field, $_1/2$ = 5 MHz. The transparency condition   z = 2.3306 is fulfilled (adapted
from [73])

deviation from the   z-axis of only 0.07 . At half of the rf period (t = / $_{rf}$), M
has moved back to the   z-axis and the e ective field is again $_e$ = $_1$. During
the second half of the rf period,   M describes a corresponding trajectory with
negative x-values, so that after the full rf period   M is again oriented along
the z-axis. Thus, the tip of the magnetization vector   M describes a figure of
eight in the very close vicinity of the   z-axis. If the transparency condition is
not fulfilled, the trajectory is no longer a closed curve, and the magnetization
moves on a toggling path towards the   xy-plane, as is shown in Fig. 25b for
the first five rf periods and   z = 1.

For strong mw fields, as used in pulse EPR experiments, the third-order
contribution $c_0^{(3)}$ (24) to the e ective field amplitude can no longer be neglected. This reduces the value of   z for which transparency is observed. For
example, for an mw field with amplitude $_1/2$ = 5 MHz, corresponding to a
pulse with a length of 100 ns, transparency occurs at   z = 2.3306 instead of
z = 2.4048. The tip of M describes again a figure of eight, but with larger
deviations from the   z-axis (Fig. 25c).

Bichromatic pulses can be used as a tool to experimentally control the
transition amplitude. This also includes the full suppression of the interaction
of the mw field with the spin system. For example, during a resonant mw
pulse a free evolution period can   artificially be created by an rf pulse with
a field amplitude that fulfills the first transparency condition. In this way an
electron spin echo can be observed using a   single mw pulse [76].

The pulse sequence for such a one-pulse echo experiment is shown in
Fig. 26a. During the mw pulse of length t_{mw} , the rf pulse of length $_1$ is
turned on at time t_{p1} and turned o at time $t_{mw} - t_{p2}$, with $2 t_{p1} = t_{p2}$ and

(a)

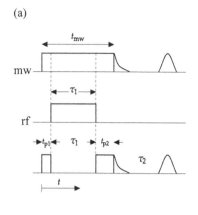

(b)

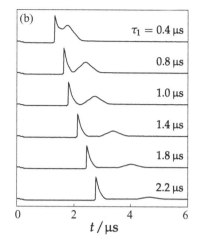

$\tau_1 = 0.4\,\mu s$

$0.8\,\mu s$

$1.0\,\mu s$

$1.4\,\mu s$

$1.8\,\mu s$

$2.2\,\mu s$

$t/\mu s$

Fig. 26. One-pulse echo experiment with a   -photon-induced free evolution period on a coal sample, $_{mw}/2$ = 9.626 GHz, $_{rf}/2$ = 15 MHz, B_0 = 341.1 mT. (a) Pulse sequence with t_{p1} = 300 ns, t_{p2} '0 ns, t_{mw} = 1.3–3.1 µs, $_1$ = 0.4–2.2 µs. The rf pulse applied during time $_1$ fulfills the first transparency condition for the single-photon transition. (b) Experimental time traces showing FIDs and echoes for di erent $_1$ values (adapted from [73])

$_1 = t_{mw} - t_{p1} - t_{p2}$. The mw field amplitude is chosen such that the nominal flip angle is $_1 = /2$ during t_{p1} and $_2 = $ during t_{p2}.

The pulse scheme thus corresponds to a two-pulse echo sequence, ($/2$) − $_1 -$ ( ) − $_2 -$ (echo), with a   -photon-induced free evolution period   of time $_1$, and a usual free evolution period of time $_2 = _1$. During time $_1$ the spin coherence evolves as it would do during a free evolution period, apart from a phase shift that depends on   z, the rf phase, and the length of the bichromatic pulse.

Experimental time traces of such one-pulse echo experiments are shown in Fig. 26b for di erent $_1$ values. The signal that follows the mw pulse consists of an FID and an electron spin echo at time $_2 = _1$, as in a conventional two-pulse echo experiment.

The Bichromatic Pulse as a Substitute for a Second mw Frequency

The   -photon-induced transparency phenomenon can be applied, for example, in pulse EPR experiments that require two mw frequencies $_{mw,1}$ and $_{mw,2}$. A bichromatic pulse, which is transparent for the mw frequency $_{mw} = _{mw,1}$, may then be used as a substitute for the second mw frequency $_{mw,2}$ [77]. When the transparency condition is fulfilled for the center band at $_{mw}$, the bichromatic pulse only drives e ciently two- and three-photon transitions with resonance frequencies $_{mw} \pm _{rf}$ and $_{mw} \pm 2 _{rf}$.

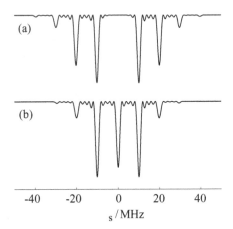

Fig. 27.   Simulation of the polarization patterns of an $S = 1/2$ spin system with an inhomogeneous linewidth $_{inh} =$ after a bichromatic pulse with a length of 500 ns and a radio frequency of 10 MHz, as a function of $_s$. (a) Transparency condition fulfilled, $z = 2.4$. The intensity of the   center band is zero. (b) Transparency condition not fulfilled, $z = 1.5$ (adapted from [77])

This is illustrated in Fig. 27, which shows simulations of the M_z magnetization of an $S = 1/2$ spin system with an inhomogeneous linewidth $_{inh} =$ after a bichromatic   pulse of length 500 ns, as a function of the resonance o -set $_s$. If the transparency condition is fulfilled ($z = 2.4$), only spins with resonance frequencies $_{mw} \pm$ $_{rf}$, $_{mw} \pm 2$ $_{rf}$, and $_{mw} \pm 3$ $_{rf}$ are excited (Fig. 27a). The contribution of higher-order multiple-photon transitions can be neglected.

A bichromatic pulse of proper length, fulfilling the transparency condition, can thus e ciently be used for broadband excitation with frequencies di erent from the mw frequency. If the transparency condition is not fulfilled, transitions at $_{mw}$ and $_{mw} + k$ $_{rf}$ are simultaneously excited. This is shown in Fig. 27b for $z = 1.5$. For the transverse magnetizations M_x and M_y an analogous behavior is found, reflecting the excitation at frequencies $_{mw} + k$ $_{rf}$ by the bichromatic pulse.

• Stimulated Soft ESEEM

Stimulated Soft Electron Spin Echo Envelope Modulation (SS-ESEEM) is a pulse EPR experiment that is used to determine weak hyperfine and nuclear quadrupole interactions in paramagnetic species in solids [78]. The pulse sequence of SS-ESEEM shown in Fig. 28a makes use of two mw frequencies.

The first pulse with flip angle $/2$ and frequency $_{mw,1}$ creates electron coherence on an allowed (forbidden) transition. The second pulse with flip angle   and frequency $_{mw,2}$ is resonant with a forbidden (allowed) transition and transfers the electron coherence to nuclear coherence, which then evolves

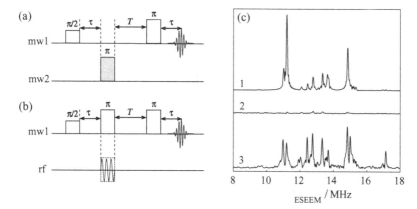

Fig. 28. Stimulated soft ESEEM. (a, b) Pulse sequences for SS-ESEEM (a), and bichromatic-pulse SS-ESEEM (b). (c) Comparison of three-pulse ESEEM and bichromatic-pulse SS-ESEEM experiments on a single crystal of ^{63}Cu(II)-doped Zn(picolinate) $_2$, arbitrary orientation, T K. 1: Three-pulse ESEEM spectrum obtained with mw pulses with a length of 20 ns. 2: Corresponding spectrum obtained with mw pulses with a length of 400 ns. 3: Bichromatic-pulse SS-ESEEM spectra obtained with pulses with a length of 400 ns and a radio frequency of 14 MHz. Plots (2) and (3) use the same scale (adapted from [77]).

during the following free evolution period of time   T. The third pulse with flip angle   and frequency $_{mw,1}$ transfers the nuclear coherence back to forbidden (allowed) electron coherence, which refocuses to a coherence-transfer echo. In this experiment, time   T is incremented, and the modulation of the echo amplitude is monitored. SS-ESEEM experiments provide the same information as three-pulse ESEEM. They have, however, the advantages to be free of blind spots, and to require an mw power which is about two orders of magnitudes lower than for the standard three-pulse ESEEM experiment.

In the bichromatic version of SS-ESEEM, the second pulse with frequency $_{mw,2}$ is replaced by a bichromatic pulse (Fig. 28b) [77]. The general features of SS-ESEEM are discussed elsewhere [78], here we concentrate on the characteristics of the experiment with a bichromatic pulse.

The potential of bichromatic-pulse SS-ESEEM is demonstrated on a single crystal of ^{63}Cu(II)-doped Zn(picolinate) $_2$. All the spectra are measured at a temperature of 20 K with an arbitrary crystal orientation. The three-pulse ESEEM spectrum shown in Fig. 28c, trace 1, is obtained with short non-selective mw pulses with a length of 20 ns, and consists of several proton peaks close to the proton Larmor frequency of 12.7 MHz.

Trace 2 and 3, demonstrate the superiority of SS-ESEEM over three-pulse ESEEM, when long low-power   / 2 and   pulses with a length of 400 ns are used. Under this condition, three-pulse ESEEM is no longer operative, but SS-ESEEM still is. Trace 2 shows the three-pulse ESEEM spectrum. The excitation bandwidth of a pulse with a length of 400 ns is by far too small to simultaneously excite allowed and forbidden transitions. Trace 3 shows the bichromatic-pulse SS-ESEEM spectrum measured with two mw pulses and a bichromatic pulse with a radio frequency of 14 MHz. The improvement in the signal/noise ratio is tremendous.

The line positions in three-pulse ESEEM and the bichromatic-pulse SS-ESEEM spectrum are the same. Since SS-ESEEM does not su er from blind spots, some of the lines are more intense in the bichromatic-pulse experiment than in three-pulse ESEEM. The line intensities are di erent in the two spectra, because the selective bichromatic pulse excites a narrow band of frequencies close to $\omega_{mw} + k\omega_{rf}$. Bichromatic-pulse SS-ESEEM is easy to implement and does not require a precise tuning of the bichromatic pulse to the transparency condition.

- Double Electron-Electron Resonance

Another experiment that makes use of two mw frequencies is double electron-electron resonance (DEER) [79], which has recently been extended to four pulses [80]. DEER may be used for the determination of distances between two spins A and B in solids by measuring the dipole coupling between two unpaired electrons.

The pulse sequence for four-pulse DEER is shown in Fig. 29a. It consists of three pulses with frequency $\omega_{mw,1}$ and one pulse with frequency $\omega_{mw,2}$. The first pulse with flip angle $\pi/2$ and frequency $\omega_{mw,1}$ creates electron coherence of the A-spins. The second pulse with flip angle π and frequency $\omega_{mw,1}$ refocuses this coherence to an electron spin echo, which is formed at time τ_1 after

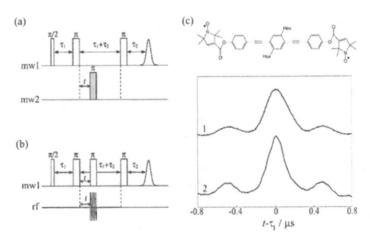

Fig. 29. Four-pulse DEER. (a, b) Pulse sequence for DEER (a), and bichromatic-pulse DEER (b). (c) Comparison of DEER time-traces of the nitroxide biradical CAS 312624-83-21 shown on top. 1: Standard four-pulse DEER with two mw frequencies, $(\omega_{mw,1} - \omega_{mw,2})/2\pi$ MHz. 2: Four-pulse DEER with a bichromatic pulse, $\omega_{rf}/2\pi$ MHz and z 2.4 (adapted from [77])

the pulse, and the third pulse with flip angle   and frequency $_{mw,1}$ refocuses the coherence to a refocused echo. At time   t after the second pulse an mw pulse with flip angle   and frequency $_{mw,2}$ is applied to the B-spins. The effect of this pulse on the refocused echo intensity is recorded by varying time t between 0 and 2 $_1$ ($_1$ < $_2$). The pulse with frequency $_{mw,2}$ can again be replaced by a bichromatic pulse with mw frequency $_{mw}$ = $_{mw,1}$ and radio frequency $_{rf}$ (Fig. 29b).

Figure 29c shows experimental DEER time traces obtained with the two-frequency experiment (trace 1) and with bichromatic-pulse DEER (trace 2). The time traces are very similar, and the dipolar coupling obtained from the corresponding spectra agrees well with results reported earlier [80].

In the case of DEER, it is important that the bichromatic pulse fulfills the transparency condition. Otherwise the residual excitation of A-spins at frequency $_{mw}$ destroys the relevant signal, since A-spins and B-spins with resonance frequencies $_{mw}$ and $_{mw}$ + k $_{rf}$ are both flipped simultaneously. A bichromatic pulse of at least one period of the radio frequency has to be used in bichromatic-pulse DEER. This leads to a loss in signal intensity compared to standard four-pulse DEER. On the other hand, in bichromatic-pulse DEER a larger number of spins with frequencies $_{mw}$ + k $_{rf}$ are simultaneously excited, which again improves the signal intensity.

The examples of SS-ESEEM and DEER demonstrate that bichromatic pulses can successfully be used to substitute pulses with a second mw frequency.

In this Chapter we described the different types of multiple-photon transitions that can be observed and used in EPR, from both a theoretical and an experimental point of view. We especially emphasized on the multiple-photon transitions induced by an additional longitudinal oscillating field. These transitions play a major role in traditional field-modulated cw EPR and are responsible for its derivative-like lineshapes. Moreover,   -photon-induced transparency based on this type of transitions has proved to be a very interesting and useful phenomenon. One area of applications refers to two-frequency pulse EPR experiments such as DEER, Soft ESEEM and others. In the future, we will study the multiple-photon phenomena in more detail and screen their potential for practical applications.

A Second Quantization

A.1 Hamiltonian of the Photon Field

For a fully quantum-mechanical description of the interaction between matter and an electro-magnetic radiation field the oscillating magnetic fields have to be replaced by a set of quantized harmonic oscillators [3, 81], this is called second quantization. The Hamiltonian of a spin system in a radiation field is then given by

$$H = H_S + H_R + H_i , \qquad (35)$$

where H_S is the Hamiltonian of the undisturbed system, H_R describes the radiation field, and H_i describes the magnetic dipole coupling between spin system and radiation field.

The radiation field is represented by a superposition of an infinite number of transverse electro-magnetic waves, propagating in all possible directions. The different modes, distinguished by their propagation vectors k with $|k| = {}_k c$, are orthogonal and do not interact with each other. Every mode has two independent orthogonal polarizations   . The polarization of the magnetic field component of a mode is given by the vector $_k$. For a circular field it is more convenient to describe the polarization with the unit vectors $_{+1} = -(_x + i_y)/\overline{2}$ (right circular polarization), $_{-1} = (_x - i_y)/\overline{2}$ (left circular polarization) and $_0 = _z$ (linear polarization), instead of using the basis system of Cartesian unit vectors, $\{_x, _y, _z\}$.

Using the correspondence principle the Hamiltonian of the radiation field is found to be

$$H_R = \sum_k {}_k \left(a_k^\dagger a_k + \frac{1}{2} 1 \right) , \qquad (36)$$

with the photon   annihilation   and creation   operators a and $a^\dagger$. The combined operator $N_k = {}_k {}^\dagger a_k$ is called the photon number   operator, since it gives the number n_k of photon quanta $_k$ present in an eigenstate $|n_k \rangle_k$ of mode k ,

$$N_k |n_k \rangle_k = n_k |n_k \rangle_k . \qquad (37)$$

Consequently, the eigenstates of the operator N_k are called number states .

The polarization vector of a radiation mode does enter in the interaction Hamiltonian H_i, which describes the coupling between the magnetic field and the magnetic dipole of the spin,

$$H_i = -\frac{1}{}B^T\mu = -\frac{1}{} \sum_k B_k^T \mu . \qquad (38)$$

The operator of the quantized magnetic field is given by

$$B(r) = i \sum_k {}_k \sqrt{\frac{_k \mu_0}{2V}} \left(a_k e^{ik\cdot r} - a_k^\dagger e^{-ik\cdot r} \right) . \qquad (39)$$

For low frequencies and correspondingly long wavelengths, as used in NMR and EPR spectroscopy, one finds $k^T r$ 1, so that in good approximation $\exp(ik \cdot r) = 1$, over all positions   r within the sample volume. In this   dipolar approximation   the interaction between the radiation field and the magnetic dipole is reduced to the energy of a dipole in a dipolar field. The interaction Hamiltonian of a certain mode k can then be written as

$$H_i = - \sqrt{\frac{_k \mu_0}{2V}} i \, _k^T \left(a_k + a_k^\dagger \right) S , \qquad (40)$$

with the square root describing the magnetic field produced by a single photon with frequency ω_k in a cavity of volume V. For a linear field mode oriented parallel to the j-axis ($j = x, y, z$) the interaction Hamiltonian is given by

$$H_i = \kappa \left(a_k + a_k^\dagger \right) S_j , \qquad (41)$$

with an effective coupling factor κ. For a right circular field mode, $\epsilon_k = \mathbf{i}_{+1} = (i, -1, 0)/\sqrt{2}$, we find for the corresponding Hamiltonian

$$H_i = \kappa \left(a_k S^+ + a_k^\dagger S^- \right) . \qquad (42)$$

Since these coupling elements connect spin states with $\Delta m = 1$, a right circular field consists of photons with helicity $m_J = 1$, i.e., σ^+-photons. The first term in (42) with the photon annihilation operator a_k and the raising operator S^+ connects an initial state ($| \ldots$ state + photon) with a final state ($| \ldots$ state, photon absorbed (annihilated)). The second term $a_k^\dagger S^-$, on the other hand, describes the emission (creation) of a photon. For a left circular field, the situation is reversed.

A.2 Coherent States

The quantum-electrodynamical equivalents to classical radiation fields are coherent states [82, 83, 2]. A coherent state $|\alpha\rangle$ is an eigenfunction of the annihilation operator with a complex eigenvalue α, ($a|\alpha\rangle = \alpha|\alpha\rangle$), leading to the definition

$$|\alpha\rangle = e^{-\frac{1}{2}|\alpha|^2} \sum_{n=0}^{\infty} \frac{\alpha^n}{\sqrt{n!}} |n\rangle . \qquad (43)$$

The populations of the different number states follow a Poisson distribution

$$|\langle n|\alpha\rangle|^2 = e^{-|\alpha|^2} \frac{|\alpha|^{2n}}{n!} \qquad (44)$$

with an average photon number $N = |\alpha|^2$.

A typical semiclassical interaction Hamiltonian of a linear radiation field, oriented perpendicular to the static magnetic field, is

$$H_{i,\mathrm{semicl}} = 2 \omega_1 \cos(\omega t) S_x . \qquad (45)$$

By definition it is equal to the corresponding quantized interaction Hamiltonian in (41), $H_{i,\mathrm{quant}} = H_{i,\mathrm{semicl}}$, which leads to a coupling factor of

$$\kappa = \frac{\omega_1}{N^{1/2}} . \qquad (46)$$

The fully quantized Hamiltonian finally results in

$$H = H_S + H_R + H_{i,\text{quant}}$$

$$= \omega_S S_z + \omega \left(a^\dagger a + \frac{1}{2} 1 \right) + \frac{\omega_1}{N^{1/2}} (a^\dagger + a) S_x , \qquad (47)$$

with a function space spanned by $|m_S, n\rangle = |m_S\rangle \otimes |n\rangle$. Physically, this can be understood as spin states being dressed by a certain number of photons. The resulting states are thus called dressed states [84].

For low frequencies and high power as used in magnetic resonance experiments $\langle N \rangle$ is extremely large, which allows further simplifications, as is shown in the following section.

A.3 Floquet Theory

For very high average photon numbers $\langle N \rangle$ the Poisson distribution in (43) approximates a Gaussian distribution with very narrow relative linewidth. Almost exclusively states with $n \approx \langle N \rangle$ are populated. For these the coupling terms $\langle m_S, n | H_{i,\text{quant}} | m_S - 1, n + 1 \rangle$ in (47) are in very good approximation equal to $\frac{1}{2} \omega_1$. The quantized field Hamiltonian can then be replaced by a Hamiltonian in the semiclassical Floquet space [15, 74, 85],

$$H_F \approx H - \omega \left(\langle N \rangle + \frac{1}{2} \right) 1 . \qquad (48)$$

The advantages of the Floquet Hamiltonian are the reduced number of considered photons (only differences of photon numbers are of interest) and the simpler mathematical treatment due to the periodicity of eigenfunctions and eigenvalues.

Floquet theory was introduced in magnetic resonance by Shirley [15] to describe transition probabilities of nuclear spins in an atomic beam, subject to a linear rf field. Later it was applied to analyze saturation-transfer EPR spectra, by calculating the steady-state solution of a stochastic Liouville equation under the influence of radiation and modulation fields [86, 87, 88, 89]. Nowadays Floquet theory is used, for example, for the description of multiple-photon processes in NMR [16] and EPR [46, 17, 32], and for explaining experimental observations in magic angle spinning (MAS) NMR [74, 85].

Mathematically, the Floquet formalism is based on a solution of linear differential equations with periodic coefficients [90]. A semiclassical periodically time-dependent Hamiltonian $H(t) = H_0 + H_1(t)$ in the eigenbasis of H_0, with fundamental frequency $\omega/2\pi$ and period $t_p = 2\pi/\omega$, is expanded to a Fourier series,

$$\langle p | H(t) | q \rangle = \sum_{n=-\infty}^{\infty} \langle p | h^n | q \rangle e^{in\omega t} , \qquad (49)$$

where h^n is the nth Fourier component of $H(t)$. A Floquet Hamiltonian is defined as a Hermitian matrix of infinite dimension in Floquet space

$$p, n|H_F|q, m = p|h^{n-m}|q + n_{nm} pq ,$$
(50)

where $|p, n$ represents a spin function $|p$ dressed by n quanta with frequency
. The evolution of the density operator in Floquet space is described in analogy to the Hilbert space by

$$_F(t) = U_F(t; t_0) _F(t_0)U_F^{-1}(t; t_0) ,$$
(51)

with the propagator

$$U_F(t; t_0) = e^{-iH_F(t-t_0)} .$$
(52)

For the calculation of the evolution of the spin system in Hilbert space either the density operator or the propagator has to be transformed back to Hilbert space.

For time-dependent Hamiltonians with more than one basic frequency the Floquet theory is expanded to the   many-mode Floquet theory [91, 92]. When transient e ects are investigated the evolution of the density operator in Hilbert space can be calculated numerically [93].

References

1. J. Margerie, J. Brossel: C. R. Acad. Sc.   241 , 373 (1955)
2. C. Cohen-Tannoudji: Optical pumping and Interaction of Atoms with the Electromagnetic Field. In: Carg`ese Lectures in Physics , Vol. 2, ed by M. L'evy
   (Gordon and Beach, New York 1968) p 347
3. C. Cohen-Tannoudji, J. Dupont-Roc, G. Grynberg: Atom-Photon Interactions
   (Wiley, New York 1992)
4. N.B. Delone, V.P. Krainov: Multiphoton processes in atoms , 2nd edn, (Springer-
   Verlag, Berlin 2000)
5. E.A. Donley, R. Marquardt, M. Quack, J. Stohner, I. Thanopulos, E.-U.
   Wallenborn: Mol. Phys.   99 , 1275 (2001)
6. S. Haroche: Ann. Phys.   6, 189, 327 (1971)
7. J.-M. Winter: Ann. Phys.   4, 745 (1959)
8. M. G"oppert-Mayer: Ann. Phys. (Leipzig)   5, 273 (1931)
9. F.H.M. Faisal: Theory of multiphoton processes   (Plenum Pres, New York 1986)
10. D.G. Gold, E.L. Hahn: Phys. Rev. A   16 , 324 (1977)
11. R. Boscaino, G. Messina: Physica C   138 , 179 (1986)
12. P. Bucci, P. Cavaliere, S. Santucci: J. Chem. Phys.   52 , 4041 (1970)
13. F. Chiarini, M. Martinelli, L. Pardi, S. Santucci: Phys. Rev. B   12 , 847 (1975)
14. F. Persico, G. Vetri: Phys. Rev. B   8, 3512 (1973)
15. J.H. Shirley: Phys. Rev.   138 , B979 (1965)
16. Y. Zur, M.H. Levitt, S. Vega: J. Chem. Phys.   78 , 5293 (1983)
17. M. Jele'n, W. Froncisz: J. Chem. Phys.   109 , 9272 (1998)
18. A. Schweiger, G. Jeschke: Principles of Pulse Electron Paramagnetic Resonance
    (Oxford University Press, Oxford 2001)
19. C.A. Hutchison, B.M. Weinstock: J. Chem. Phys.   32 , 56 (1960)
20. J. Forrer, A. Schweiger, N. Berchten, H.H. G" unthard: J. Phys. E: Sci. Instrum.
    14 , 565 (1981)

21. T.T. Chang: Phys. Rev. **136**, 1413 (1964)
22. D. Backs, R. St" osser, M.Lieberenz: Phys. Stat. Sol. B **131**, 291 (1985)
23. D.H.L. A. Carrington, T.A. Miller, J.S. Hyde: J. Chem. Phys. **47**, 4859 (1967)
24. D. Gourier, D. Simons, D. Vivien, N. Ruelle, M.P. Thi: Phys. Stat. Sol. B **180**, 223 (1993)
25. N.S. Dalal, A. Manoogian: Phys. Rev. Lett. **39**, 1573 (1977)
26. P.P. Sorokin, I.L. Gelles, W.W. Swith: Phys. Rev. **112**, 1513 (1958)
27. S. Yatsiv: Phys. Rev. **113**, 1522 (1959)
28. A.S.M. Rudin, H.H. G" unthard: J. Magn. Reson. **51**, 278 (1983)
29. J.W. Orton, P. Auzins, J.E. Wertz: Phys. Rev. **119**, 1691 (1960)
30. P. Tiedemann, R.N. Schindler: J. Chem. Phys. **54**, 797 (1971)
31. J.W. Orton, P. Auzins, J.E. Wertz: Phys. Rev. Lett. **4**, 128 (1960)
32. I. Gromov, A. Schweiger: J. Magn. Reson. **146**, 110 (2000)
33. B. Clerjaud, A. Gelineau: Phys. Rev. Lett. **48**, 40 (1982)
34. J.M. Winter: Le Journal de Physique et le Radium **19**, 802 (1958)
35. R. Boscaino, I. Ciccarello, C. Cusumano, M.W.P. Strandberg: Phys. Rev. B **3**, 2675 (1971)
36. R. Boscaino, F.M. Gelardi, G. Messina: Solid State Comm. **46**, 747 (1983)
37. R. Boscaino, F.M. Gelardi, G. Messina: Phys. Lett. **97A**, 413 (1983)
38. R. Boscaino, F.M. Gelardi, G. Messina: Phys. Rev. A **28**, 495 (1983)
39. S. Agnello, R. Boscaino, M. Cannas, F. Gelardi: Phys. Rev. B **64**, 174423 (2001)
40. R. Boscaino, F.M. Gelardi: Phys. Rev. B **46**, 14550 (1992)
41. G. Bimbo, R. Boscaino, M. Cannas, F.M. Gelardi, R.N. Shakhmuratov: J. Phys.: Condens. Matter **15**, 4215 (2003)
42. J.S. Hyde: Multiquantum EPR. In: Foundation of Modern EPR , ed by G.E. Eaton, S.S. Eaton, and K.M. Salikhov (World Scientific, Singapore 2001) p 741
43. J.S. Hyde, P.B. Sczaniecki, W. Froncisz: J. Chem. Soc. Faraday Trans. I **85**, 3901 (1989)
44. P.B. Sczaniecki, J.S. Hyde, W. Froncisz: J. Chem. Phys. **93**, 3891 (1990)
45. P.B. Sczaniecki, J.S. Hyde: J. Chem. Phys. **94**, 5907 (1991)
46. H.S. Mchaourab, J.S. Hyde: J. Chem. Phys. **98**, 1786 (1993)
47. R.A. Strangeway, H.S. Mchaourab, J.R. Luglio, W. Froncisz, J.S. Hyde: Rev. Sci. Instrum. **66**, 4516 (1995)
48. G. Goelman, D.B. Zax, S. Vega: J. Chem. Phys. **87**, 31 (1987)
49. E.M. Krauss, S. Vega: Phys. Rev. A **34**, 333 (1986)
50. P. Bucci, M. Martinelli, S. Santucci: J. Chem. Phys. **53**, 4524 (1970)
51. P. Bucci, S. Santucci: Phys. Rev. A **2**, 1105 (1970)
52. M. van Opbergen, N. Dam, A.F. Linskens, J. Reuss, B. Sartakov: J. Chem. Phys. **104**, 3438 (1996)
53. H.S. Mchaourab, T.C. Christidis, W. Froncisz, P.B. Sczaniecki, J.S. Hyde: J. Magn. Reson. **92**, 429 (1991)
54. H.S. Mchaourab, T.C. Christidis, J.S. Hyde: J. Chem. Phys. **99**, 4975 (1993)
55. H.S. Mchaourab, S. Pfenninger, W.E. Antholine, C.C. Felix, J.S. Hyde, P.H. Kroneck: Biophys. J. **64**, 1576 (1993)
56. J.S. Hyde, M. Pasenkiewicz-Gierula, A. Jesmanowicz, W.E. Antholine: Appl. Magn. Reson. **1**, 483 (1990)
57. J. Burget, M. Odehnal, V. Pet˘ ríček, J. Sàcha, L. Trlifaj: Czech. J. Phys. B **11**, 719 (1961)
58. T. Hashi: J. Phys. Soc. Japan **16**, 1243 (1961)

59. P. Lehtovuori, H. Joela: J. Phys. Chem. A   106 , 3061 (2002)
60. R. Gabillard, B. Ponchel: C. R. Acad. Sc.   254 , 2727 (1962)
61. R. Karplus: Phys. Rev.   73 , 1027 (1948)
62. B. Smaller: Phys. Rev.   83 , 812 (1951)
63. K. Halbach: Helv. Phys. Acta   29 , 37 (1956)
64. H. Primas: Helv. Phys. Acta   31 , 17 (1958)
65. W.A. Anderson: Magnetic field modulation for high resolution NMR. In: NMR and EPR spectroscopy, Varian's 3rd annual workshop on NMR and EPR   (Pergamon Press, Oxford 1960) p 180
66. J.D. Macomber, J.S. Waugh: Phys. Rev.   140 , A1494 (1965)
67. O. Haworth, R.E. Richards: Prog. NMR Spec.   1 , 1 (1966)
68. H. Wahlquist: J. Chem. Phys.   35 , 1708 (1961)
69. C.P. Poole: Electron Spin resonance: A comprehensive treatise of experimental techniques (Wiley, New York 1983)
70. P.A. Berger, H.H. G¨unthard: J. Appl. Math. and Phys.   13 , 310 (1962)
71. A.G. Redfield: Phys. Rev.   98 , 1787 (1955)
72. M. K¨alin, I. Gromov, A. Schweiger: J. Magn. Reson.   160 , 166 (2003)
73. M. K¨alin: Multiple Photon Processes in Electron Paramagnetic Resonance Spectroscopy. PhD Thesis, ETH Z¨ urich, No. 15142, Z¨urich (2003) http://e-collection.ethbib.ethz.ch/cgi-bin/show.pl?type=diss&nr=15142.
74. S. Vega: Floquet theory. In: Encyclopedia of Nuclear Magnetic Resonance   , ed by D.M. Grant and R.K. Harris (Wiley, Chichester UK 1996) p 2011
75. F. Bloch, A. Siegert: Phys. Rev.   57 , 522 (1940)
76. M. K¨alin, I. Gromov, A. Schweiger: Phys. Rev. A   69 , 033809 (2004)
77. M. Fedin, M. K¨ alin, I. Gromov, A. Schweiger: J. Chem. Phys.   120 , 1361 (2004)
78. E.J. Hustedt, A. Schweiger, R.R. Ernst: J. Chem. Phys.   96 , 4954 (1992)
79. A.D. Milov, K.M. Salikhov, M.D. Shirov: Sov. Phys. Solid State   24 , 565 (1981)
80. M. Pannier, S. Veit, A. Godt, G. Jeschke, H.W. Spiess: J. Magn. Reson.   142 , 331 (2000)
81. M. Weissbluth: Photon-Atom Interactions (Academic Press, Boston 1989)
82. R.J. Glauber: Phys. Rev.   131 , 2766 (1963)
83. N. Polonsky, C. Cohen-Tannoudji: J. Physique   26 , 409 (1965)
84. C. Cohen-Tannoudji, S. Reynaud: J. Phys. B   10 , 345 (1977)
85. T.O. Levante, M. Baldus, B.H. Meier, R.R. Ernst: Mol. Phys.   86 , 1195 (1995)
86. B.H. Robinson, J.-L. Monge, L.A. Dalton, L.R. Dalton, A.L. Kwiram: Chem. Phys. Lett.   28 , 169 (1974)
87. J.H. Freed, G.V. Bruno, C.F. Polnaszek: J. Phys. Chem.   75 , 3385 (1971)
88. L.R. Dalton, B.H. Robinson, L.A. Dalton, P. Co ey: Saturation Transfer Spectroscopy. In: Advances in Magnetic Resonance , Vol. 8, ed by J.S. Waugh (Academic Press, New York 1976) p 149
89. D.J. Schneider, J.H. Freed: Spin Relaxation and Motional Dynamics. In: Advances in Chemical Physics: Lasers, Molecules, and Methods , Vol. 73, ed by J.O. Hirschfelder, R.E. Wyatt, and R.D. Coalson (Wiley-Interscience, New York 1986) p 387
90. M.G. Floquet: Ann. Ecole Norm. Sup.   12 , 47 (1883)
91. T.-S. Ho, S.-I. Chu: J. Phys. B   17 , 2101 (1984)
92. S.-I. Chu: Adv. Atom. Mol. Phys.   21 , 197 (1985)
93. M. K¨alin, A. Schweiger: J. Chem. Phys.   115 , 10863 (2001)

Multi-Frequency EPR Study
of Metallo-Endofullerenes

Klaus-Peter Dinse [1] and Tatsuhisa Kato [2]

[1] Physical Chemistry III, Darmstadt University of Technology, Petersenstrasse 20
64287 Darmstadt, Germany
dinse@chemie.tu-darmstadt.de

[2] Institute for Molecular Science, Myodaiji, Okazaki 444-8585, Japan, (present
address: Department of Chemistry, Josai University, 1-1 Keyakidai, Sakado
350-0295, Japan)
rik@josai.ac.jp

Abstract. Immediately after the observation that fullerenes can act as cages for persistent trapping of ions or atoms, Electron Paramagnetic Resonance (EPR) was used for the investigation of these compounds. In contrast to group 15 atoms entrapped in C_{60} or C_{70}, which are nearly perfectly decoupled from the cage and which at time average are found at the center of the cage, it was observed that single lanthanide ions are localized at specific binding sites of the internal carbon surface. Apparently this is no longer true for encased ion clusters like La_2^{6+} or Sc_3N^{6+}, which are exploring the full inner space of the fullerenes in such a way that the inherent symmetry of the cage is retained. Although simple 2-pulse sequences have been used to elucidate ion and cage motions in the single ion case, more advanced techniques like pulsed electron-nuclear double resonance (ENDOR) as well as two-dimensional EPR had to be invoked for the study of the cluster compounds. Performing experiments at 95 GHz, one is benefiting from superior spectrometer sensitivity as well as from the inherent capability for improved orientation selection. In this contribution results about $Sc_3N@C_{80}^-$, $La@C_{82}$, and $La_2@C_{80}$ compounds are compiled. It was the purpose of our investigation to obtain information about the amount of spin and charge transfer in these cage compounds. Using DFT methods, a reasonable description of the observed spin Hamilton parameters could be obtained.

1 Introduction

Since trace amounts of metallo-endofullerenes (MEF) were detected more than a decade ago in 1991, scientific interest was focused on two problems, i.e., the possibility of ion localization at a specific binding site and the amount of charge transfer from the encased metal ion to the cage [1, 2]. To clarify both points would be a prerequisite for a deeper understanding of electronic properties of MEF. With respect to localization, major progress was made after pure substances in macroscopic quantities became available several years

K.-P. Dinse and T. Kato: Multi-Frequency EPR Study of Metallo-Endofullerenes , Lect. Notes
Phys. **684** , 185–207 (2006)
www.springerlink.com

later. This enabled direct structure determination via synchrotron radiation powder di raction techniques, the analysis not only yielding the binding site but also allowing to identify the cage topology [3]. Furthermore, under favorable conditions, the oxidation state of the ion (or cluster) could also be determined by performing a combined Rietveld/Maximum Entropy analysis of the X-ray data. Compared to X-ray di raction, application of less direct magnetic resonance methods like NMR and EPR for structure determination nevertheless has advantages given by the fact that such investigations can be performed using highly diluted material. In most cases, a very wide temperature range can also be investigated, thus probing the binding potential and testing the influence of solvents and solid matrices. Even more important is the fact that magnetic resonance methods are uniquely capable to determine the spin multiplicity of the compound, thus enabling to explore details of the molecular wave function of the compound. Knowledge of the e ective electronic spin for instance is of utmost importance for a complete understanding of the weakly coupled electronic system, generated by attaching the metal ion to the carbon cage. Envisioning the MEF as examples of internal charge transfer complexes, sign and size of the resulting exchange coupling between cage and ion will determine the properties of the ground state of the coupled system. It should be noted, however, that the assumption of a purely ionic interaction is too simplistic, because photoelectron spectroscopy of La@C$_{82}$ for instance has revealed that there is a finite electron density in 5 d levels of La [4]. Finally, for magnetic resonance, an additional degree of freedom is given by the possibility to study dilute solutions of compounds that can be reduced or oxidized by chemical or electrochemical methods, thus allowing to probe charge redistribution in the status of electron excess or deficiency.

The localization problem is best stated by describing the potential of the inner fullerene surface seen by the encased atom or ion after intra-molecular charge transfer. Depending on the topology of this potential, either localization or quasi-free motion is possible. Clearly, the binding potential will depend on fullerene topology (i.e., carbon number and cage symmetry) as well as on the identity of the metal ion, implying that both extremes of localized particle and free motion can be realized depending on temperature.

In order to distinguish between both cases, magnetic resonance techniques can be invoked. Using magnetic resonance, static as well as dynamic terms of the e ective spin Hamiltonian can in principle be determined from line positions and line widths data, respectively. For this purpose, a line shape analysis of individual spectral transitions is performed using a model of thermally activated hopping between di erent binding sites. If the correlation time of this process is fast on the time scale of typical frequency di erences in the spectrum, averaging of spectroscopic properties of di erent sites occurs. Molecular tumbling on the other hand leads to characteristic line broadening, which can be analyzed in terms of various second-rank tensor interactions. Changing the temperature, the correlation time as well as the variance of the time-modulated interactions can be deduced. By this method, it is in principle possible to distinguish between internal hopping of the ion and a tumbling motion of the whole molecule with the ion rigidly attached to its docking position [5].

Although quite powerful for the fast correlation limit, magnetic resonance investigations based on such a line shape analysis are limited to a temperature range for which liquid solvents are available. Even more restrictive is the condition that a fully resolved and assigned spectrum must be available. Meeting these conditions, a qualitative picture emerges from a fit of individual lines using Lorentzians of varying widths, thus revealing the concerted action of the anisotropic g-matrix, the electron-nuclear magnetic dipole-dipole, and the nuclear quadrupole interactions. Additionally the principal elements of these second-rank tensors can quantitatively be deduced, although an axes assignment is generally not possible. At lower temperatures, for which localization is expected, spectral analysis of the rigid-limit spectrum allows to determine these tensor elements directly, thus providing information about the binding site of the localized ion. Combining results from a high temperature dynamic study with rigid limit data, it is possible to detect the anticipated transition from a temperature-activated large-scale internal motion to rigid attachment from an analysis of the hyperfine (hf) tensor elements.

The ground state of a coupled spin system is advantageously defined by its spin multiplicity. In particular, if MEF are considered to consist of encased Lanthanide ions with an empty valence shell after charge transfer and a partially filled cage molecular orbital (MO), overlap of 4f orbitals with carbon centered orbitals will be small even at short distance. As a result, the exchange part of the Coulomb interaction will probably be small, leading to close-lying spin multiplets.

Information about the effective electronic spin S_{eff} of the compound can be determined from the analysis of the EPR spectrum. If S_{eff} 1, second and higher rank electronic spin-spin tensor interactions generally dominate the EPR spectrum. Evaluation of these Zero-Field-Splitting (ZFS) terms usually allows a qualitative characterization. Multi-frequency EPR data are in most cases required to discriminate between integer and half-integer spin systems. A simple method to distinguish between different S_{eff} 2 spin systems with additional fourth-rank tensor components is unfortunately not available for disordered samples.

In neutral MEF, a subtle balance of electron affinity of the cage and ionization potential of the encaged atom governs charge distribution between ion and cage. Depending on the energy level structure of both subcomponents, accommodation of an additional electron, for instance provided by reduction to the mono-negative ion, cannot be predicted unambiguously. The question arose, whether the oxidation state of the ion could be "switched" by reduction, thus allowing changing its effective spin multiplicity. Such reversal of charge transfer would be counter-intuitive, however, if modelling the MEF as Faraday cage system. Very early, the surprisingly small isotropic hyperfine interactions (hfi) of encased ions observed in $Sc@C_{82}$ [7, 6], $Y@C_{82}$ [9, 8], and

La@C_{82} [2] were taken as evidence for a complete charge (and spin) transfer of all three metal valence electrons to the cage. A convenient way to probe a possible spin redistribution under reduction or oxidation is therefore given by probing the isotropic Fermi contact interaction of the ion.

In the following, we present recent results obtained by multi-frequency EPR investigations of three different MEF La@C$_{82}$, La$_2$@C$_{80}$, and Gd@C$_{82}$. These compounds nicely exemplify the above stated problems, viz. localization, exchange coupling, and charge redistribution under reduction of the compound.

The system La@C$_{82}$ (I) was chosen for this study because of conflicting conclusions published by various groups. X-ray studies by the group of Takata indicated large amplitude motion around a well-defined site (quasi-localization) at room temperature [10, 11]. An early theoretical study of Andreoni and Curioni [12] came to the conclusion that the La ion could be localized at two different positions, one of them allowing large amplitude motion. As was shown later [10, 13], the C$_{82}$ topoisomer investigated in their first study had the wrong cage symmetry (C_2 instead of C_{2v}). In a later study by the same authors [14], the proper C_{2v} topoisomer was also investigated, apparently exhibiting a unique binding site with strong localization in addition to a second local minimum. The hypothesis of different binding sites was in contradiction to our finding that even at room temperature molecular tumbling controls the correlation time of hf interactions seen by the ion [5]. The observation of resolved ^{13}C satellites in the EPR spectrum confirmed our model but would still allow large amplitude motion about a defined equilibrium position, which would be in agreement with Takata's room temperature result.

Apart from C$_{60}$, only C$_{80}$ can be used to realize confinement with the highest possible symmetry I_h. As an example, diamagnetic La$_2$@C$_{80}$ was studied by NMR in solution, indicating that the encased cluster is not restricted to any specific binding site, thus preserving the I_h symmetry on time average as seen by ^{13}C nuclei of the cage [15]. Using qualitative arguments derived from MO theory, six electrons can be accommodated by the 4-fold degenerate Highest Occupied Molecular Orbital (HOMO) of C$_{80}$, suggesting an oxidation state of +6 for the cluster. This hypothesis was recently confirmed by powder diffraction data [16]. In case of the mono-negative ion the ground state of the compound could either be represented either as La$_2^{6+}$:C$_{80}^{7-}$ or as La$_2^{5+}$:C$_{80}^{6-}$, resulting in a drastic change of the La hfi. In the former case, the "spin-less" configuration with small hfi of the cluster would be preserved under reduction, whereas the latter configuration would be characterized by a large La hfi, because the additional electron would have to occupy 5d or 6s orbitals.

2 La@C$_{82}$ – a Case Study of Ion Localization

2.1 Experimental

Conventional arc vaporization of a rod composed of graphite and La$_2$O$_3$ in a helium atmosphere was used to prepare a fullerene soot containing La@C$_{82}$. After soxhlet extraction, the major topoisomer La@C$_{82}$(I) was separated using multi-stage HPLC techniques as described in [17]. Diluted solutions of the pure 98 percent compound were degassed on a high vacuum line and subsequently sealed in 4 mm or 0.9 mm o.d. quartz tubes appropriate for EPR investigations at 9.5 and 94 GHz, respectively. EPR spectra were taken with a BRUKER ELEXSYS 680 spectrometer at 9.4 GHz (X-band) and 94 GHz (W-band). W-band ENDOR experiments were performed using a home-built ENDOR probe head [18]. The microwave pulse power of 5 mW provided by the commercial micr owave bridge was boosted to a level of 100 mW using a multi-stage micr owave amplifier [19]. The microwave field amp litude at the sample was approximately 0.25 mT, as determined by the / 2 pulse length of 35 ns.

2.2 Results and Discussion

EPR Spectra

Caused by the rather low C$_{2v}$ symmetry of the C$_{82}$ cage of the major topoisomer of La@C$_{82}$(I), the spin Hamiltonian parameters of the paramagnetic compound will exhibit the full possible anisotropy of second-rank tensor interactions. For this reason, the g-matrix as well as electron-nuclear dipolar hfi and electric field gradient (EFG) at the La site will be characterized by three di erent components in their respective eigen frames. Isotropic averaging by molecular tumbling let one recover the trace of these interactions. Data thus derived can be used for a convenient identification of the various compounds which is important to discriminate, for instance, incorporation of La in C$_{82}$ cages of di erent topology [20]. Fluctuating values of spin Hamiltonian parameters induced either by molecular tumbling or by internal mobility of the encased ion will in general lead to characteristic relaxation processes. Under favorable conditions, an analysis of relaxation-induced line broadening of resolved hf components (hfc) of the EPR spectra can be utilized to discriminate between various possible relaxation processes and this information can be used to derive a detailed model of cage and ion mobility. Solution spectra of La@C$_{82}$ are well resolved, not only showing the dominant hfi of the La nucleus ($I = 7/2$) but also showing hfi with ^{13}C nuclei of the cage (in natural abundance).

In frozen solution, however, no hf information can be extracted neither from X- nor from W-band EPR powder spectra, probably because g-matrix and hfi tensor eigen frames are not co-linear. Nevertheless, the g-matrix

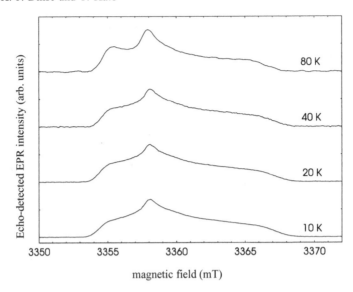

Fig. 1.   95 GHz (W-band) echo-detected EPR spectra of La@C$_{82}$ (I) as a function of temperature

anisotropy is su ciently large to lead to a powder pattern in W-band, which is characteristic for a rhombohedral g-matrix as expected.

The individual subcomponent line width masked at low temperatures by inhomogeneous spectral broadening can be recovered in principle by simple 2-pulse echo experiment. In the typical case of a paramagnetic sample of high dilution, the de-phasing time T_2 of a particular EPR transition is of the order of a few microseconds resulting in a homogeneous line width of less than 1 MHz. Because this value is much less than the spectral range of the powder spectrum, selective excitation by micr owave pulse seq uences is possible. Considering that the e ective spectral width of microwave pulses used for W-band ENDOR is less than 5 MHz, orientation selection is possible. In Fig. 1, a series of low temperature EPR spectra is depicted. As can be noted, spectral features are still changing even below 80 K.

Similar observations have already been reported in [21]. Because spectra are obtained using a simple 2-pulse echo sequence with fixed pulse delay, apparent intensity variations could result from an orientation-dependent spin de-phasing time T_2. This could be taken as evidence for residual cage mobility. In addition, the total spectral width is reduced noticeably comparing 80 K with 10 K spectra as is shown in Fig. 2. This change in g matrix anisotropy probably indicates structural changes, which occur also in this temperature interval. Note that in Fig. 2, field modulation was used to emphasize canonical positions in the spectrum.

A completely di erent EPR spectrum is obtained when observing La@C$_{82}$ in liquid solution. Hfi with ^{139}La nuclei (100 percent natural abundance) is

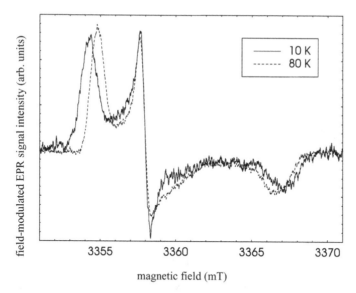

Fig. 2.   Continuous wave (cw) EPR spectra of La@C$_{82}$ (I) at 10 K and 80 K

apparent from 8 equidistant lines. Because of very narrow EPR lines, it is also possible to resolve additionally various hf coupling constants (hfcc) with ^{13}C nuclei of the cage (see Fig. 3). Assuming equal spin de-phasing rates for all La hfc, spin statistics predict 8 major lines of equal intensity. The observed characteristic deviation from equal intensities can be evaluated and used to determine the dominant EPR relaxation mechanisms. A consistent picture had emerged when fitting the widths of individual La hfc in the liquid phase temperature range of 200 to 300 K. A detailed analysis was performed by two research groups, showing that fluctuations of the EFG at the La site in combination with electron-nuclear dipole-dipole interaction, both caused by molecular tumbling, are the major sources of electron spin relaxation. Considering that ^{13}C hfi could also be observed over a wide temperature range, it could be concluded that the ion is rigidly attached to the inner surface of the carbon cage. Any additional internal hopping of the ion to a di erent attachment site would drastically influence the ^{13}C hfi, because the spin density distribution on the cage would change if the ion would be attached to another site, thus modifying the ^{13}C hfcc. As a result, the widths of these ^{13}C satellites would drastically di er from those of the main components.

In addition to the isotropic part of La hfi (a_{iso} (^{139}La) = 3 .22 MHz), which can be directly deduced form the EPR spectrum, the line width analysis resulted in an estimate of the principal component of the traceless dipole-dipole interaction A_{zz} (^{139}La) = 6 .5 MHz in its eigen frame denoted by z'. For this analysis, axial symmetry had to be assumed, and no axis assignment with respect to the molecular frame is possible. An estimate of the nuclear quadruple

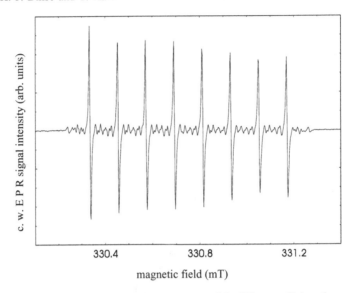

Fig. 3.   Room temperature EPR spectrum of La@C$_{82}$ (I) in toluene

coupling constant e ^{2}Qq/ (2I (2I − 1)h) = 0 .85 MHz was also obtained. Because nuclear quadrupole interaction (nqi) is traceless, this interaction does not lead to first order line splittings in the EPR spectrum. Any attempt to confirm this analysis therefore had to invoke electron-nuclear double resonance techniques.

ENDOR Spectra

Apart from simplifying crowded EPR spectra, ENDOR gives the opportunity to detect EFG using the electric quadrupole moment Q of $I > 1/2$ nuclei as local sensor. This can be seen by calculating ENDOR frequencies from eigenvalues of the spin Hamiltonian

$$H /h = \nu_e - \nu_n + S\,AI + I\,QI , \qquad (1)$$

under the condition of $m_s = 0$ and $m_I \pm 1$. Here, in usual notation, the electron and nuclear Larmor frequencies are indicated as ν_e and ν_n, respectively. The dipolar hf coupling tensor (including its isotropic part) is denoted as A, the traceless quadrupole coupling tensor as Q (all values given in frequency units). Its principal value Q_{zz} is related to the quadrupole coupling constant e ^{2}Qq/h by $Q_{zz} = e^2qQ/ (2I (2I − 1)h)$. The presence of a non-vanishing nqi leads to two characteristic multiplets of 2 I lines, each centered at $| \nu_n + A_{zz} m_s |$.

$$\nu_{m_I+1 ,m_I} = | \nu_n + A_{zz} (,)m_s + 3 / 2 Q_{zz} (,)(2m_I + 1) | . \qquad (2)$$

In (2), the nuclear spin quantum number m_I is ranging from $-|I|$ to $|I|- 1$. For simplicity only allowed NMR transitions are considered which might be too restrictive if e cient nuclear spin mixing results from non collinear nqi. In both electron spin sublevels, the same characteristic quadrupolar splitting of $3Q_{zz}$ between adjacent lines is predicted. The orientational dependence of all interactions is denoted by the angles   and   , relating the eigen axes of the interactions (given as "primed" coordinates) to the laboratory z axis, which in turn is defined by the magnetic field direction. The orientation dependence of both hf interactions causes a "powder-like" ENDOR line shape, in general precluding the determination of the coupling constants. In contrast, if e cient orientation selection can be obtained by frequency-selective excitation of spin packets in the EPR spectrum, a resolved "single crystal-like" ENDOR spectrum is predicted. Su cient orientation selection can for instance be obtained if electron Zeeman anisotropy dominates the EPR spectrum, thus leading to an unique set of   and   values for a given field setting. In our case, because of C_{2v} symmetry of the compound, which is preserved as site symmetry at the ion, there will be a common eigen axes system of the g matrix and La hfi and nqi tensors. Selective excitation at the edge of the EPR absorption spectrum is thus predicted to yield a narrow-line ENDOR spectrum, which can be analyzed in terms of one of the eigenvalues of the hfi and nqi tensors.

In case of $La@C_{82}$ (I), electron spin de-phasing times of approximately $5 \mu s$ were observed, nearly independent of temperature. Spin-lattice relaxation rates were strongly temperature dependent and reached 5 ms at 10 K. Under these conditions, pulsed ENDOR spectra could be recorded using a   - T- /2- -   sequence ($t_{/2} = 40$ ns, $T = 180 \mu s$,   = 1 μs), nuclear spin transitions being excited during interval   T after the inverting   pulse. ENDOR signals were extremely weak and could only be detected after more than 10 hours accumulation time. The resulting spectrum is depicted in Fig. 4.

In agreement with (2), equidistant transitions are observed, centered about the nuclear Zeeman frequency $_n$ (^{139}La) = 20 .4 MHz, the low-frequency multiplet, however, being barely detectable. As is indicated by the simulation, the expected seven-line pattern of both ms multiplets, separated by A_{zz} , overlap partially. Clear identification of transitions of the high-frequency multiplet nevertheless allows the unambiguous determination of all relevant parameters, viz. $|A_{zz}| = 5.9(2)$ MHz, $a_{iso} = 4.1(2)$ MHz, and $|Q_{zz}| = 0.44(2)$ MHz. Comparing these values with the data set determined in liquid solution, viz. $|A_{zz}| = 5.8(2)$ MHz, and $a_{iso} = 3.2(2)$ MHz, it can be stated that the dipolar hfi is nearly invariant over the full temperature range.

Evaluation of the nqi data is more complicated because no axis assignment of the solution value e $^2qQ/2I(2I - 1)h = 0.85(1)$ MHz can be made. Assuming that z is also the principal axis of the nqi, the   z orientation-selected low temperature quadrupole coupling $|Q_{zz}| = 0.44$ MHz corresponds to the solution value 0.85 MHz, thus indicating a significant reduction. Experiments performed under di erent orientation selection conditions indicate, however, a larger quadruple splitting, better in agreement with the solution value [22]. Further pulsed experiments have to be performed at still higher Larmor frequencies with inherent better orientation selection to clarify this point.

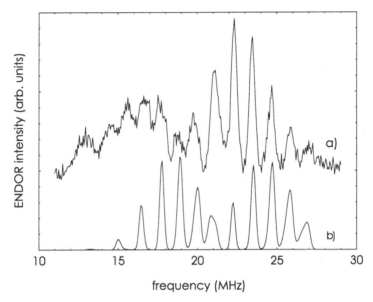

Fig. 4. (a) W-band pulsed ENDOR spectrum of La@C$_{82}$(I) measured at 10 K at
B$_0$ = 3371 .6 mT. The spectrum was accumulated when exciting the high-field edge
of the EPR spectrum. (b) Simulated spectrum used to identify the individual line
positions

HYSCORE Spectra

As was observed quite early, hfi can lead to pronounced deviations from an
exponential decay of 2-pulse electron spin echo decays. The so called electron
spin echo envelope modulation (ESEEM) results from the interference of electron spin coherences generated by the micr owave pulses and can be used to
extract hfi data otherwise only available by double resonance techniques like
ENDOR [23]. Later an elegant 2-dimensional version termed Hyperfine Sublevel Correlation Spectroscopy (HYSCORE) was proposed, by which nuclear
spin dependent frequency di erences in di erent electron spin sublevels can
be correlated [24]. This technique is now widely applied in the study of disordered samples and is particularly useful as the method complementary to
ENDOR. This 2-dimensional technique is particularly useful in disentangling
spectral features in ENDOR spectra originating from di erent nuclear spins,
a condition frequently met in X-band ENDOR.

The observation of deep ESEEM e ects using a ^{13}C-labelled sample of
La@C$_{82}$(I) prompted us to use HYSCORE to obtain additional information
about La nqi. In contrast to the attempt to use e cient orientation selection
to obtain frequency resolved transitions within the nuclear spin manifold as
described above, nearly perfect sampling of all orientations was attempted
here. The reason for this strategy is as follows: Powder broadening of nuclear
spin transitions occurs because of a varying orientation of the magnetic field

"seen" by the nuclear spin with respect to its eigenframe, for instance, defined by the local EFG. If the external field is compensated by the hyperfine field, the resulting nuclear spin transitions resemble a pure quadrupole resonance spectrum ("zero field NQR") [25]. The possibility to vary the external field using multi-frequency EPR and thus to "tune" into the perfect cancellation condition is advantageously used to determine ^{14}N nqi in biological samples. Clearly, compensation is only possible in one of the electron spin manifolds because of the dependence of the hyperfine field on the electron spin magnetic quantum number m_S. A minimum number of narrow lines therefore is only obtained if averaging over the full orientation sphere can be performed, thus wiping out transitions originating from one of the electron spin manifolds. This requires exciting the full spectrum. Being limited in frequency width of the microwave exc itation pulses, this condition cannot be met in X-band. Using ^{13}C labelling, however, the spectral broadening by strong ^{13}C hfi leads to e ective orientation averaging even under semi-selective excitation conditions.

A representative HYSCORE spectrum of the major isomer obtained at 80 K is shown in Fig. 5. Projections of the 2D plot on both frequency axes are also shown. The resulting plots correspond to spectra which would be obtained performing a one-dimensional ENDOR experiment. Because of spectral overlap in the range from 1 to 6 MHz originating from ^{13}C and ^{139}La ENDOR transitions, no clear assignment would be possible if only ENDOR data are available. In the 2D plot, however, three major features can be noted, one is a diagonal peak at 14.7 MHz, another is a pair of ridges which runs perpendicular to the frequency diagonal with a crossing point at 3.8 MHz, the other is a strong diagonal peak at 2.9 MHz. The first peak at 14.7 MHz corresponds to the free proton Zeeman frequency and it arises from "distant" protons of spurious solvent molecules. The observed ridges, crossing the diagonal at 3.8 MHz, can be assigned to arise from ^{13}C dipolar hfi with various non-equivalent nuclei of the cage. The prominent diagonal peak at 2.9 MHz is identical with the line obtained by Fourier transformation of the deep ESEEM modulation observed in both 2-pulsed and 3-pulsed spin echo decays. Although the nuclear Zeeman frequency of ^{139}La at the chosen field of 2.1 MHz is close to this value, a transition at the La nuclear Zeeman frequency cannot be observed unless lanthanum would be situated at a site of at least cubic symmetry, which is not available in our case. We therefore have to assign this prominent narrow line feature to a quadrupole transition in e ective zero field between La nuclear spin levels, which could be classified in first order (neglecting mixing of m_I states by non-axial EFG components) as $m_I = \pm 1/2 \pm 3/2, \pm 3/2 \pm 5/2$, or $\pm 5/2 \pm 7/2$, respectively. It should be noted, that this unambiguous assignment is possible only because in the 2D frequency plot of HYSCORE the partially overlapping ENDOR transitions from ^{13}C nuclei are separated and clearly identified.

In the axial approximation, transitions at 3, 6, and 9 in units of $e^2qQ/42h$ are predicted. Unfortunately, because of limited signal-to-noise and overlap with ^{13}C lines we were not successful in identifying unambiguously the full set

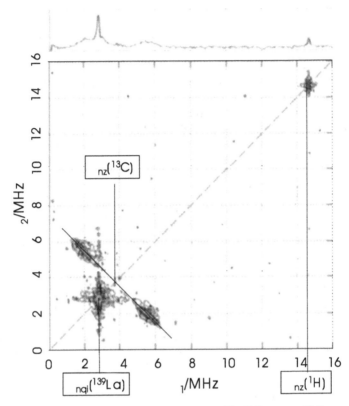

Fig. 5.   X-band HYSCORE spectrum of La@C$_{82}$ at 80 K

of three quadrupole transitions, which would have enabled the determination of the asymmetry parameter also. Lacking this additional information, only an order of magnitude interpretation can be given: Assuming that the observed prominent peak at 2.9 MHz corresponds to one of the transitions given above, quadrupole coupling constants e^{2}qQ/2I(2I − 1)h would be obtained as 0.96, 0.64, and 0.32 MHz, respectively. These values are well in the range of $|Q_{zz}|$ = 0.44 MHz obtained from the ENDOR experiment.

2.3 La@C$_{82}$ – Resum´e

The analysis of EPR and ENDOR data obtained at different electron Larmor frequencies and in different phases leads to the conclusion that no significant change occurs with respect to the binding site of the encased ion. Apparently, the room temperature large-scale motion of the encased ion observed freezes at low temperatures, and because of averaging of hfi on the time scale of the EPR experiment no drastic change of dipolar and quadrupolar hfi is observed. These interactions are sensitive measures of the structure of the binding site because they are determined by the local spin and charge distribution. The detection of hfi in disordered samples was possible only by using orientation selection in the 94 GHz EPR spectrum, which is expanded with respect to magnetic field by g-anisotropy. It would be a challenge to reproduce these results of spin and charge density distribution at the Lanthanum ion by quantum-chemical modelling. This will be essential for a better understanding of the electronic structure of the complex.

3 La$_2$@C$_{80}^-$ Radical Anion – Evidence for Reduction of the Encased Cluster

3.1 Experimental

La$_2$@C$_{80}$ was prepared and separated by the method reported in [15, 26]. Mono anions of La$_2$@C$_{80}$ were produced by chemical reduction using 1, 5-diazabicycloundecene (DBU) as well as by electrochemical reduction. The electrochemical reduction was performed in well-dried o-dichlorobenzene (ODCB) with electrochemical grade tetra-n-butylammonium perchlorate (TBAP). EPR spectra of La$_2$@C$_{80}^-$ mono anions obtained by these two reduction processes were recorded by EPR using a microwave freq uency of 95 GHz (W-band) and of 9.5 GHz (X-band). Anions prepared by both methods gave identical X-band EPR spectra in the temperature range from 3 K to 295 K.

3.2 Results and Discussion

EPR Analysis

Having shown by NMR that neutral La$_2$@C$_{80}$ has an electronic singlet spin ground state [15], the question arose how the mono anion would accommodate an additional electron, provided by chemical or electrochemical reduction. As stated above, the ground state of the negative ion could either be represented in first approximation as [La$_2^{6+}$:C$_{80}^{7-}$] or as [La$_2^{5+}$:C$_{80}^{6-}$]. In this notation, localization of the additional electron either on the cage or at the confined cluster is evident. In the former case, a very small La hfi is predicted because of vacant La valence orbitals and a completely empty 4 f shell. In contrast, the latter configuration would be characterized by a large La hfi, because the additional electron would have to occupy La 5 d or 6s valence orbitals. First evidence of the possibility of "in cage reduction" was provided by the EPR spectrum of the mono-anion of Sc$_3$N@C$_{80}$, characterized by Sc hfi, one order of magnitude larger [27] than observed for Sc$_3$@C$_{84}$ [6]. Spin polarization was proposed as reason for the small Sc hfi in Sc$_3$@C$_{84}$. The increase in hfi of the mono anion was explained by assuming electron localization at the cluster in reduced Sc$_3$@C$_{84}$, synonymous for "in cage reduction".

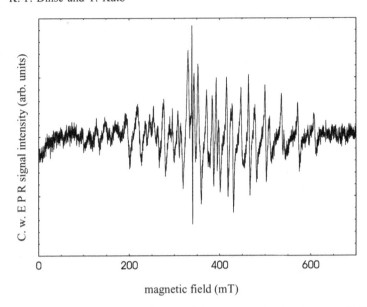

C. w. E P R signal intensity (arb. units)

magnetic field (mT)

Fig. 6.   X-band EPR spectrum of La $_2$@C$_{80}$ mono anion in toluene at 235 K

In Fig. 6, an EPR spectrum of La $_2$@C$_{80}$ mono anions in solution is depicted. In first order perturbation theory, a simple equidistant 15 line spectrum would be expected for this spin system consisting of $S = 1/2$ coupled to two equivalent $I = 7/2$ nuclei. The observed spectrum clearly cannot be explained even qualitatively within this scheme. The reason being twofold: First, the La hfi is so large that the usual first-order truncated spin Hamiltonian

$$H/h = {}_eS_z - {}_nI_z + a_{iso}S_zI_z , \tag{3}$$

obtained by omitting non-secular terms $a_{iso}(S_+I_- + S_-I_+)$ cannot be used. These terms are lifting the frequency degeneracy of allowed EPR transitions being characterized by a total value of M_I. The magnitude of frequency shifts of the center of each of these M_I multiplets is of the order of $a_{iso}^2/{}_e$ which amounts to 4 mT in our case (vide infra). Second, nuclear spin dependent line width variations might broaden some transitions beyond detectability.

For spectral simulation, the spectrum of two equivalent nuclear spins can easily be constructed by using a coupled representation of total nuclear spin, in our case ranging from $I_{tot} = 14$ to 0. Only within these nuclear spin subspaces, EPR transitions are allowed, giving a total of 64 transitions. Because of the large hfi constant, the spectrum extends over a very wide field range thus varying the frequency shifts considerably. As a result, even in the simulated spectrum no simple feature evolves. However, as is shown in Fig. 7, the high field edge of the spectrum is least disturbed, allowing to extract $a_{iso} = 38.6$ mT from the separation of outmost hf components by direct comparison with the experimental data.

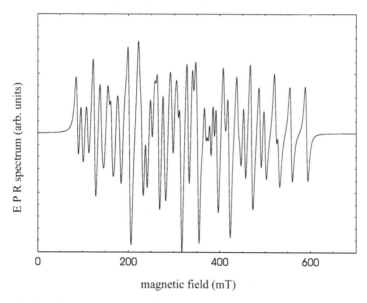

Fig. 7.   Simulated X-band EPR spectrum of La $_2$@C$_{80}$ mono anion (g = 1 .984, a$_{iso}$ = 38 .6 mT, $_e$ = 9 .4 GHz.)

Modelling La hfi

The measured isotropic La hfi constant of 38.6 mT corresponding to 1.03 GHz clearly cannot be rationalized as arising from spin polarization e ects. In LaO, a large La hf constant of 3.89 GHz has been observed, the absence of a noticeable anisotropy being taken as evidence for dominant spin density in the 6s orbital of the metal [28]. This assumption is consistent with the observation of a hfcc of 8.1 GHz of the free La $^{2+}$ ion. Taking the free ion value as reference for La hfi arising from a singly occupied 6 s orbital, in first approximation a local 6s spin density of approximately 0.25 can be deduced for both metal atoms of the internal La $_2^{5+}$ cluster.

Apart from this qualitative argument, Density Functional Theory (DFT)-type calculations were performed to support the hypothesis of internal re-duction. For the calculation a fixed geometry had to be assumed, although at room temperature the internal cluster is highly mobile, preserving the   I$_h$ symmetry of the cage on the time scale of EPR. Preserving the inversion symmetry, the La $_2$ cluster was placed along a C $_3$ axis through the center of 6-membered carbon rings. The results of the DFT calculations, which were performed using the   B3LYP   parametrization and 6-31G* and Hay/Wadt ba-sis sets for carbon and lanthanum, respectively, are best compiled showing the change in charge density of the compound under reduction. In agreement with the qualitative argument given above, charge density is piled up at the cluster as seen in Fig. 8. Correspondingly, the spin density is also concentrated at La. Calculated La hfcc are much smaller than the observed value, however, which

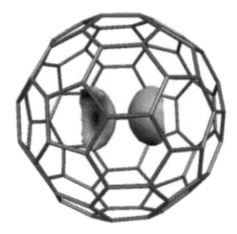

Fig. 8.   Calculated charge di erence between neutral La $_2$@C$_{80}$ and its mono anion. For details of the calculation see text

is not too surprising considering the fixed core basis set used for La, which does not allow spin polarization of the inner shell orbitals to occur.

La$_2$@C$_{80}^-$ Radical Anion – Resum´ e

The analysis of the EPR spectrum of mono anions of La $_2$@C$_{80}$ gives evidence for the rare case of "in cage reduction" for this compound. In analogy to mono anions of Sc$_3$N@C$_{80}$, the additional electron cannot be accommodated in the closed 4-fold degenerate HOMO of the carbon cage but rather resides on the internal cluster. The observed very large La hfcc of 1.03 GHz indicates that 6s orbitals of the metal cluster are mainly populated. The predicted small anisotropy of the hfi was confirmed by the observation of a well resolved EPR spectrum in solid solution, which could be fully analyzed invoking spectra taken at 94 GHz as well as at 9.4 GHz [29].

4 Gd@C$_{82}$ – Determination of Exchange Coupling Between Ion and Cage

4.1 Experimental

A highly purified sample of the major isomer of Gd@C $_{82}$(I)(C $_{2v}$) was pre-pared by applying various HPLC steps as described above [20, 30]. Liquid solution ESR spectra using   CS$_2$ or tri-chlorobenzene (TCB) as solvent as well as powder spectra of 100 percent Gd@C $_{82}$ were obtained using a Bruker E500 (X-band) and a E680 (W-band) spectrometer. Temperature was controlled by a helium flow cryostat set at 4.0 K for the X-band ESR measurement and at 20 K and 4.0 K for W-band experiments.

4.2 Results and Discussion

EPR Analysis

Having identified lanthanides as class of atoms which can easily be encapsulated in fullerenes it was obvious to search for effects originating from different orbital and spin momenta resulting from varying number of $4f$ electrons. Lanthanum, gadolinium and lutetium are special because of their vanishing orbital momentum. In contrast to lanthanum and lutetium, each of spin multiplicity one, gadolinium with its half filled $4f$ shell has an octet spin multiplicity. Here we assume complete charge transfer of three valence electrons for each of the elements. As a result, La@C$_{82}$ and Lu@C$_{82}$ exhibit EPR properties characteristic of effective electronic spin $S_{eff} = 1/2$. In case of Gd@C$_{82}$, however, the spin multiplicity of the ground state will depend on the details of coupling between ion and cage. Taking $S_{cage} = 1/2$ and $S_{ion} = 7/2$, $S_{eff} = 4$ or 3 can be realized.

The energy difference between both states can be described by an effective spin Hamiltonian

$$H = H_1 + H_2 - 2J S_1 S_2 . \tag{4}$$

In (4) contributions from the isolated constituents (viz., ion and cage) are denoted by H_1 and H_2, respectively, the coupling being approximated by assuming an isotropic exchange term. In this description the analogy to singlet/triplet splittings in organic molecules and charge transfer complexes is evident. For a rough estimate of J in our case the following arguments can be used: In typical small organic molecules the energy difference between the first excited singlet and the lowest triplet state is of the order of $1\,eV$, in charge transfer (CT) complexes this value is reduced by at least one order of magnitude, giving values for J ranging from $+4000\,cm^{-1}$ to $+100\,cm^{-1}$. The positive sign of J indicates that the triplet state is situated below the corresponding singlet level as is usually found in small organic molecules. The decrease in absolute value of J under CT conditions can be attributed to charge localization at different centers with resulting decrease in orbital overlap. For weakly coupled CT complexes even both signs are observed. In MEF like Gd@C$_{82}$ with large charge transfer and localization of the metal ion at the inner surface, a rather large electronic exchange term could result. However, because of an empty valence shell after transfer of three electrons, only the charge distribution of compact $4f$ electrons has to be considered, which might lead to a drastic reduction of J. In brief, highly localized atomic orbitals (AO) in combination with small charge separation distances prevailing in MEF preclude a reliable prediction of sign and size of J in this case.

Determination of the ground state spin multiplicity would be sufficient to deduce the sign of J. Although it is rather straightforward to discriminate between integer and non integer spin states by EPR, the predicted difference in EPR spectra of $S_{eff} = 3$ and $S_{eff} = 4$ spin systems is not large if for instance the same fine structure splitting parameter is used. In Fig. 9 fictitious

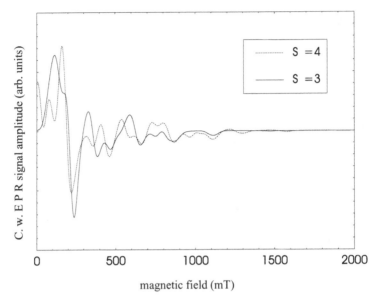

Fig. 9.   Simulated X-band EPR spectra of fictitious   S = 3 and   S = 4 spin systems

X-band powder spectra are depicted, calculated with an identical ZFS tensor. Here, for simplicity we have not included higher forth-rank terms in the spin hamiltonian which could contribute for both spin systems. Diagonal elements $D_{ii} = [-1.56 - 2.64\ 4.20]$ GHz and $g_{ii} = [1.99\ 1.99\ 2.00]$ were chosen, which are in the range of those determined for the Gd@C$_{82}$ compound (see below). The spectra are calculated by summing over all transitions between eigenstates of a truncated spin Hamiltonian

$$H = \mu_e B^T g S + S^T D S \ . \tag{5}$$

In (5) the index T indicates transposition.

The situation is slightly better if spectra are taken at W-band, i.e., by increasing the Zeeman term by one order of magnitude. Simulated spectra using the same parameter set as in Fig. 9 are shown in Fig. 10. Obviously, if the spin system exhibits a non-axial g matrix and ZFS tensor, both cases cannot be discriminated unambiguously without further knowledge.

It should be noted, however, that the assumption of an identical ZFS tensor for both $S_{eff} = 3$ and $S_{eff} = 4$ multiplets is not justified. A small exchange term J implies a weakly coupled spin system consisting of   S = 7/2 and S = 1/2 constituents. Denoting the principal element of   D as $\tilde{D}$, under weak coupling condition the resulting $\tilde{D}$ can easily be calculated in the coupled representation by invoking the relevant Clebsch-Gordan coeﬃcients giving

$$\tilde{D}^{(S=3)} = 5/4\tilde{D}^{(Gd)} - 1/8\tilde{D}^{cage} \ , \tag{6}$$

$$\tilde{D}^{(S=4)} = 3/4\tilde{D}^{(Gd)} + 1/8\tilde{D}^{cage} \ . \tag{7}$$

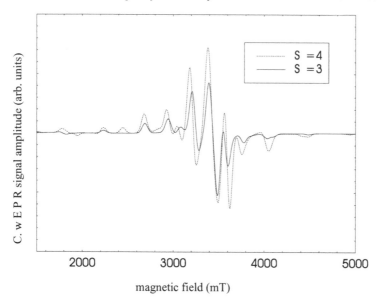

Fig. 10.   Simulated W-band EPR spectra of fictitious $S_{eff} = 3$ and $S_{eff} = 4$ spin systems

In this approximation one finds an exclusive contribution from the metal ion to the ZFS tensor   D because of a vanishing $\tilde{D}$ of the S = 1 / 2 subcomponent. The respective $\tilde{D}$ values of both spin multiplets are predicted to scale in the ratio 5 : 3.

Anticipating that   J might be rather small and in the range of a few wavenumbers only, EPR spectra taken at low temperature can in principal then be used to determine the ground state spin multiplicity and the energy di erence to the upper electronic spin multiplet by observing the temperature dependence of line intensities and the appearance of additional lines because of increasing population of the upper levels in thermal equilibrium. Because of the predicted rather large change in $\tilde{D}$, incipient thermal population of the "excited" spin manifold will lead to much larger e ects in the EPR spectrum than shown in Figs. 9, 10.

In case of close-lying spin multiplets the interpretation of EPR spectra becomes also complicated because no a priori information about relaxation rates neither between states of di erent e ective spin nor within a spin multiplet is provided. It can be argued, however, that spin selection rules will restrict matrix-induced transitions between the upper and lower spin sublevels compared to transitions between states of identical e ective spin. In this case no dominant lifetime broadening of excited state levels is expected and therefore no di erent line shape parameters have to be assumed for spectral simulations of transitions within both spin multiplets. Anticipating that line broadening by orientational disorder prevents resolution of individual spin packets, an

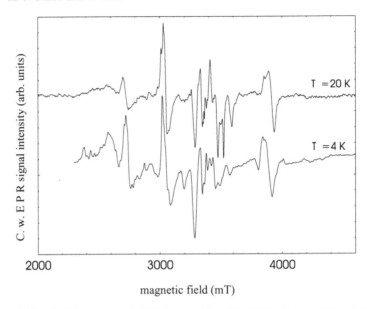

Fig. 11. W-band EPR spectra of Gd@C $_{82}$ diluted in TCB taken at 4 K and 20 K, respectively

identical line width parameter can also be used for transitions connecting both spin multiplets.

Spectra taken in the temperature range of 4 K and 20 K showed a noticeable change of spectral features, as shown in Fig. 11. The appearance of additional lines cannot be easily explained unless thermal population of a close-lying electronic state occurs. As was reported elsewhere [31], this observation can be explained by assuming a negative exchange term $J = -1.8\,\mathrm{cm}^{-1}$, corresponding to $S_{\mathrm{eff}} = 3$ as ground state, situated 14 .4 cm^{-1} below the $S_{\mathrm{eff}} = 4$ excited state. Furthermore, there was evidence that the exchange coupling is not completely isotropic. Such a traceless contribution could either arise from dipole-dipole interaction of electronic spin distributions at cage and ion or from anisotropic overlap between subcomponent orbitals. There is no way to discriminate between both possibilities by EPR. In Fig. 12 simulated spectra are shown derived from the eigenvalues of the following spin Hamiltonian

$$H = \mu_e B^T g_1 S_1 + \mu_e B^T g_2 S_2 + S_1^T D_1 S_1 - 2J S_1^T S_2 + S_1^T D_{12} S_2 . \quad (8)$$

In (8), hfi contributions are omitted because no data could be extracted from the spectra. Components attributed to the metal ion and the cage are indexed as 1 and 2, respectively. Only for the $S = 7/2$ subcomponent corresponding to the metal ion a ZFS term D has to be included. Numerical values for diagonal elements of all matrices were obtained by a spectral fit and are compiled in Table 1. Although no very good agreement between experimental and simulated spectra was obtained, the general trend is well reproduced.

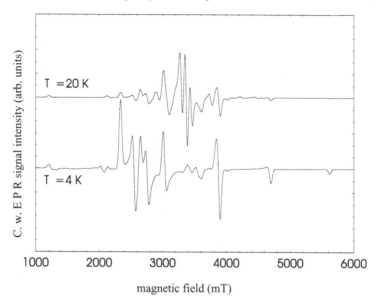

Fig. 12.   Temperature dependence of W-band EPR spectra of Gd@C$_{82}$ (simulated)

Table 1.   ZFS (in GHz), g-matrix elements, and isotropic coupling constant J = − 54 GHz used for the simulation of the EPR spectrum of   Gd@C$_{82}$

$D_{xx}^{(1)}$	$D_{yy}^{(1)}$	$D_{zz}^{(1)}$	$g_{xx}^{(1)}$	$g_{yy}^{(1)}$	$g_{zz}^{(1)}$
− 2.364	− 2.784	+5 .151	2.0090	2.0100	1.9775
$D_{xx}^{(12)}$	$D_{yy}^{(12)}$	$D_{zz}^{(12)}$	$g_{xx}^{(2)}$	$g_{yy}^{(2)}$	$g_{zz}^{(2)}$
+6 .471	+8 .031	− 14.499	2.1050	2.0970	2.0570

Most important, not even qualitative agreement can be obtained if the sign of the exchange coupling constant   J is reversed.

Exchange Interaction in Gd@C$_{82}$ Unravelled

Apparently the compound Gd@C$_{82}$ is an interesting model system for a class of compounds exhibiting weak Coulomb exchange. In such a case Coulomb interaction between spin-separated subcomponents results in energy separation of spin multiplets with the e ect that thermal population of "excited" levels not only leads to noticeable spectral changes but also in a non-standard temperature dependence of the magnetic susceptibility. If the compound is investigated as "pure" substance, i.e., if Coulomb interaction is also admitted between adjacent molecules for instance in closely packed crystalline topology, the resulting e ective spin will also be influenced by the electronic band structure. "Anti-ferromagnetic" coupling could lead to complete spin compensation of cage spins, leaving the remaining "shielded" $S = 7/2$ spins of the metal ions which form a super paramagnet. Switching from an integer spin system to a dipolar coupled half-integer situation by studying pure $Gd@C_{82}$ instead of a dilute solid solution was indeed observed [31].

5 Conclusion

The examples presented give an impression about the rich variety of problems related to ground state properties of these new molecular structures with varying degree of electronic interaction between well defined subcomponents. Application of advanced EPR methods like pulsed EPR and pulsed ENDOR as well as the capability to use electronic Zeeman energies comparable with ZFS interaction is mandatory for successful characterization of MEF. A clear understanding of the transition between localization observed in $Gd@C_{82}$ and $La@C_{82}$ and quasi-free internal rotation as seen in $La_2@C_{80}$ anions probably requires development of new parametrization for DFT calculations. Reliable predictions of redox potentials, and more general about chemical reactivity also await improvement of current quantum chemical methods.

Acknowledgement

In this report results obtained by various graduate students (P. Schweitzer, M. Rübsam, P. Jakes, S. Okubo) and post-docs (Dr. N. Weiden, Dr. H. Matsuoka, Dr. K. Furukawa) are presented. Without their dedicated work this study could not have been completed. Fruitful and long lasting collaborations with Prof. H. Shinohara and Prof. T. Akasaka were essential for this interdisciplinary research. Financial support by the Deutsche Forschungsgemeinschaft under various grants is gratefully acknowledged. German/Japanese collaboration was supported by a visiting professor fellowship and travel grants of the Institute for Molecular Science, Okazaki. We thank Dr. S. Stoll (ETH Z"urich) for providing the EasySpin software package.

References

1. Y. Chai, T. Guo, C. Lin, R. Haufler, L.P.F. Chibante, J. Fure, L. Wang, J.M. Alford, R.J. E. Smalley: Phys. Chem. 95, 7564 (1991)
2. R.D. Johnson, M.S. de Vries, J.R. Salem, D.S. Bethune, C.S. Yannoni: Nature 355, 239 (1992)
3. For a recent review see: Endofullerenes, A New Family of Carbon Clusters, Akasaka, T. and Nagase, S., Eds. (Kluwer Academic Publishers, Dordrecht 2002)
4. B. Kessler, A. Bringer, S. Cramm, C. Schlebusch, W. Eberhardt, S. Suzuki, Y. Achiba, F. Esch, M. Barnaba, D. Cocco: Phys. Rev. Lett. 79, 2289 (1997)

5. M. Rübsam, P. Schweitzer, K.-P. Dinse: J. Phys. Chem.   100, 19310 (1996)
6. H. Shinohara, H. Sato, M. Ohkohchi, Y. Ando, T. Kodama, T. Shida, T. Kato, Y. Saito: Nature   357, 52 (1992)
7. C.S. Yannoni, M. Holnkis, M.S. de Vries, D. Bethune, J.R. Salem, M.S. Crowder, R.D. Johnson: Science   256, 1191 (1992)
8. H. Shinohara, H. Sato, Y. Saito, M. Ohkohchi, Y. Ando: J. Phys. Chem.   96, 3571 (1992)
9. J.H. Weaver, Y. Chai, G.H. Kroll, C. Jin, T.R. Ohno, R.E. Haufler, T. Guo, J.M. Alford, J. Conceicao, L.P.F. Chibante, A. Jain, G. Palmer, R.E. Smalley: Chem. Phys. Lett.   190, 460 (1992)
10. E. Nishibori, M. Takata, M. Sakata, H. Tanaka, M. Hasegawa, H. Shinohara: Chem. Phys. Lett.   330, 497 (2000)
11. M. Takata, E. Nishibori, M. Sakata: Endofullerenes, A New Family of Carbon Clusters , T. Akasaka and S. Nagase, Eds. (Kluwer Academic Publishers, Dordrecht 2002)
12. W. Andreoni, A. Curioni: Phys. Rev. Lett.   77, 9606 (1996)
13. T. Akasaka, T. Wakahara, S. Nagase, K. Kobayashi, M. W" alchli, K. Yamamoto, M. Kondo, S. Shirakura, S. Okubo, Y. Maeda, T. Kato, M. Kako, Y. Nakadaira, R. Nagahata, X. Gao, E. van Caemelbecke, K.M.J. Kadish: J. Am. Chem. Soc. 122, 9316 (2000)
14. W. Andreoni and A. Curioni: Appl. Phys. A   66, 299 (1998)
15. T. Akasaka, S. Nagase, K. Kobayashi, M. W" alchli, K. Yamamoto, H. Funasaka, M. Kako, T. Hoshino, T. Erata: Angew. Chem. Int. Ed. Engl.   36, 1643 (1997)
16. E. Nishibori, M. Takata, M. Sakata, A. Taninaka, H. Shinohara: Angew. Chem. Int. Ed.   40, 2998 (2001)
17. K. Yamamoto, H. Funasaka, T. Takahashi, T. Akasaka: J. Phys. Chem.   98, 2008 (1994)
18. N. Weiden, B. Goedde, H. K" aß, K.-P. Dinse, M. Rohrer: Phys. Rev. Lett.   85, 1544 (2000)
19. The add-on unit was produced by W. Krymov.
20. S. Okubo, T. Kato, M. Inakuma, H. Shinohara: New Diamond and Frontier Carbon Technology   11, 285 (2001)
21. S. Knorr, A. Grupp, M. Mehring, U. Kirbach, A. Bartl, L. Dunsch: Appl. Phys. A 66, 257 (1998)
22. N. Weiden, T. Kato, K.-P. Dinse: J. Phys. Chem. B   108, 9469 (2004)
23. W.B. Mims: Phys. Rev. B   5, 2409 (1972)
24. P. H"ofer: J. Magn. Reson. A   144, 77 (1994)
25. H.L. Flanagan, D. Singel: J. Chem. Phys.   87, 5606 (1987)
26. T. Suzuki, Y. Maruyama, T. Kato, K. Kikuchi, Y. Achiba, K. Kobayashi, S. Nagase: Angew. Chem. Int. Ed. Engl.   34, 1094 (1995)
27. P. Jakes and K.-P. Dinse: J. Amer. Chem. Soc.   123, 8854 (2001)
28. W. Weltner, Jr., D. McLeod, P.H. Kasai: J. Chem. Phys.   46, 3172 (1967)
29. T. Kato, S. Okubo, H. Matsuoka, K.-P. Dinse in: Fullerenes and Nanotubes: The Building Blocks of Next Generation Nanodevices , Proc. of the Intern. Symp. on Fullerenes, Nanotubes, and Carbon Nanoclusters (Paris, France 2003); D.M. Guldi, P.V. Kamat, F. D'Souza, Eds: Vol. 13,   The Electrochemical Society   (Pennington, NJ 2003)
30. S. Okubo, T. Kato: Appl. Magn. Reson.   23, 481 (2003)
31. K. Furukawa, S. Okubo, H. Kato, H. Shinohara, T. Kato: J. Phys. Chem. A 107, 10933 (2003)

Beyond Electrons in a Box: Nanoparticles of Silver, Platinum and Rhodium

J.J. van der Klink

Institut de Physique des Nanostructures, EPFL, 1015 Lausanne, Switzerland
jacques.vanderklink@epfl.ch

Abstract. This chapter presents a review of NMR in small particles of three elemental metals: the "odd electron" Ag and Rh, and the "even electron" Pt. These experiments on elements close to each other in the periodic system demonstrate the versatility of the NMR technique, detecting several properties that are not captured by the simplest "electrons in a box" approximation. The NMR of Ag particles shows Bardeen-Friedel oscillations in the spatial density of Fermi-level electrons. The case of Pt is dominated by a strong spatial variation in the local density of states of d-like electrons, and Rh shows at low temperatures signs of incipient antiferromagnetism. As an introduction, a summary of relevant NMR theory is presented, and its application to the NMR of bulk Na metal.

1 Introduction

The framework provided by the early explanation of the NMR shift in metals in terms of a hyperfine coupling and a Pauli-type of susceptibility [1], is still solidly upright today. But although the estimates for these quantities at the time were good enough to show the essential correctness of the theory, it was only much later, with the advent of methods developed by Kohn, Hohenberg and Sham, that a su cient theoretical and numerical power became available to yield nearly quantitative comparisons between theory and experiment. It is somewhat unfortunate that in the meantime a vast body of experimental data had been accumulated (for a compilation, see [2]), without an adequate theoretical framework for their interpretation. While the last extensive treatment of metal NMR in book form appeared in 1970 [3], the first modern collection of "Calculated Electronic Properties of Metals" was published only eight years later [4], and the first review paper that attempted to discuss metal NMR somewhat comprehensively in the light of such theoretical results dates from 2000 [5]. As we will see, some of these calculated results are astonishingly close to what are now considered as the best experimental data for the simple metals. For transition metals the available experimental data are not complete enough to provide a full test of the numerical results, although a

1983 calculation specific to Pt [6] has given an amazing agreement with the experimental value for the so-called core polarization hyperfine field, a quantity first discussed, both experimentally and theoretically, by Yafet, Jaccarino and colleagues around 1964 [7].

This hyperfine field is one of three phenomena not covered in the very first descriptions of metal NMR; in fact it was expected at the time [1, p. 115] that no shift would be associated with the paramagnetism from   d electrons. The two others are the Stoner enhancement of the static uniform susceptibility   ¯, as compared to the value predicted by simple Pauli theory, and the Moriya desenhancement of the spin-lattice relaxation rate T_1^{-1} with respect to its Korringa value. All three phenomena are consequences of the exchange enhancement of the nonlocal complex paramagnetic susceptibility. Our present ideas on this point [8] are based on theories concerning the inhomogeneous electron gas published around 1975 [9, 10]. Formally, this susceptibility is a zero-field groundstate property that can be calculated in a linear response theory; in practice the core polarization fields are more easily obtained by explicitly including a magnetic field in the Hamiltonian.

In most metals the temperature dependence of the susceptibility is only weak, and therefore   ¯ can be calculated as a zero-temperature property. However, some strongly enhanced metals, of which Pd is the archetypical example [11], show a marked (and not necessarily monotonic) temperature dependence of ¯. This phenomenon can be described by spin fluctuation theory [12]. The basic idea is that a thermal fluctuation sets up an "internal field", similar to that in Stoner theory, which couples to the other fluctuation modes. In this way, there is a temperature-dependent magnetic energy in zero applied field, associated with the thermal fluctuations of the local spin density around a zero average. This theory is purely thermodynamical and considers the homogeneous electron gas: it yields expressions for the low-frequency wavevector-dependent susceptibility, but not for the corresponding hyperfine fields.

The first NMR experiments on nonbulk metal used thin filaments of lead, embedded in porous glass [13]. The results were interpreted in terms of surface e ects. That discussion appeared just a year after an influential theoretical paper predicting a quantum size e ect for the susceptibility of small metal particles [14]. From simple free electron theory for   N electrons in a small "box", one expects the spacing between energy levels to be of the order E_f/N , where E_f is the Fermi energy of the corresponding bulk solid. At temperatures kT E_f/N , the susceptibility is expected to deviate from its bulk value. Furthermore, for an element with an odd number of electrons per atom there should appear a Curie-like susceptibility for those "boxes" in the sample (supposed to consist of many particles of similar, but not identical, size) that contain an odd number of atoms. Several groups have searched for such e ects in samples of small copper particles, with moderate success. A possible reason is that at temperatures low enough for the discrete energy levels to be felt, the usual T_1 mechanism (electron spin flip– flops) disappears because the requirement of energy conservation cannot be satisfied, so that the signal is very easily saturated.

This paper presents a review of the NMR in small particles of three elements: the "odd electron" Ag and Rh, and the "even electron" Pt. As a preliminary, the next Sect. 2 introduces the notation and mentions results for some bulk metals. The NMR of Ag particles shows the Bardeen-Friedel oscillations in the spatial density of Fermi-level electrons, and their "washing-out" at higher temperatures. The case of Pt is dominated by a strong variation in the local density of states of d-like electrons (the "exponential-healing" phenomenon) and Rh shows at low temperatures signs of incipient antiferromagnetism. These experiments on elements close to each other in the periodic system demonstrate the versatility of NMR as a technique, detecting several properties that are not captured by an "electrons in a box" approximation. Perhaps the most evident of such properties is the size-dependent structural change detected in small particles of Tc by ^{99}Tc NMR [15]. That nucleus has spin 9/2, and therefore the hexagonal structure of the bulk leads to a NMR spectrum of about a MHz wide (in a 7 T field), whereas small particles (average diameter 2 .3 nm, but with a rather wide size distribution) give a single narrow line, about 1–5 "kHz wide. This clearly shows that the small Tc particles have a cubic structure.

2 Bulk Metals

2.1 Theory

The theory of the NMR of paramagnetic metals [5, 8] is based on the local density approximation of the density functional theory for the low-frequency limit of the complex nonlocal spin susceptibility of the inhomogeneous electron gas (r, r ;), which gives the response in r to a magnetic field with frequency applied in point r [9, 10, 16, 17, 18]. That susceptibility can be written in the form of an integral equation:

$$(, + R ;) = \quad _P(, + R ;)$$
$$+ \quad _P(, _1 + R ;) (n(_1 + R))$$
$$(_1 + R , + R ;) d _1, \tag{1}$$

where (n(r)) is related to a second derivative of the exchange-correlation energy, and is (in the local-density approximation) a function only of the charge density n at r. The quantity $_P(, + R ;)$ is the "noninteracting" Pauli susceptibility. The vectors and $_1$ are in the unit cell at the origin, and R and R are lattice vectors. N is the number of unit cells.

The dependence of   on N lattice vectors in real space can be replaced by a dependence on N vectors q in the reciprocal lattice through the Bloch Fourier transform:

$$\tilde{}(, ;q ;) = \sum_{=1}^{N} e^{iq \cdot R} (, + R ;) . \tag{2}$$

The $(n(r))$ that appears in (1) creates simultaneously the Stoner enhancement of the susceptibility, the core-polarization hyperfine fields, and the desenhancement factor in the spin-lattice relaxation rate [8]. In terms of $\tilde{}$ (the real part of $\tilde{}$) the static uniform susceptibility is

$$\overline{} = {}^{-1} \int_{cell} \tilde{}(, ;0;0) d \, d \, , \tag{3}$$

where   is the volume of the unit cell. For some metals, the spin-only static susceptibility has been measured from the intensity of the ESR absorption. The ESR signal can be calibrated by using the same setup for an NMR measurement [1] on the same sample, since the nuclear susceptibility can easily . be calculated. Such experiments were performed by Schumacher and Slichter [19].

The (isotropic part of the) Knight shift   K is

$$K( ) = \frac{2}{3} \int \tilde{}(, ;0;0) d \, , \tag{4}$$

where the nucleus under consideration has a relative position   in the unit cell.

By definition, the hyperfine field B_h is simply related to the ratio of the Knight shift and the uniform susceptibility through the dimensionless quantity

$$\frac{B_h( )}{\mu_0 \mu_B} = \frac{K( )}{\overline{}} , \tag{5}$$

but we will see later, in (9), that this simple definition of "the" hyperfine field is not always physically meaningful, and that it can be more judicious e.g. to attribute different hyperfine fields to s-like and to d-like electrons.

The general expression for the relaxation rate of a nucleus in   .   is

$$S(T_1( )T)^{-1} = \frac{\mu_0}{4} \frac{4\mu_B}{3}^2 \frac{2}{N} \sum_{=1}^{N} \frac{\tilde{}(, ;q ; s - {}_I)}{(s - {}_I)} , \tag{6}$$

where $S = (2 \mu_B)^2/(4 \, k {}^2)$, and s and I are the electronic and nuclear Larmor frequencies. The imaginary part of the susceptibility   is an odd function of frequency, and linear for small values of   . The right-hand side of (6) is then frequency-independent and can be evaluated in the limit of vanishing frequencies. For metals   is (nearly) temperature-independent, in which case $T_1^{-1} \propto T$, as was first found by Heitler and Teller in 1936 [20].

For completeness we mention the expression for the scalar coupling between two nuclei, at positions $\mathbf{1}$ and $\mathbf{2}$ in the unit cell, mediated by the electron spin susceptibility:

$$J(\mathbf{1}, \mathbf{2}) = \mu_0 (2/3)^2 \, \mathbf{1} \, \mathbf{2} N^{-1} \sum_{=1}^{N} \tilde{} (\mathbf{1}, \mathbf{2}; q; 0). \tag{7}$$

This coupling was experimentally detected and theoretically explained by Bloembergen and Rowland [21]; it has also been discussed by Ruderman and Kittel [22].

The origin of the Knight shift is the contact interaction, which is "symmetric" in nuclear and electronic spins. The corresponding shift of the ESR line, the Overhauser shift O, is proportional to the thermal average of the nuclear spin magnetization $\mathbf{I} = \mathbf{n} H_0$, with $\mathbf{n}$ the nuclear magnetic susceptibility, just as the Knight shift is proportional to the electron spin susceptibility. If the (5) is meaningful, we will therefore have that $O = (\mathbf{n}/\bar{})K$. The interesting thing is that at practical temperatures the nuclei form a magnetically non-interacting system with an easily computed (Langevin) susceptibility, so that a measurement of O can be interpreted as a measurement of B_{hf}.

2.2 Practical Equations

Starting from (1), the static susceptibility, (3), the Knight shift, (4), and the relaxation rate, (6), for transition metals can be decomposed approximately into sums of s- and d-like contributions [8, 23], to which an additional orbital term [24] must be added:

$$\frac{}{} = \mu_0 \mu_B^2 {}^{-1} \frac{D_s(E_f)}{1 - {}_s} + \frac{D_d(E_f)}{1 - {}_d} + {}_{orb} = {}_s + {}_d + {}_{orb} \tag{8}$$

$$K = \frac{}{\mu_0 \mu_B} {}_s B_{hf,s} + {}_d B_{hf,d} + {}_{orb} B_{hf,orb} = K_s + K_d + K_{orb} \tag{9}$$

$$S(T_1 T)^{-1} = k_s K_s^2 + k_d K_d^2 R_d + (\mu_B D_d B_{hf,orb})^2 R_{orb}, \tag{10}$$

where   is the atomic volume; D_s and D_d are densities of states at the Fermi energy E_f; ${}_s$ and ${}_d$ Stoner enhancement factors that are related to exchange integrals $I_{s,d} = {}_{s,d}/D_{s,d}(E_f)$; the B_{hf} e ective hyperfine fields (including core polarization); k_s and k_d Moriya desenhancement factors; and R_d and R_{orb} reduction factors related to the decomposition of D_d into contributions of di erent symmetry $D_{t_{2g}}$ and D_{t_e}. The three densities of states $D_s(E_f)$, $D_{t_{2g}}(E_f)$ and $D_{e_g}(E_f)$ can be found from band structure calculations, and the d-like hyperfine field can sometimes be determined by experiment. The ${}_l$ ($l = s, d$) are treated as fittable parameters. It is usually assumed that k_l can be calculated from some l-independent function of the Stoner parameter

$k(\ )$, thus $k_l = k(\ _l)$. We have often used the Shaw–Warren result [25], that can be fitted to

$$k_{SW}(\ ) = (1 - \ )(1 + 1/4\ ), \tag{11}$$

while the $k(\ )$ relation of the original Moriya equation [26, 27] can be approximately represented as

$$k_M(\ ) = (1 - \ )(1 + 5/3\ ^2). \tag{12}$$

For strongly enhanced paramagnets, $\quad K$ and/or $T_1 T$ may become temperature dependent through spin fluctuations. In Sect. 5.2 we summarize how these temperature dependences are expected to di er for ferromagnetic and antiferromagnetic enhancements. For noninteracting electrons $\quad = 0$, and for the case of s-electron metals we retrieve the Korringa relation [28]

$$S(T_1 T)^{-1} = K^2, \tag{13}$$

that has been independently derived by Slichter [1]. As we will see shortly, even for Na metal the relation (13) is only very approximately fulfilled.

2.3 Na Metal

The properties of the alkali metals are often discussed with reference to a free electron gas with the same electron density. If the volume per electron (and therefore also per atom) is $\quad$ and the electron density parameter $\quad r_s$ is defined through $\quad \frac{4}{3}\ r_s^3$ then the density of states at the Fermi energy (number of states per atom and per Joule) in the free electron gas (feg) is

$$D_{feg}(E_f) = 1.869 \times 10^{17}(r_s/a_0)^2, \tag{14}$$

where a_0 is the Bohr radius. The susceptibility is

$$_{feg} = 32.65 \times 10^{-6}(r_s/a_0)^{-1} \tag{15}$$

and has no dimension, although it is often referred to as the "volume" susceptibility. The corresponding cgs value is obtained by dividing the right-hand side by 4 . The cgs susceptibility in emu mol $^{-1}$ results from a further multiplication by the molar volume $V_m = N_A$, and therefore has the dimension of a volume, and units cm ^{3} .

In the early papers [1] the Knight shift $\quad K_{feg}$ was written as the product of a susceptibility as obtained above and the intensity at the nucleus of a Fermi-energy single-electron wavefunction, $|\ (0)|^2$, normalized in the volume

$$\begin{aligned}
K_{feg} &= \frac{_{feg}}{4}\ \frac{8}{3}\ |\ (0)|^2 \\
&= \frac{_{feg}}{4}\ \frac{4}{\mu_0}\ \frac{B_{hf}}{\mu_B} \tag{16} \\
&= \mu_B\, D_{feg}(E_f) B_{hf},
\end{aligned}$$

where the second line introduces the equivalent hyperfine field B_{hf}, with units Tesla (T). The corresponding cgs equations are obtained by replacing the factors $\chi_{feg}/4\pi$ by χ_{cgs}/N_A , where χ_{cgs} is the susceptibility in emu mol^{-1}, and further replacing μ_0 by 4π. The resulting hyperfine field is in Gauss (G), and $1\,T = 10^4\,G$.

In the case of Na, the "best" experimental value for the spin susceptibility at low temperature (4.2 K) is thought to be $(13.63 \pm 0.14) \times 10^{-6}$ [29], and the value found for the hyperfine field by Overhauser shift measurements is $B_{hf} = 24.76 \pm 1.2\,T$ [30]. The molar volume of sodium at low temperatures is $V_m = 22.72\,cm^3$. These ESR-based values of susceptibility and hyperfine field predict a value for the Knight shift $K = (1.09 \pm 0.05) \times 10^{-3}$, in good agreement with the accepted value $K = 1.07 \times 10^{-3}$ at 10 K. The value of B_{hf} corresponds to $|\psi(0)|^2 = 120 \, \Omega^{-1}$, which shows clearly how the wavefunctions that are important for the contact shift pile up near the nucleus.

For $r_s/a_0 = 3.93$, appropriate for Na, (14) gives a density of states $D_{feg} = 2.89 \times 10^{18}\,J^{-1}$, or 6.3 states per atom and per Rydberg (1 Ry = 2.1795×10^{-18} J) in good agreement with the value $6.12\,Ry^{-1}$ from band structure calculations [4]. Therefore $\chi_{feg} = 8.31 \times 10^{-6}$ is close to the value from the band structure calculation in the Pauli approximation, but both are clearly different from the experimental $\chi_{observed} = 13.63 \times 10^{-6}$. The Stoner exchange integral I, as determined from the $\chi_{observed}$ and the calculated $D(E_f)$ is $I = 66.5\,mRy$, whereas the directly calculated value is $I = 67.8\,mRy$. The calculated value of the direct hyperfine field (without core polarization) is $B_{hf} = 31\,T$; the agreement with experiment (24.8 T) is not perfect, but nevertheless rather impressive. From the calculated density of states, exchange integral and hyperfine field we find $K = 1.38 \times 10^{-3}$, which is 30% higher than the experimental value. All these theoretical results are from the collection of calculated properties of metals [4].

Associated with the Stoner enhancement of the uniform susceptibility is a disagreement between the T_1T value calculated from the Korringa relation (13), 3.29 s K, and the experimental value $5.1 \pm 0.3\,s\,K$. The Shaw–Warren model, (11), gives $k(0.407) \approx 0.653$, close to the value of 0.64 observed in Na. All this shows that susceptibility enhancement is important even in such a "simple" metal as sodium, and that computational methods as already available in 1978 [4] can yield good agreement with experiment.

The free electron gas model has a very smooth density of states around the Fermi energy, and the only temperature effect is thermal expansion, which changes r_s/a_0. The same parameter describes pressure effects. The volume dependences of the susceptibility, (15), and of the Knight shift, (17), are given by

$$\frac{d\ln\chi_{feg}}{d\ln\Omega} = -1/3 \tag{17}$$

$$\frac{d\ln K_{feg}}{d\ln\Omega} = \frac{d\ln\chi_{feg}}{d\ln\Omega} + 1 + \frac{d\ln|\psi(0)|^2}{d\ln\Omega}. \tag{18}$$

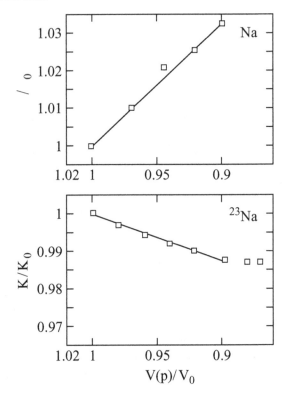

Fig. 1.   Volume dependence of the spin susceptibility in Na metal, measured by the CESR method and of the ^{23}Na Knight shift. All quantities relative to room temperature and atmospheric pressure. [After Kushida et al., [34] and Benedek et al., [31]. c 1976, 1958 by the American Physical Society]

Clearly, changes of volume can affect both the uniform susceptibility and the hyperfine field, and both variations have been observed for Na, see Fig. 1. The change of susceptibility with volume at constant temperature (4.2 K) has been measured by the CESR method [31] for volume reductions to (p)/ (0) = 0 .9, and the quantity in (17) has been found as − 0.34 ± 0.03. The ^{23}Na pressure-dependent NMR data [31, 32, 33] in the same range of relative volume give a slope of +0 .13 ± 0.02 for (18), so that d ln | (0)|^2/ d ln − 0.54. If the changes in volume scale at the level of the wavefunctions, such that | (0)|^2 remains constant, then d ln | (0)|^2/ d ln = − 1. As a kind of opposite case, the changes in volume might affect the wavefunction far away from the nucleus, while | (0)|^2 stays constant: d ln | (0)|^2/ d ln = 0. In the free electron gas approximation we may thus expect for the volume dependence of the Knight shift

$$- 1/ 3 \quad \frac{d \ln K_{feg}}{d \ln} \quad 2/ 3 . \tag{19}$$

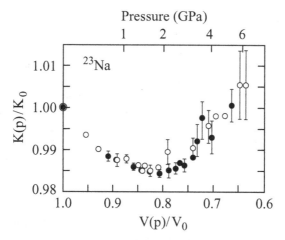

Fig. 2.   Volume dependence of the relative ^{23}Na Knight shift in Na metal measured at 295 K in a diamond anvil cell, achieving very large volume reductions. Two sets of data, taken in slightly di erent cells. [After Kluthe et al., [33].   c 1996 by the American Physical Society]

In sodium metal, this slope of the Knight shift changes sign at a volume reduction $V(p)$ $0.8V(0)$, see Fig. 2, and becomes approximately -0.1. There are no corresponding data for the susceptibility, but if we continue to use the value of -0.34, then $d \ln | (0)|^2/ d \ln$ -0.76, closer to the value -1 expected for $| (0)|^2$ constant. In this interpretation of the NMR data, initially the value of the wavefunction far   away from the nucleus changes more than the value close to it; but at higher pressure the "compression" becomes more uniform.

For Li, the best value for the spin susceptibility is $(27$ $.4 \pm 0.1) \times 10^{-6}$ [29], and the hyperfine field from the Overhauser shift is 5 $.39 \pm 0.2$ T [35]. The molar volume is 12.8 cm ^{3}. The Knight shift calculated from these data is $(2.69 \pm 0.1) \times 10^{-4}$, while the experimental value is 2 $.6 \times 10^{-4}$. The calculated direct hyperfine field (without core polarization) is 6.9 T and the calculated $K = 4 .33 \times 10^{-4}$. These satisfactory agreements hide in fact a conceptual problem concerning the meaning of the hyperfine fields. Several bandstructure calculations that project out the partial densities of states of   s-, p- and d-symmetries, see e.g. [36], agree that for Li about 3/4 of the density of states at the Fermi energy corresponds to   p-states, that have zero amplitude at the nucleus; for Na the value is about 2/5. A CESR experiment, and therefore the Overhauser experiment also, sees all electrons at the Fermi energy, as shown by the good agreement between measured and calculated susceptibilities: the measured hyperfine field is an average value over all these electrons.

We now face a delicate question on the principle of a Knight shift calcula-tion, and even more so for the relaxation rates. Should we consider only the s-like susceptibility, its enhancement, and the (high) purely   s-like hyperfine field, or rather the average susceptibility, its enhancement and the (lower) average hyperfine field? The reasonable agreement between calculated properties and the observed $^-$ and K for two alkali metals, Li and Na, with rather different band structures speaks in favor "average sp"-approach over the "pure s-type" models. A disadvantage of the average sp-approach is that its hyperfine fields cannot really be considered as atomic properties, and can, strictly speaking, not be compared easily between unit cells in di erent environments. Nevertheless, we will assume the existence of two site-independent hyperfine fields when using (9) for the interpretation of small-particle ^{195}Pt NMR. The expression for the Knight shift is a sum of terms linear in the densities of states and hyperfine fields, and can be interpreted as representing an average, weighted by lifetime. The relaxation rate contains a sum of squares, and its interpretation as an average requires that on some relevant time scale the d electrons are distinguishable from the s (or rather sp) electrons. The relevant time scale is of the order of the time it takes for an electron to fly through the Wigner-Seitz cell [37]. Formally, the requirement for the separation into these two contributions is that cross-enhancement between d- and sp-susceptibilities be negligible [8].

3 Small Silver Particles: Bardeen-Friedel Oscillations

For bulk silver metal, $V_m = 10.27\,cm^3$, $K = 5.22 \times 10^{-3}$ and $(T_1 T)^{-1} = 0.111\,(s\,K)^{-1}$. Using (11) and the calculated $D(E_f) = 3.67\,Ry^{-1}$ yields I = 153 mRy and $^- = 24.3 \times 10^{-6}$ (in dimensionless SI units). Between room temperature and the liquid phase at 1360 K the molar volume increases by nearly 13%. The Knight shift of ^{109}Ag plotted as a function of molar volume is continuous across the melting transition, Fig. 3, but the desenhancement factor k( ) shows a discontinuous increase in relaxation rate in the liquid, Fig. 4 [38]. Over this range $d\ln K/d\ln = 0.873 \pm 0.004$; and additional data down to 25 K also fall very well on this plot [39]. This value is large compared to that for Na, and falls outside the expected range in the free electron gas, (19).

The apparent jump in k( ), Fig. 4, might be due to additional contributions to the relaxation rate in the liquid phase, e.g. through atomic motion. On the other hand, there might be a real increase in the relaxation rate, still described by (6), but not correctly parameterized by (11). It has been argued [40] that disorder enhances the dynamic susceptibility (related to the relaxation rate), while it does not a ect the static susceptibility (connected to the shift) by much. The mechanism can be understood by considering the relaxation process as an example of Abragam's [37] "scalar relaxation of the second kind" [5]. This is a form of BPP theory in the extreme narrowing limit, so that the relaxation rate is proportional to a mean-squared interaction multiplied by a relevant correlation time. It is plausible to take this time proportional to the hopping time for electronic conduction: increasing the disorder tends to

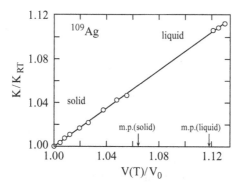

Fig. 3.   Relative shift of ^{109}Ag in Ag metal as a function of volume. Here the experimental parameter is temperature (rather than pressure). The plot is continuous across the melting transition. [After El-Hanany et al., [38].   c 1974 by the American Physical Society]

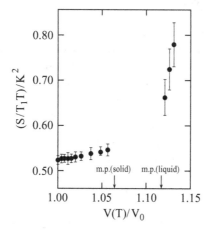

Fig. 4.   The Korringa ratio ($S/T_1T)/K^2$ for ^{109}Ag in Ag metal as a function of volume. The relaxation rate shows a discontinuous increase in the liquid phase. [After El-Hanany et al., [38].   c 1974 by the American Physical Society]

localize the electron, increases the hopping time, and therefore the spin-lattice relaxation rate. The mean-squared interaction, proportional to K^2, remains una ected by the disorder.

The number of NMR studies on small silver particles is quite limited, which is rather surprising when one thinks of the advantage that ^{109}Ag has over ^{63}Cu in being a spin 1/2, thereby avoiding possible problems with quadrupole broadening of the signals. In bulk silver, the full linewidth at half maximum (FWHM) H is of the order of 0.75 G, which in a 7 T field gives   H/H = 11 ppm, compared to K = 5200 ppm. For copper the corresponding figures are H/H   = 120 ppm, K = 2500 ppm. It therefore seemed that studies on small silver particles might resolve some of the ambiguities left in attempts to detect the quantum size e ect [14] by ^{63}Cu NMR [41, 42, 43, 44]. Three groups have reported ^{109}Ag NMR [45, 46, 47] in systems of supported particles: but two of them found it impossible to detect signals in samples where all particles were smaller than 50 nm.

In practice, it is not easy to avoid the presence of such large particles: the reactivity of silver (think e.g. of its use in photographic processes) makes the preparation of small particles of this metal rather di cult, and often leads to asymmetrical or even bimodal size distributions [45]. We have prepared some samples by the colloidal route, using surfactants as protecting agents, and deposing the particles on alumina, titania or silica carriers. The sample with the smallest particles [47] had an average particle diameter determined from X-ray line broadening analysis (XLBA) of 13 nm. We have very little information on the particle size distributions, because of contrast problems in transmission electron microscopy (TEM). The available TEM pictures suggest that nearly all samples contained at least some large particles of the order of 50 nm.

The ^{109}Ag NMR in all our samples showed a symmetric line, detectable by Fourier transform methods but clearly broadened with respect to a bulk signal. The line position and the spin-lattice relaxation time were those of bulk silver. Most experiments were performed at 20 K in 7 or 8 T fields, but occasionally temperatures down to 15 and up to 400 K were used. For quantitative analysis of the line broadening, the observed signals were fitted to a sum of two Lorentzians, centered at the same frequency, but with di erent width. While this was done simply for convenience, it was also found this way that the lineshape for di erent samples could be very di erent. This is thought to be related to details in the particle size distributions (e.g. bimodality), and makes it very di cult to characterize the linewidth by a single number. Two di erent measures were used: the first simply the FWHM of the fitted line; the other the ratio (M_2^3/M_4)$^{1/2}$, with M_2 and M_4 the second and fourth moment for a truncated Lorentzian (it is well known that the moments of a full Lorentzian diverge). If the two measures gave a very di erent ranking for the same spectrum in the particle diameter vs. linewidth ordering, the data on that sample were discarded. Nevertheless, there remained considerable scatter in the plot of linewidth vs. particle diameter d, although it is pretty sure that the variation is somewhere between proportional to d^{-1} and d^{-2}, see Fig. 5.

Given the particle sizes and the temperature at which the experiments are performed, the quantum size e ects can be excluded as the reason for the particle-size dependent line broadening. The proposed explanation is related to one of the first discussions of size e ects in the NMR of metals, where the broadening of the ^{207}Pb NMR line in filaments of lead, obtained by the impregnation of porous glass, was ascribed [13] to surface-induced spatial variations in the density of Fermi-level electrons. The associated charge oscillations in the free electron gas have been mentioned by Bardeen [48], and go asymptotically as $(2 k_f x)^{-2} \cos(2k_f x)$ where x is the distance from the surface

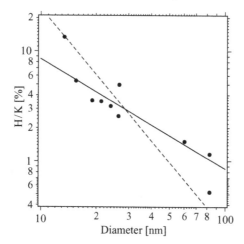

Fig. 5.   Linewidths (fwhm) in units of the bulk metal shift for [109] Ag in small silver particles as a function of particle diameter. The lines are simple guides to the eye, showing slopes d^{-1} (solid) and d^{-2} (dashed). Data at 20 K. [After Bercier et al., [47]. c 1993 by the Institute of Physics]

and k_f the Fermi wavevector. The asymptotic variation of the susceptibility goes as

$$(x) \quad \frac{\sin(2k_f x)}{2k_f x} \quad \text{for } x \quad k_f^{-1}, \qquad (20)$$

and therefore decays slower than the charge perturbation [49]. The free electron gas has no crystal lattice, and therefore it is not easy to compare its characteristic distance, the inverse of the Fermi wavevector, with a distance in a crystal lattice. Anyway, it has been proposed [47] that these Bardeen-Friedel oscillations imply that in the surface region some sites have higher and other sites have lower local susceptibilities than the average, thereby leading to NMR line broadening. Strictly speaking of course the decaying oscillation cannot give a perfectly symmetric broadening, but there remains a clear difference with the "exponential-healing" model to be discussed in Sect. 4 that results in a one-sided broadening.

A very coarse estimate of the broadening can be made on basis of the free electron model. The calculation of the position-dependent susceptibility [49] (with the free electron gas parameter set for silver) indicates that the relative shift of the surface atoms with respect to the bulk might be as large as ± 0.15, of the order of the linewidth of the smallest sample at low temperatures. It is then calculated that the local susceptibility oscillations decay to an amplitude that corresponds to the bulk linewidth over a distance of ten atomic layers. In the largest-size sample studied, this region contains about one tenth of all atoms; in the smallest-size sample it is about half of all atoms. This model suggests that for samples with average particle diameters above 5.7 nm (twenty

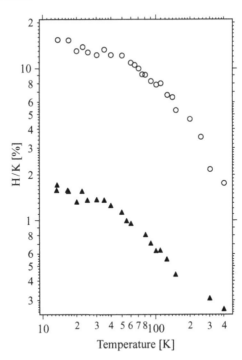

Fig. 6.   Linewidths (FWHM) in units of the bulk metal shift for   [109] Ag in small silver particles as a function of temperature. Typical particle diameter 1.3 nm (top) and 8.0 nm (bottom). [After Bercier et al., [47].   c 1993 by the Institute of Physics]

atomic layers) the linewidth should decrease as d^{-1}. The scatter in the data of Fig. 5 is then attributed to di erences in the particle size distributions.

There is a remarkable variation of the observed linewidth with temperature, Fig. 6. To a good approximation, the lineshapes observed at 20 K and at 80 K can be brought into coincidence by scaling of the shift axis (and, of course, of the amplitude). The scaling factor varies somewhat from one sample to the next, but is about 1.6, clearly di erent from the factor 4 that one would expect for broadening by paramagnetic impurities in the dilute limit. (The lineshapes scale linearly with applied field between 4 T and 8 T). When the temperature is increased further, the linewidth decreases more rapidly. The temperature dependence does not have the characteristics of thermally activated motional narrowing, e.g. by self-di usion.

The proposed explanation is based on the vibrational motion of the surface. This can be an important e ect in small particles, where increased vibration amplitudes are seen in the Debye-Waller factor [50] and in a lowering of the melting temperature [51]. The motion diminishes the sharpness of the boundary of the electron gas, and thus diminishes the amplitude of the Bardeen oscillations in the model of [49]. A more dynamic image of this process can be obtained by considering the local hyperfine field acting on some arbitrarily selected nucleus inside the particle. When the surface vibrates, the Bardeen oscillations and therefore the local field on this nucleus vary in time. When the variation is rapid compared with the total range of NMR frequencies of the nucleus during the course of one vibration period, there will be "motional narrowing" of the NMR signal, but this time not due to motion of the nucleus itself, but of the surface of the particle.

The phenomenon of size- and temperature-dependent magnetic broadening of NMR lines also seen by others in copper [42, 43], and perhaps in lead and tin [52, 53] (but not in d-band metals like Pt) might well arise from intrinsic properties of small particles, although it has been usual to ascribe it to, often poorly defined, extrinsic sources.

4 Small Platinum Particles: Exponential Healing

The NMR parameters for bulk Pt, according to (8) to (10), are given in Table 1 [8, 23]. The values of $B_{hf,s}$, I_s and I_d have simply been fitted. $B_{hf,d}$ can be measured from the correlation between Knight shift and susceptibility, with temperature as parameter [23]. The values of $B_{hf,orb}$, $_{orb}$ and $_{dia}$ are estimated [23]. There is no term in the Knight shift or the relaxation that corresponds to the diamagnetic part of the susceptibility. The shift associated with diamagnetism of core electrons is included in the definition of the zero of the shift scale. The Landau diamagnetism creates a demagnetizing field, leading to overall shifts of a few tens of ppm (the numerical value of the Landau diamagnetic susceptibility). For small particle NMR, we keep all parameters fixed at the bulk value, except the densities of state. We neglect the loss of cubic symmetry near surface sites (where the R-factors can be expected to be different) as well as the site to site variation of the orbital Knight shift (assuming that such variations are small compared to variations in the spin Knight shift).

Table 1.   Fitted values of partial contributions to the susceptibility, the Knight shift and the relaxation rate of ^{195}Pt in platinum metal. The next four rows give the parameters used: hyperfine fields B_h, reduction factors R, exchange integrals I and densities of states $D(E_f)$. $V_m = 9.10\,cm^3$ From [5], with corrections

	s	d	orb	dia	sum	exp.	
	22.2	289	13.8	− 41	284	284	× 10⁻⁶
K	7.8	− 44.2	2.0		− 34.4	− 34.4	× 10⁻³
$(T_1 T)^{-1}$	7.2	19.9	6.35		33.4	33.4	$s^{-1}K^{-1}$
B_{hf}	270	− 118	110				T
R		0.208	0.389				
I	98	37.7					mRy
$D(E_f)$	4.08	20.4					Ry⁻¹

While the Ag lineshape described in Sect. 3 could be described with concepts from the free electron gas, this is clearly not the case for the   d-band metal Pt, as shown e.g. by an interesting series of NMR experiments on   [195]Pt in bulk platinum alloyed with very small quantities of other transition metals [54]. In these spectra "satellite" lines appear (see Fig. 7), due to   [195]Pt in sites close to an alloying impurity. The di erence in shift between a satellite line and the bulk NMR line was taken as a direct measure for a di erence in local susceptibility, the hyperfine fields being considered as site-independent. This spatial distribution of the susceptibility on Pt sites around the solute is related to the change in the bulk susceptibility of similar (more concentrated) alloys. To fit the data, an exponential decay of the change in susceptibility with distance from the impurity, rather than the damped sinusoid of (20), was assumed. This assumption is empirical, it has no fundamental justification. The characteristic length of the decay is between 0.5 and 0.6 lattice constants (0.392 nm), depending on solute. The spatial decay of the local susceptibility corresponds to a decay in the local density of states at the Fermi energy: the experimental data show that a measurable change in this quantity can extend as far as the third neighbor Pt shell.

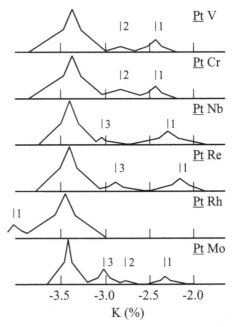

Fig. 7.   Sketches of the main (near $K = -0.035$) and satellite   [195]Pt lines in Pt metal slightly alloyed with other transition metals. The assignments to first, second and third neighbors (numbered arrows) are based on simultaneous analysis of all data. [After Inoue et al., [54].   c 1978 by the Physical Society of Japan]

A similar exponential decay has been assumed to explain the ^{195}Pt linewidth in small metal particles [55], with the particle surface now playing the part of the impurity in the so-called layer model. In this model, the atoms in a small-particle sample are divided into groups belonging to dierent atomic layers: the surface layer, the subsurface layer and so on. To find the fraction of atoms in each group, the particle size histograms obtained by electron microscopy are interpreted in terms of fcc cubooctahedra.

We assume that the dierent sites in a given layer, Fig. 8, of a cubooctahedral particle are suciently similar that the resonance frequencies of all nuclei in the same layer are relatively close to each other on the scale of the total spectrum width. The superposition of NMR signals from all nuclei in a given layer is supposed to result in a Gaussian, completely characterized by the position of its maximum in the spectrum, by its width (of the order of a MHz) and by its integral. The integral must be proportional to the relative number of atoms in the corresponding layer, given in Fig. 8. For the position of the maximum as a function of layer number, we impose an exponential decay (see (21) below), similar to the behavior in the vicinity of an impurity in very dilute alloys, Fig. 7. Finally, the width of the Gaussian is considered a freely fittable parameter, but for not too dierent samples (like the case of Fig. 9) it is assumed to be sample-independent. The maximum of the Gaussian peak corresponding to the nth layer is assumed to occur at a Knight shift K_n (K_0 is the Knight shift on the surface, K_∞ that in the infinite solid) obeying the relation:

$$K_n - K_\infty = (K_0 - K_\infty) e^{-n/m} , \qquad (21)$$

where the dimensionless constant m represents the "healing length" for the Knight shift, expressed in units of a layer thickness (0.23 nm). According to

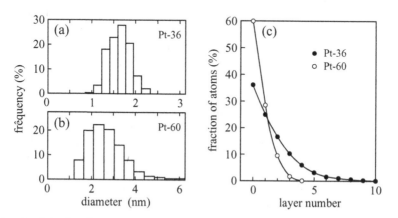

Fig. 8.   Particle size distributions and layer statistics. (a,b) size histograms, giving the percentage of particles in dierent diameter classes, for two samples of Pt particles on titania TiO$_2$. (c) distribution of atoms over the layers of the NMR layer model. Layer 0 is the surface. [After Bucher et al., [56].   c 1989 by Elsevier Science]

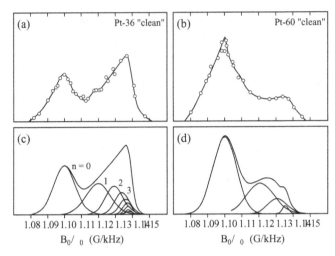

Fig. 9. [195] Pt NMR spectrum and layer statistics. (a,b) point-by-point spectra under clean-surface conditions for the samples in Fig. 8. (c,d) fits by a superposition of Gaussians that represent the NMR line of a given layer. [After Bucher et al., [56]. c 1989 by Elsevier Science]

this assumption and to the data in Fig. 8c, the NMR spectrum of Fig. 9b should consist mainly of a superposition of three Gaussian peaks with relative areas 0.60, 0.29 and 0.09. The spectrum of Fig. 9a contains these same Gaussians (having the same positions in the spectrum and the same widths), but now with relative areas 0.36, 0.25 and 0.17, and several more Gaussian peaks. Fits according to this principle are shown in Fig. 9c and Fig. 9d. They correspond to $K_0 = 0$ and m = 1.35. The agreement between fitted and experimental spectra is not perfect, but su cient to demonstrate the usefulness of the NMR layer model. The fitted subsurface (n = 1) peak of the clean sample falls approximately halfway between the surface and bulk resonances. This is in very good agreement with a five-layer slab calculation [6] and shows that more than half of the spectrum contains information from the surface region.

The NMR layer model simply considers the layer-to-layer variation of the NMR shift. It might perhaps be more reasonable to look instead at the local density of states at E_f (LDOS). The necessary additional experimental information can be obtained from the spin-lattice relaxation data in the following way. It is easily seen from (9) that at a single resonance frequency (fixed value of K) one might find signals from nuclei with many di erent combinations of s- and d-like LDOS. Each such combination would give rise to the same K , but a di erent T_1. Therefore, generally, the spin-lattice relaxation curves measured at a certain resonance frequency could be nonexponential. The spin-lattice relaxation mechanism by conduction electrons requires that such nonexponential decay curves obey "time-temperature scaling", because the spin-lattice relaxation rate T_1^{-1} for each individual nucleus is proportional to temperature T, so that $T_1 T = C$ [20]. In a relaxation experiment at a given resonance position and temperature, we measure a series of recovered signal amplitudes A_i as a function of the relaxation interval τ_i between the initial saturation pulses and the inspection pulse(s). When the A_i are normalized by the fully relaxed amplitude, the Bloch equation for a single site is

$$1 - A_i = e^{-\tau_i/T_1} \qquad (22)$$

and using the Heitler–Teller relation, one has

$$1 - A_i = e^{-\tau_i T/C} . \qquad (23)$$

If the relaxation curve is a sum of N different exponentials (corresponding to different sites), there are N different constants C; but a collection of curves taken at different values of T will collapse onto a single curve when plotted as normalized A_i vs. $\tau_i T$, as shown in Fig. 10 [57]. It is usually impossible to determine a value of N from the experimental data. We find that most relaxation curves can be described by a sum of two exponentials, with temperature-independent amplitude ratios. While it is clear that in such a case nuclei in at least two different environments resonate at the frequency under consideration, it is of course impossible to demonstrate that there are not more than two environments. Therefore one should be very cautious in interpreting the amplitude ratio of the two exponential decays as a ratio of "site occupations".

The time-temperature scaling in Fig. 10 is seen to persist down to at least 22 K. These particles have (on the average) 60% of their atoms exposed on the surface. For comparison, a cubooctahedron of 309 atoms (2.1 nm equivalent diameter) has 162 atoms in the surface, 92 in the first subsurface layer, and 42, 12 and 1 atoms in the next layers. It is therefore somewhat astonishing that no hint of discreteness of the electron energy spectrum is visible in Fig. 10. However, such a "loss of metallic character" has been found for platinum in zeolite of similar or slightly higher dispersion. The platinum particles are inside the (sometimes partially damaged) supercages of the zeolite matrix. Samples of this dispersion show virtually no signal at the bulk resonance position, since in terms of the NMR layer model less than 1% of the atoms are below layer 2 (the sub-subsurface layer).

In three Pt/zeolite systems the [195]Pt spin-lattice relaxation at the position $1.10\,\mathrm{G\,kHz}^{-1}$ was studied between 22 and 250 K, see Fig. 11. In the following discussion we focus mainly on the sample "64-clean". From 250 K down to 80 K the relaxation curves can very well be normalized by time/temperature scaling. Below 80 K only a temperature-dependent fraction of the experimental relaxation curve can still be normalized this way. At 22 K, only about 1/3 of the nuclei resonating at this position still show the "metallic" spin-lattice relaxation. The scaling behavior of the relaxation could be studied below 83 K at one spectral position, $1.091\,\mathrm{G\,kHz}^{-1}$, where only surface atoms should resonate (the layer model supposes that the maximum of the surface

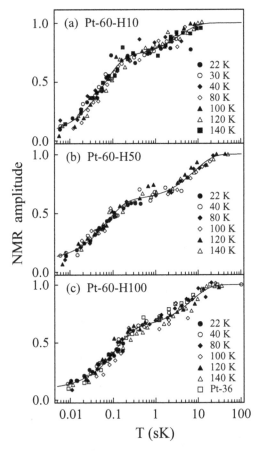

Fig. 10. Time-temperature scaling for nonexponential relaxation curves in a series of ^{195}Pt saturation-recovery experiments. The time points   are multiplied by the temperature of the experiment (see key to symbols) and the individual equilibrium signal amplitudes are scaled to the same value. The sample is Pt/TiO$_2$ of dispersion 0.6 under several hydrogen coverages (0.1, 0.5 and 1.0 monolayer). The squares in (c) show data at 110 K for another Pt/TiO$_2$ catalyst of dispersion 0.36. [After Tong et al., [57]. c 1994 by the American Chemical Society]

peak is at 1.10 G kHz^{-1}, but there is some signal from the subsurface layer at this frequency), and another, 1.110 G kHz^{-1}, where the contribution from the subsurface layer should be more important. At both frequencies, roughly the same 1/3 of the signal has metallic relaxation at 22 K. This suggests that there are two classes of particles: those that have at 22 K metallic relaxation on all sites, and those that have no metallic relaxation. It is then reasonable to think that the metallic particles must be the larger ones, and from the size distribution it is calculated that their diameter is 1.6 nm or more.

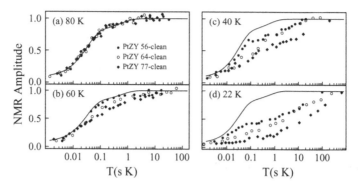

Fig. 11. Time-temperature-scaled relaxation data for three clean-surface Pt/zeolite samples, taken at spectral position 1.100 GHz^{-1}. The full curve in all panels is a double exponential fit to the 80 K data [After Tong et al., [58].   c 1995 by the American Physical Society]

The di erence between relaxation behavior in oxide-supported and in zeolite-encaged particles may be related to the fact that particles sit on the surface of the oxide, but are enclosed in the matrix of the zeolite. The latter probably fit rather snugly inside the cavity created by the local partial collapse of the aluminosilicate network, while the surfaces of the oxide-supported particles are relatively unconstrained. Free particles can execute "breathing" vibrations, whereby their volume varies in time. This is a phonon-like motion, and the electron distribution can adjust very rapidly on this timescale, so as to be at any moment in equilibrium with the instantaneous configuration of the nuclei.

As a simplification we consider the particle as an elastic continuum to describe the vibrations, and as a free electron gas inside this vibrating membrane to calculate an e ective density of states. When under influence of the vibration the radius of the particle varies between $R + \frac{1}{2} R$ and $R - \frac{1}{2} R$, the electron density parameter will vary between $r_s + \frac{1}{2} r_s$ and $r_s - \frac{1}{2} r_s$, with $r_s/r_s = R/R$ and $R = N^{1/3} r_s$. The variation of $D_{feg}(E_f)$ is given by (14). The usual estimate of the level splitting in a particle containing N electrons is $= 2/ND(E_f)$. Washing-out of the level structure will then occur if the variation in Fermi energy that corresponds to the variation in r_s is larger than the level splitting :

$$E_f \quad r_s - \frac{1}{2} r_s \quad - E_f \quad r_s + \frac{1}{2} r_s \quad 2/ND (E_f) , \qquad (24)$$

which leads to the condition $R/r_s \quad (2/3)N^{-2/3}$. For $N = 100$, the requirement is that the amplitude of the surface motion is about 3% of an atomic radius; in bulk gold the atomic rms displacements are estimated to satisfy the requirement above 100 K; and it is well known that the amplitude of thermal motion in a surface is larger than that in the bulk.

Such a vibration mechanism could therefore explain the absence of discrete-level e ects above 20 K in particles with essentially "free" surfaces, such as those anchored on oxides. On the other hand, the "encaged" particles in zeolites are mechanically clamped, so that higher temperature is needed to have su cient amplitude of these breathing modes to average out the discrete level structure.

5 Small Rhodium Particles: Incipient Antiferromagnetism

In pure bulk metals, magnetism is found only in the 3   d row of the periodic system, from Cr to Ni. Non-bulk forms of 4   d and 5 d metals can also be magnetic, either due to surface e ects as in films [59] or because of size e ects, as in atomic clusters [60].

As shown in Sect. 4, the local susceptibility in the surface layer of platinum is lower than that of the bulk, while the bulk value is retrieved about 3 atomic layers deep. At least qualitatively, this agrees with calculations for a five-layer platinum film [6]. The calculated Stoner enhancement for bulk rhodium is rather small [16], but the calculated surface-LDOS is larger the bulk value [61, 62]. The experimental situation is not quite clear [63], but suggests that the Rh(100) surface shows superparamagnetism or some extremely unstable two-dimensionsional ferromagnetic order. In rhodium clusters of up to roughly a hundred atoms, size-dependent magnetism has been found both experimentally [64] and in calculations [62].

The incipient magnetic structure of a paramagnetic system is best described by its wavevector dependent susceptibility $\tilde{}$ (q). In this respect, calculations give distinct results for bulk Pd and bulk Rh [65, 66, 67]. The $\tilde{}$ (q) in Pd peaks strongly at zero wavevector and decays monotonously for larger q, which is characteristic for an incipient ferromagnet. The calculated   q = 0 enhancement in Rh is clearly weaker than that in Pd; but there is a rather strong secondary maximum that corresponds to a tendency towards antiferromagnetic ordering.

We have used NMR of [103] Rh of small supported rhodium particles to study some of these issues [68]. As a prerequisite for the data analysis, we give the bulk [103] Rh NMR parameters and present a brief review of spin fluctuation theory as applied to NMR.

5.1 NMR of Bulk Rh

The Knight shift of bulk Rh is nearly temperature independent up to 100 K and then decreases (becomes less positive) going to room temperature. We find the metal resonance at low temperatures at 1   .34373(5) MHz T $^{-1}$, in excellent agreement with the literature value of 1   .34374(3) MHz T $^{-1}$ [69]. But because we use another reference frequency [5], we give the corresponding Knight shift

Table 2.   Fitted values of partial contributions to the susceptibility, the Knight shift and the relaxation rate of ^{103} Rh in rhodium metal. The next rows give the parameters used: hyperfine fields B_{hf}, reduction factors R, exchange integrals I and densities of states $D(E_f)$. The molar volume $V_m = 8.30\,cm^3$. For cgs in $\mu emu\,mole^{-1}$ multiply the table body entries by $V_m/(4\pi)$.

	s	d	orb	dia	sum	exp.	
	2.5	130.6	71.2	− 59.7	144.6	143.8	$\times 10^{-6}$
K	0.60	− 5.25	8.42		3.76	3.75	$\times 10^{-3}$
$(T_1 T)^{-1}$	1.3	11.6	96.9		109.8	111.5	$\times 10^{-3}\,s^{-1}\,K^{-1}$
B_{hf}	200	− 34	100				T
R		0.20	0.44				
I	0	28					mRy
$D(E_f)$	0.7	18.0			18.7		Ry^{-1}

as $K = 3.75 \times 10^{-3}$, instead of 4.3×10^{-3}. At 280 K, we have $K = 3.57 \times 10^{-3}$. Between 15 and 200 K, we find the relaxation rate linear in temperature, with $T_1 T = 8.97(5)\,s\,K$, in agreement with the earlier result 9 s K at helium temperatures [69].

These low-temperature NMR data for bulk Rh can be fitted by the usual (8) to (11), see Table 2. In this fit, the diamagnetic susceptibility of Rh was set equal to that of Ag (− 39.4 μemu mole^{-1} in cgs units), as calculated from the total experimental susceptibility of silver (− 19.5 μemu mole^{-1}) and its spin susceptibility as determined by ^{109} Ag NMR [5] (19.9 μemu mole^{-1}, see Sect. 3). The reduction factors R were set to equal occupancy for all five types of d-orbitals, and a term 1/25 was added to R_{orb} to account for dipolar relaxation [24]. Since $D_s(E_f)$ is low anyway, the corresponding exchange integral $I_s = \alpha_s/D_s(E_f)$ was set to zero; for the same reason the s-like hyperfine field was simply set to a plausible value, the fitted value for Pd [5]. The d-like hyperfine field was constrained to be between zero and the value determined experimentally for Pd [11]. The orbital susceptibility and hyperfine field were constrained to be comparable to calculated values [70]. The d-like exchange integral I_d was constrained to yield a moderate susceptibility enhancement. The fitted enhancement factor is 2.02, in reasonable agreement with the calculated value of 1.79 [16]. The parameter values reproduced in Table 2 [68] are somewhat different from those in the preliminary analysis proposed in [71]. The latter unfortunately contains numerical errors and is therefore inconsistent.

5.2 Spin Fluctuations, K and T_1

For strongly enhanced paramagnets, K and/or $T_1 T$ may become temperature dependent through spin fluctuations. Below we summarize how these temperature dependences are expected to differ for ferromagnetic and antiferromagnetic enhancements. These theories have been developed for the homogeneous electron gas [12, Ch. 5], and we will give the relevant quantities an index h.

Paramagnets with Ferromagnetic Spin Fluctuations

These are systems that show a strong temperature dependence of the uniform static susceptibility, but that nevertheless remain paramagnetic. The paramagnetic phase of low- T_C ferromagnets can be described by the same theory. In addition to an exchange parameter χ_h, spin fluctuation theory introduces a function $\lambda(T)$, such that the temperature dependent susceptibilities in a ferromagnetically enhanced paramagnet are given by

$$\tilde{\chi}_h(q,T) = \frac{1}{\chi_h}\,\frac{\tilde{\chi}_{P,h}(q)}{(\lambda(T)+1)\,\tilde{\chi}_{P,h}(0) - \tilde{\chi}_{P,h}(q)}. \tag{25}$$

The Stoner enhancement factor is related to the exchange parameter and the $q = 0$ static susceptibility through

$$\chi_h = \chi_h\,\tilde{\chi}_{P,h}(0;\lambda = 0) \tag{26}$$

and to the low-temperature limit of $\lambda(T)$ by

$$\lim_{T \to 0}\lambda(T) = \frac{1 - \chi_h}{\chi_h}. \tag{27}$$

The inverse of the static uniform susceptibility, as in (8), is

$$\chi_h^{-1}(T) = \chi^{-1}\lambda_h(T). \tag{28}$$

At low temperatures the usual Stoner enhanced susceptibility is retrieved because of (27). At higher temperatures the experimental Curie-Weiss behaviour implies that

$$\lambda(T) \propto (T - T_C) \tag{29}$$

and the cases of interest to us are close to the limit $T_C \to 0$.

According to (28) the Knight shift becomes temperature dependent, and from (6) and (25) the spin-lattice relaxation rate is given by:

$$S(T_1 T)_h^{-1} = \frac{2(\mu_B \chi_{hf})^2}{\mu_0\,(2\pi)^3}$$

$$\int \frac{\tilde{\chi}_{P,h}(q;\lambda)\,\lambda(T)+1}{\chi_h}\left[\frac{1}{\lambda(T)+1 - F(q)}\right]^2 dq \tag{30}$$

$$S(T_1 T)_{P,h}^{-1}\,\frac{\lambda(T)+1}{\chi_h}\left\langle\left[\frac{1}{\lambda(T)+1 - F(q)}\right]^2\right\rangle_{FS} \tag{31}$$

$$S(T_1 T)_{P,h}^{-1}\,\frac{1 + \frac{5}{3}(\lambda(T)+1)^{-2}}{\chi_{P,h}}\,\lambda_h(T), \tag{32}$$

where $F(q)$ is the Lindhard function. The angular brackets in (31) indicate an average over vectors q that connect two points on the spherical Fermi

surface. The expression in the numerator of (32) comes from the approximation in (12). It is usually assumed that the main temperature dependence of
$T_1 T$ is contained in the factor $\chi_h(T)$, which for an incipient ferromagnet is
$1/T$. The result is a T_1 independent of temperature, and a Knight shift
$K \sim \chi_h(T) \sim 1/T$. The low-temperature limit of (27) introduces the Moriya
desenhancement factor of (12) into (32):

$$(T_1 T)_h^{-1} = (T_1 T)_{P,h}^{-1} k_M(\alpha_h) . \tag{33}$$

Paramagnets with Antiferromagnetic Spin Fluctuations

In antiferromagnetic systems, the static paramagnetic susceptibility has its
maximum at an ordering vector $Q \ne 0$. A moderate tendency towards antiferromagnetism has been found in calculations for bulk rhodium, where a
secondary maximum in $\tilde{\chi}(q)$ appears at a nonzero wavevector [66], the absolute maximum remaining at $q = 0$. In the NMR of the paramagnetic state
of antiferromagnets a situation can arise where T_1 is dominated by $\tilde{\chi}(Q)$,
whereas the Knight shift is determined by $\tilde{\chi}(0)$. In that case, there exists no
equivalent of the $k(\alpha)$ relation as in (11) or (12), since the χ in K refers to
$q = 0$, and that in T_1 to Q.

There is no closed expression, analogous to (30), for the spin-lattice relaxation in antiferromagnetically enhanced paramagnets. Still the temperature dependence of $T_1 T$ can be estimated in the following way. The starting
assumption is that, because of the enhancement, $\tilde{\chi}$ is strongest around a
wavevector Q different from zero, and that around that wavevector it can be
expanded as

$$\frac{\tilde{\chi}(Q + q; \omega)}{\omega} = \frac{C}{\kappa_Q(\eta_Q(T) + A q^2)^2} , \tag{34}$$

where C and A are coefficients of the expansion, and κ_Q and $\eta_Q(T)$ are
analogous to the quantities in ferromagnetic fluctuations theory. The sum
in (6) is converted into an integral over a spherical volume equivalent to a
Brillouin zone centered at Q, of radius q_B with $q_B^3 = 6\pi^2/\Omega$, so that $(T_1 T)^{-1}$
is proportional to

$$\int_{q=0}^{q_B} \frac{4\pi q^2}{\kappa_Q(\eta_Q(T) + A q^2)^2} dq = \frac{\pi^2}{A \kappa_Q (A \eta_Q(T))^{1/2}} \quad \text{for } \eta_Q \to 0 \tag{35}$$

$$= 2\pi q_B^3 \kappa_Q \tilde{\chi}_P^2(Q) \quad \text{for } \eta_Q \to 0 . \tag{36}$$

In the weak-enhancement limit, $\eta_Q \to 0$, the $T_1 T$ is independent of T, as
expected. In the limit of a strong antiferromagnet, $\eta_Q \to 0$, we have

$$T_1 T \propto \eta(T) \propto T - T_N \tag{37}$$

and for an incipient antiferromagnet, $T_N \to 0$, the resulting relaxation rate is
proportional to the square root of temperature.

The detection of antiferromagnetic enhancement by NMR is usually based on qualitative arguments. At low temperature the Knight shift is temperature independent, whereas $T_1 T$ is not. The experimental relaxation rate $T_1^{-1}(T)$ is fitted to

$$T_1^{-1} = aT + bT^{1/2} .\tag{38}$$

The term T is attributed to the term $B_{hf,orb}$ in (10). It is usually impossible to give a quantitative interpretation of the value of b.

5.3 ^{103} Rh Spectra of Small Rhodium Particles

We have studied four samples of small Rh particles by NMR. A first sample, denoted Rh/TiO$_2$, has an average diameter $d = 3.6$ nm, dispersion 26%. A second one, Rh/PVP, has $d = 3.0$ nm, dispersion 37%. A third sample was prepared on RP3 alumina (Rhˆ one-Poulenc) with $d = 2.6$ nm and dispersion 40%. The fourth sample was on GSF400 alumina (Rhˆ one-Poulenc) with 4% loading. Its size distribution is difficult to measure by TEM; we have used small-angle X-ray scattering to estimate the average diameter as 1.5 nm and the dispersion as 54%. The ^{103} Rh spectra for these four small-particle samples are shown in Fig. 12. They are all approximately centered at the bulk

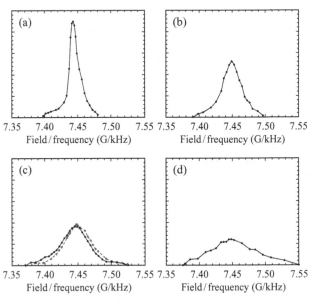

Fig. 12.   Point by point ^{103} Rh NMR spectra for clean surface Rh particles of different sizes on different supports, taken at 80 K. The spectra are normalized to the same area. The samples are (a) Rh/TiO$_2$, (b) Rh/PVP, (c) Rh/RP3, (d) Rh/GSF400. The dotted line in (c) shows the effect of chemisorption of a monolayer of hydrogen. On this scale, the width of the bulk Rh resonance is comparable to a line thickness. [After Burnet et al., [68].   c 2002 by the Institute of Physics]

resonance position, and broaden rather symmetrically when the particle size decreases. Chemisorption of hydrogen has a measurable, but rather small effect, Fig. 12(c). These spectral characteristics are markedly different from those of the small platinum particles in Sect. 4.

There are two reasons for this different behaviour: one in the NMR of Rh metal, the other in the electronic structure of the Rh surface. According to Table 2, the Knight shift and the relaxation rate of bulk Rh are dominated by the orbital parts, with additional contributions from the spins of the d-like electrons, and negligible s-like parts. This is very different from Pt (compare Table 1), where the d-like spin part dominates, and the orbital parts are nearly negligible. In small particles of Pt the magnetic behaviour can be reasonably well described by considering only the site to site variation of the d-like spin part, and as it happens this part of the susceptibility is measurably smaller on surface sites than in the bulk. From calculations for a rhodium slab [61] it is expected that the site to site variation of the ^{103} Rh spin Knight shift is comparatively small: therefore the spectral shapes in Fig. 12 arise from competition between (positive) orbital shifts and (negative) spin shifts. On some surface sites the net result is negative, on others positive; therefore no distinct surface signal can be found in the spectra. From calculations for the fcc(111) surfaces for Pd and Rh [72] it is found that hydrogen has markedly less influence on the $D(E_f)$ at surface sites on rhodium than on palladium. It is believed that in this respect Pt behaves as Pd, which would explain that the ^{195} Pt spectral shape shows large qualitative changes upon hydrogen adsorption, whereas the ^{103} Rh spectrum in Fig. 12(c) only shows a shift (by about 14 kHz or 740 ppm, upfield).

5.4 Spin-Lattice Relaxation of ^{103} Rh in Small Particles

In ^{195} Pt NMR of clean surface particles the relaxation rate at a given resonance position is independent of support or particle size [73] (with the exception of zeolite carriers). Therefore Pt nuclei in different samples that resonate at the same spectral position are in very similar environments. Such a simple identification of atomic environment with spectral position cannot be made for Rh particles: at the same resonance position, we can find very different relaxation rates.

In the sample with the largest particles, Rh/TiO$_2$, the value of $T_1 T$ is roughly the same at 80 K and 20 K for several spectral positions, see Fig. 13(a,b). Assume for the sake of argument that the drawn line in Fig. 13(b) correctly represents the spectral variation of the $T_1 T$ product. It is nowhere larger than the bulk value and, the contribution of $D_s(E_f)$ to the relaxation being negligible, (10) says that there can be no atomic sites with a local value for the density of states $D_d(E_f)$ smaller than the bulk value. The symmetry of the NMR spectrum and of the sketched $T_1 T$ curve imply that there must also be a site to site variation of the orbital susceptibility $_{orb}$. The parts (c) and (d) of Fig. 13 show the relative changes across the spectrum of $D_d(E_f)$ and $_{orb}$, assuming all other parameters in Table 2 to be constant.

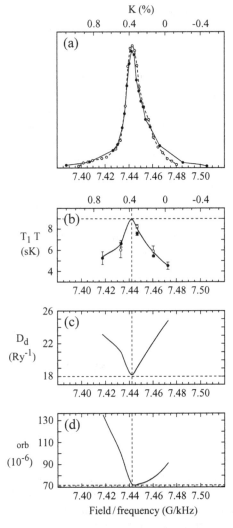

Fig. 13. The NMR spectrum (a) and value of $T_1 T$ for several spectral positions (b) in sample Rh/TiO$_2$, at 20 K (full circles) and 80 K (open circles). The line drawn in (b) is meant to suggest that $T_1 T$ is independent of temperature, and that it is close to the bulk value (straight dashed lines) in the centre of the spectrum. The drawn line in (b) can be represented by the variations of $D_d(E_f)$ and of $_{orb}$ shown in (c) and (d). [After Burnet et al., [68]. c 2002 by the Institute of Physics]

At the low field end of the spectrum for somewhat smaller particles, Rh/PVP, the relaxation rate is less than it is in the bulk, see Fig. 14(a,b). Since the orbital relaxation rate should be unaffected by magnetic fluctuations, this immediately says that at least in part of this sample the LDOS is lower than the $D(E_f)_{bulk}$. There is a clear low-temperature enhancement of

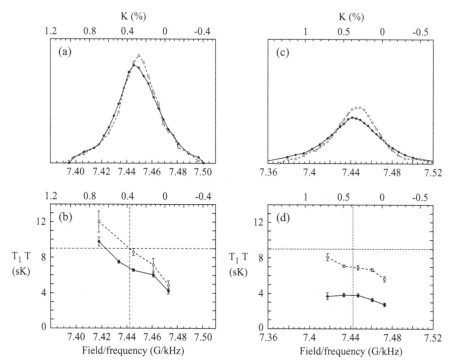

Fig. 14.   The NMR spectrum (a) and value of $T_1 T$ for several spectral positions (b) in sample Rh/PVP, at 20 K (full circles) and 80 K (open circles). The lines in (b) are drawn only to connect points at the same temperature. The temperature e ect on the relaxation (but not on the spectrum) is much larger than in Fig. 13. At the low field end in (b) the relaxation is slower than in bulk Rh (straight dashed lines). Similar data for sample Rh/RP3 are shown in (c) and (d). Here the site to site variation is rather small, but the temperature e ect is large. [After Burnet et al., [68]. c 2002 by the Institute of Physics]

the relaxation rate over most of the spectral width, while the spectral shape in Fig. 14(a) hardly varies.

For still smaller particles, the spectral shape remains nearly independent of temperature, but the $T_1 T$ products vary strongly, see Fig. 14(c,d). At fixed temperature, the site to site variation of the relaxation is relatively small, since T_1 is nearly constant across the spectrum. From an interpolation at the bulk resonance position we have $T_1^2 T = 0.63 \pm 0.03 \, s^2 K$ for the two temperatures. This is a strong indication for relaxation dominated by the $bT^{1/2}$ term in (38). The relative unimportance of the aT term implies a drop in $D(E_f)$ with respect to the bulk value, as has already been seen at the low field end in Fig. 14. Since the spectrum remains centered at the bulk position, there must be a drop in $_{orb}$ as well.

The smallest particle sample in Fig. 12(d) contains only a small weight of rhodium, and its signal is spread over a large spectral width, which makes it very difficult to perform the relaxation measurements. Some results (not shown) have been obtained at 80 K. Their large error bars range from the 80 K values in Fig. 14(b) to those in Fig. 14(d). Very qualitatively the results agree with those in the other samples, but the large error range and the lack of low-temperature data prevent further discussion.

5.5 Surface and Size Effects

The ^{103}Rh NMR of the 3.6 nm particles, Rh/TiO$_2$, is mainly determined by surface effects. Calculations for slabs exhibiting different Rh surfaces [61, 62] indeed show enhanced values of $D(E_f)$ at the surface, as suggested by the curve in Fig. 13(c). From comparison with Fig. 13(d), it follows that on sites where $D_d(E_f)$ increases, $_{orb}$ increases also.

In particles of 3.0 nm, Rh/PVP, the T_1 values at 80 K show that at least in a part of the sample the $D_d(E_f)$ is less than the bulk value. The spectrum stays symmetric and centered at K_{bulk}, which suggests that where $D_d(E_f)$ diminishes, $_{orb}$ diminishes also. We believe that this is a size- rather than a surface effect, that somehow accompanies the onset of the size-dependent antiferromagnetic relaxation enhancement. This enhancement is more clearly seen at 20 K; at 80 K it is decreased by spin fluctuations. For the particles in this sample the enhancement probably is in between the two limits of (36), so that the overall temperature dependence of $T_1 T$ does not have the structure of (38).

The very similar values of $T_1^2 T$ at the two temperatures for the bulk resonance position in Fig. 14(d) suggest that here (38) might be applicable. However, the considerable amount of experimental time that would be required has kept us from gathering enough data to do a $T_1^{-1} = aT + b(T - T_N)^{1/2}$ analysis, for which an interpretation of a and b would be difficult anyway. To illustrate this point, assume that in Fig. 14(d) the a term has half the value of the bulk relaxation, 0.05 (s K)$^{-1}$, and furthermore that we are in the $_Q$ 0 limit of (36). In that case, our two temperature points could be fitted by assuming T_N 5–10 K, and $_{orb}$ 0.7 $_{orb,bulk}$.

The only modest variation of T_1 across the spectrum in Fig. 14(d) suggests that the relaxation mechanism acts more or less in the same way on all atomic sites: it is related to a size- rather than a surface effect. Magnetism has been found in Rh clusters of up to 80 atoms, both experimentally [64] and theoretically [62]. Our particles in Rh/PVP and Rh/RP3 are larger (at least several hundreds of atoms), but the tendency towards magnetism now expresses itself through an enhancement of the antiferromagnetic ~ (Q) found in calculations for the bulk [66]. From the ^{103}Rh NMR we conclude that rhodium particles of approximately 2.6 nm diameter show incipient antiferromagnetism below 80 K. This incipient antiferromagnetism is accompanied by a lowering of the density of states at the Fermi level and of the orbital susceptibility with respect to their bulk values.

Acknowledgements

The small-particle NMR shown here has been taken from the PhD theses of Jean-Pierre Bucher, Jean-Jacques Bercier, YuYe Tong and S´ everine Burnet. Their work was supported by the Swiss National Science Foundation, recently under grant 20-53637.98.

References

1. W.D. Knight: Solid State Phys.   2, 93 (1956)
2. G.C. Carter, L.H. Bennett, D.J. Kahan: Prog. Mat. Sci.   20, 1 (1977)
3. J. Winter: Magnetic Resonance in Metals   (Oxford University Press, Oxford 1970)
4. V.L. Moruzzi, J.F. Janak, A.R. Williams: Calculated Electronic Properties of Metals (Pergamon, New York 1978)
5. J.J. van der Klink, H.B. Brom: Prog. NMR Spectrosc.   36, 89 (2000)
6. M. Weinert, A.J. Freeman: Phys. Rev. B   28, 6262 (1983)
7. A.M. Clogston, V. Jaccarino, Y. Yafet: Phys. Rev. A   134, 650 (1964)
8. J.J. van der Klink, J. Phys.: Condens. Matter   8, 1845 (1996)
9. S.H. Vosko, J.P. Perdew: Can. J. Phys.   53, 1385 (1975)
10. O. Gunnarson, J. Phys. F: Met. Phys.   6, 587 (1976)
11. J.A. Seitchik, A.C. Gossard, V. Jaccarino: Phys. Rev.   136, A1119 (1964)
12. T. Moriya: Spin Fluctuations in Itinerant Electron Magnetism   (Springer-Verlag, Berlin 1985)
13. R.J. Charles, W.A. Harrison: Phys. Rev. Lett.   11, 75 (1963)
14. R. Kubo: J. Phys. Soc. Jpn.   17, 975 (1962)
15. V.P. Tarasov, Yu.B. Muravlev, K.E. German, N.N. Popova: Doklady Physical Chemistry 377, 71 (2001)
16. J.F. Janak: Phys. Rev. B   16, 255 (1977)
17. A.R. Williams, U. von Barth: In [18]
18. S. Lundqvist, N.H. March: Theory of the Inhomogeneous Electron Gas   (Plenum, New York 1983)
19. R.T. Schumacher, C.P. Slichter: Phys. Rev.   101, 58 (1956)
20. W. Heitler, E. Teller: Proc. Roy. Soc. A   155, 629 (1936)
21. N. Bloembergen, T.J. Rowland: Phys. Rev.   97, 1679 (1955)
22. M. Ruderman, C. Kittel: Phys. Rev.   96, 99 (1954)
23. Y. Yafet, V. Jaccarino: Phys. Rev. A   133, 1630 (1964)
24. Y. Obata: J. Phys. Soc. Jpn.   18, 1020 (1963)
25. R.W. Shaw, Jr, W.W. Warren: Jr, Phys. Rev. B   3, 1562 (1971)
26. T. Moriya: J. Phys. Soc. Jpn.   18, 516 (1963)
27. A. Narath, H.T. Weaver: Phys. Rev.   175, 373 (1968)
28. J. Korringa: Physica   16, 601 (1950)
29. C. Oliver, A. Myers: J. Phys.: Condens. Matter   1, 9457 (1989)

30. Ch. Ryter: Phys. Lett.   4, 69 (1963)
31. G.B. Benedek, T. Kushida, J. Phys. Chem. Solids   5, 241 (1958)
32. R. Bertani, M. Mali, J. Roos, D. Brinkmann: J. Phys.: Condens. Matter   2, 7911 (1990)
33. S. Kluthe, R. Markendorf, M. Mali, J. Roos, D. Brinkmann: Phys. Rev. B   53, 11369 (1996)
34. T. Kushida, J.C. Murphy, M. Hanabusa: Phys. Rev. B   13, 5136 (1976)
35. Ch. Ryter: Phys. Rev. Lett.   5, 10 (1960)
36. D.A. Papaconstantopoulos: Handbook of the Bandstructure of Elemental Solids (Plenum Press, New York 1986)
37. A. Abragam: Principles of Nuclear Magnetism   (Oxford University Press, Oxford 1961/1986)
38. U. El-Hanany, M. Shaham, D. Zamir: Phys. Rev. B   10, 2343 (1974)
39. J.-J. Bercier: Thesis EPFL, Lausanne 1993
40. B.S. Shastry, E. Abrahams: Phys. Rev. Lett.   72, 1933 (1994)
41. S. Kobayashi, T. Takahashi, W. Sasaki: J. Phys. Soc. Jpn.   32, 1234 (1972)
42. M. Ido, R. Hoshino: J. Phys. Soc. Jpn.   36, 1325 (1974)
43. P. Yee, W.D. Knight: Phys. Rev. B   11, 3261 (1975)
44. M.J. Williams, P.P. Edwards, D.P. Tunstall: Faraday Discuss.   92, 199 (1991)
45. J.K. Plischke, A.J. Benesi, M.A. Vannice: J. Phys. Chem.   96, 3799 (1992)
46. V.M. Mastikhin, I.L. Mudrakovsky, S.N. Goncharova, B.S.Balzhinimaev, S.P. Noskova, V.I. Zaikovsky: React. Kinet. Catal. Lett.   48, 425 (1992)
47. J.J. Bercier, M. Jirousek, M. Graetzel, J.J. van der Klink: J. Phys.: Condens. Matter 5, L571 (1993)
48. J. Bardeen: Phys. Rev.   49, 653 (1936)
49. R.L. Kautz, B.B. Schwartz: Phys. Rev. B   14, 2017 (1976)
50. C. Solliard: Solid State Commun.   51, 947 (1984)
51. Ph. Bu at, J.P. Borel: Phys. Rev. A   13, 2287 (1976)
52. F. Wright Jr: Phys. Rev.   163, 420 (1967)
53. W.A. Hines, W.D. Knight: Phys. Rev. B   4, 893 (1971)
54. N. Inoue, T. Sugawara: J. Phys. Soc. Jpn.   45, 450 (1978)
55. H.E. Rhodes, P.-K. Wang, H.T. Stokes, C.P. Slichter, J.H. Sinfelt: Phys. Rev. B 26, 3559 (1982)
56. J.P. Bucher, J. Buttet, J.J. van der Klink, M. Graetzel: Surf. Sci.   214, 347 (1989)
57. Y.Y. Tong, J.J. van der Klink: J. Phys. Chem.   98, 11011 (1994)
58. Y.Y. Tong, D. Laub, G. Schulz-Eklo, A.J. Renouprez, J.J. van der Klink: Phys. Rev. B 52, 8407 (1995)
59. H. Dreyss´e, C. Demangeat: Surf. Sci. Rep.   28, 65 (1997)
60. J.P. Bucher, L.A. Bloomfield: Int. J. Mod. Phys. B   7, 1097 (1993)
61. A. Eichler, J. Hafner, J. Furthm¨ uller, G. Kresse: Surf. Sci   346, 300 (1996)
62. C. Barreteau, R. Guirado-L´ opez, D. Spanjaard, M.C. Desjonqu` eres, A.M. Ol´es: Phys. Rev. B   61, 7781–7794 (2000)
63. A. Goldoni, A. Baraldi, M. Barnaba, G. Comelli, S. Lizzit, G. Paolucci: Surf. Sci. 454–456, 925–929 (2000)
64. A.J. Cox, J.G. Louderback, S.E. Apsel, L.A. Bloomfield: Phys. Rev. B   49, 12295 (1994)
65. H. Winter, E. Stenzel, Z. Szotek, W.M. Temmerman: J. Phys. F. Met. Phys. 18, 485–500 (1988)

66. L.M. Sandratskii, J. K¨ ubler: J. Phys.: Condens. Matter   4, 6927–6942 (1992)
67. J.B. Staunton, J. Poulter, B. Ginatempo, E. Bruno, D.D. Johnson: Phys. Rev. B   62 , 1075–1082 (2000)
68. S. Burnet, T. Yonezawa, J.J. van der Klink: J. Phys.: Condens. Matter   14 , 7135 (2002)
69. A. Narath, H.T. Weaver: Phys. Rev. B   3, 616 (1971)
70. H. Ebert, M. Battocletti, M. Deng, H. Freyer, J. Voitl¨   ander: J. Computational Chemistry,   20, 1246 (1999)
71. P.A. Vuissoz, T. Yonezawa, D. Yang, J. Kiwi, J.J. van der Klink: Chem. Phys. Lett.   264 , 366 (1997)
72. R. L¨ober, D. Hennig: Phys. Rev. B   55 , 4761 (1997)
73. J.J. van der Klink: Adv. Cat.   44 , 1 (2000)

The Study of Mechanisms
of Superconductivity by NMR Relaxation

Dylan F. Smith and Charles P. Slichter

Department of Physics and Frederick Seitz Materials Research Laboratory
University of Illinois, Urbana, IL 61801
cslichte@uiuc.edu

Abstract. This chapter describes the use of NMR to study superconductivity. The focus is on the use of NMR measurements of spin-lattice relaxation (T_1) to provide insight into the possible mechanisms which give rise to superconductivity. The chapter begins with a review of NMR T_1 in the normal state of metals, followed by an explanation of how, according to the theory of Bardeen, Cooper, and Schrie er (BCS), the situation is modified when the metal becomes superconducting. These ideas are then applied in a discussion of several recent superconducting systems. Data presented for alkali fullerenes show that the fullerides are conventional BCS superconductors, while data for cuprate superconductors suggest a much different type of superconducting pairing. The authors conclude with a discussion of some organic materials that exhibit superconductivity or, with slight modification in structure, antiferromagnetism. In the studies of the antiferromagnetism, it is found that NMR can detect the electron spin-spin interaction discovered by Dzialoshinskii and Moriya.

1 Introduction

Since the discovery of magnetic resonance, consideration of the interaction of the nucleus with its surroundings has been a subject of great importance. Indeed, such considerations precede the discovery of NMR since the first discussion of the interaction was motivated by an investigation of the possibility of using the nuclear magnetism of a metal to produce cooling of the metal by adiabatic demagnetization. In a classic paper, Heitler and Teller [1] calculated the spin-lattice relaxation time, T_1, of paramagnetic ions and of nuclei. They concluded that in diamagnetic insulators, the nuclear relaxation would be of order 106 years, whereas for a metal it would be about 1 second at 0 .1 K. They also showed that T_1 would vary inversely with temperature in a metal.

In their paper announcing the discovery of "Resonance Absorption by Nuclear Magnetic Moments in a Solid", Purcell, Torrey, and Pound [2] point out that "a crucial question concerns the time required for the establishment of thermal equilibrium between spins and lattice". The first experiments of the

D.F. Smith and C.P. Slichter: The Study of Mechanisms of Superconductivity by NMR Relaxation , Lect. Notes Phys. 684 , 243–295 (2006)
www.springerlink.com

Purcell group after the discovery of NMR in 1945 were studies of relaxation times of protons in liquids and were reported in the classic paper known to all students of magnetic resonance as BPP [3]. It turns out that the "solid" of their first paper, para n, was not all that solid because had it been so the T_1 would have been too long for the experiment to have been successful!

So from the start it was evident that nuclear magnetic resonance could be used to study the material in which the nuclei were imbedded. In the subsequent history of NMR, one of the most important contributions of NMR was its use in the discovery of the superfluid phase of ^{3}He by Oshero , Richardson, and Lee [4].

In this paper, we describe the use of NMR to study superconductivity. This field is enormous, so we focus on a particular aspect, the use of NMR measurements of T_1 to give insight into the possible mechanisms which give rise to superconductivity, illustrating with examples of several superconducting systems. We first review the situation in the normal state of metals, then explain how, according to the theory of Bardeen, Cooper, and Schrie er (BCS) [5, 6], the situation is modified when the metal becomes superconducting. We show how NMR confirms the essential idea of their theory. Then we discuss several other systems (the alkali fullerides and the cuprates). We conclude with a discussion of some organic materials that exhibit superconductivity or, with slight modification in structure, antiferromagnetism. In the studies of antiferromagnetism, we encountered an NMR mystery. We describe it and give its explanation, showing that NMR can detect the electron spin-spin interaction discovered by Dzialoshinskii and Moriya.

2 Normal Metals

In 1949, Knight [7] discovered the famous shift, which bears his name, of resonance frequency experienced by nuclei in metals. A year later, Korringa published his paper [8] on the theory of NMR relaxation and shifts in metals which related T_1 to the Knight shift, a formula which is known as the Korringa relation,

$$T_1 T K^2 = \frac{\gamma_e^2}{\gamma_n^2} \frac{\hbar}{4 k_B} , \tag{1}$$

where γ_e and γ_n are the electron and nuclear gyromagnetic ratios, K is the Knight shift, and k_B is the Boltzmann constant.

A second relationship, the Heitler-Teller equation, says that

$$\frac{1}{T_1} \propto T . \tag{2}$$

These relationships were studied extensively and verified, for example by Norberg and Slichter [9] and by Holcomb and Norberg [10] in the alkali metals. Pines [11] pointed out that the Korringa relationship is modified if one takes into account electron-electron interactions. If these are on the whole antiferromagnetic in nature, the right hand side is reduced, if they are ferromagnetic, the right hand side is increased. The left hand side of (1) is often called the "Korringa product". It is interesting that if one can ignore the electron-electron effects, the right hand side depends only on fundamental constants and is independent of the material in which the nucleus sits.

The constant of proportionality of (2) is given in [1] and [8]. For our purposes, we may sketch its derivation as follows. Nuclear relaxation in a metal may be viewed as a scattering process in which an electron experiences a force from the nuclear magnetic moment and is scattered, flipping the nucleus in the process. If one thinks of the electron-nuclear magnetic interaction as arising from the hyperfine coupling, one may imagine that as an electron comes close to the nucleus, the two spins form a resultant total spin angular momentum vector about which they precess. When the collision is over, both spins have changed their directions in space. If the collision takes place in the presence of an external magnetic field, the spins will have changed their Zeeman energies and, since their γ's differ, the electron must also changed its kinetic energy.

If the hyperfine coupling is the Fermi contact term [1], and we treat the electrons as a free electron gas, the scattering is independent of the direction of the electron velocity. Then we can make a simple quantum mechanical theory of the T_1 as follows. We think of the nucleus as initially in a spin up state and the electron as initially in a spin down state with wave vector k. The scattering leaves the nucleus with spin down, the electron with spin up and a new wave vector k'. We describe the scattering using Fermi's Golden Rule: the probability per second of a transition of an electron from the initial state, i, to the final state, f, is

$$W_{if} = \frac{2\pi}{\hbar} |\langle i| \hat{V} |f\rangle|^2 \, \delta(E_f - E_i - \Delta E), \tag{3}$$

where $\hat{V}$ is the hyperfine coupling energy operator (the Fermi contact interaction), ΔE is the change in nuclear spin Zeeman energy, and E_i and E_f are the energies of the initial and final states of the electron. For the electrons, the initial state must be occupied and the final state empty. The probability that the initial state is occupied is $F(E_i)$, where $F(E)$ is the Fermi function of energy E. The probability that the final state is empty is $1 - F(E_f)$.

We denote the number of states within dE at energy E by $\rho(E)dE$. Then, summing over initial and final states, the total probability per second, W, of flips of the nuclear spin up and the electron spin down is

---

[1] Interaction energy of a nucleus at the origin with an electron at position r: $\hat{V} = \frac{8\pi}{3} \gamma_e \gamma_n \hbar^2 \delta(r)$

$$W = \frac{2}{\hbar} \quad |\langle i|\hat{V}|f\rangle|^2 \,(E_f - E_i - \quad E)F(E_i)(1 - F(E_f))$$

$$\rho(E_i)dE_i\,\rho(E_f)dE_f$$

$$= \frac{2}{\hbar} \quad |\langle i|\hat{V}|f\rangle|^2 F(E_i)(1 - F(E_i + \quad E))\,\rho(E_i)\,\rho(E_i + \quad E)dE_i \,. \qquad (4)$$

Now the energy difference between the initial and final electron states is just the change in the nuclear Zeeman energy. This change is very small compared to $k_B T$ (unless one is so cold that there is gigantic nuclear polarization!). Consequently we neglect the energy difference, integrate over the final energy, and get

$$W = \frac{2}{\hbar} \quad |\langle i|\hat{V}|f\rangle|^2 F(E_i)(1 - F(E_i))\,\rho(E_i)^2 dE_i \,. \qquad (5)$$

It is easy to show that

$$F(E)(1 - F(E)) = -\frac{\partial F(E)}{\partial E} \,. \qquad (6)$$

If then we can neglect any changes of $|\langle i|\hat{V}|f\rangle|$ and $\rho(E)$ over energies comparable to $k_B T$, we get

$$W = \frac{2}{\hbar}[|\langle i|\hat{V}|f\rangle|_{E_F}]^2 \,\rho(E_i)^2 k_B T \,, \qquad (7)$$

where the subscript E_F means evaluation for electrons at the Fermi energy. Equation (7) is essentially the Heitler-Teller result. The linear dependence on T represents the fact that, as a result of the exclusion principle, only electrons within about $k_B T$ of the Fermi surface can be scattered by the nuclear moment. The number of such electrons is a fraction of the total number of electrons of the order of ($k_B T/E_F$).

The linear dependence on temperature is displayed dramatically in Fig. 1 that shows data from a paper by Spokas and Slichter [12] on nuclear relaxation in aluminum.

The measurements extend from 1.1 K to 1000 K and include data also from Anderson and Redfield [13] as well as Hebel and Slichter for temperatures below 4.2 K.

3 Development and Verification of the BCS Theory

3.1 Bardeen's Early Gap Concept

Although superconductivity was discovered in 1911 [14], its explanation resisted the efforts of many outstanding physicists for many years. In 1950 Maxwell [15], and independently Reynolds, Serin, Wright, and Nesbitt [16] discovered that the superconducting transition temperature varied with the

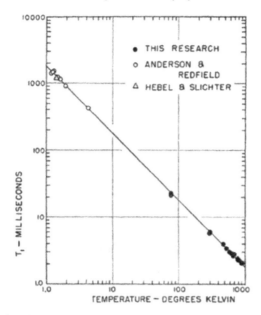

Fig. 1. Thermal relaxation time in aluminum versus the reciprocal temperature. The line describes ($T_1 T$) = 1 .85 s K. Reproduced with permission from [12]. Copyright 1959 by the American Physical Society

isotopic content of the metal, showing that in all likelihood lattice vibrations played an important role in the mechanism of superconductivity.

In 1951, John Bardeen came to the University of Illinois from Bell Laboratories. Bardeen had for many years been interested in superconductivity. The experimental discovery of the isotope e ect led him to focus on the role of the lattice vibrations. Bardeen and his postdoc, David Pines, worked out the theory of how the response of the ions to the presence of an electron changed the interaction of an electron with a second one in its vicinity [17].

Although many people associate superconductivity with the vanishing of electrical resistance, in fact the most fundamental property of a superconductor is its perfect diamagnetism: superconductors exclude magnetic fields, a property called the Meissner e ect. In 1954, Bardeen [18] proposed a very simple model that he argued could give the Meissner e ect and thus explain superconductivity. His idea was that in a superconductor there was a gap in the density of states right at the Fermi energy. The states below E_F were the same as in the normal state. States which had been above the Fermi energy were all shifted to a higher energy by an amount E_G so that the density of states vanishes between E_F and $E_F + E_G$. This density of states is shown in the middle of Fig. 2. He argued that the size of E_G would be approximately $k_B T_c$, where T_c is the temperature of the superconducting transition. One of the authors of this article, CPS, first learned of these ideas in 1954 from a

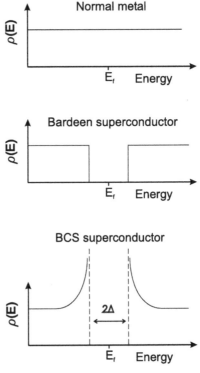

Fig. 2.   The energy dependence of the density of states for normal metals (top), the Bardeen model (middle), and the BCS model (bottom)

talk by Bardeen. Since the electrons on which Bardeen was focused were just those which give rise to the NMR T_1 of a metal, the transition to superconductivity should produce a big e ect on the nuclear T_1. CPS resolved to try to measure the nuclear T_1, but there was a problem: as a result of the Meissner e ect, superconductors exclude the magnetic fields needed to do NMR (this event predates discovery of Type II superconductors which do not exclude the magnetic field completely).

However, another consequence of the Meissner e ect is that application of a su ciently strong field suppresses the superconductivity. This critical field, H_c, is 98.4 gauss for Al metal, for which T_c is 1.172 K.

The solution to measuring T_1 in the superconducting state was to cycle the magnetic field. The cycle used by Hebel and Slichter [19, 20] is shown in Fig. 3. One sits initially at a field H_0, greater than H_c, so that the metal is in the normal state, the magnetic fields can penetrate, and a net nuclear magnetization can be established. Then one reduces the magnetic field to zero, cooling the nuclear spins (an adiabatic demagnetization), and converting the metal to its superconducting state. Since the nuclear spin temperature is now much lower than the lattice temperature, the nuclear spins warm towards

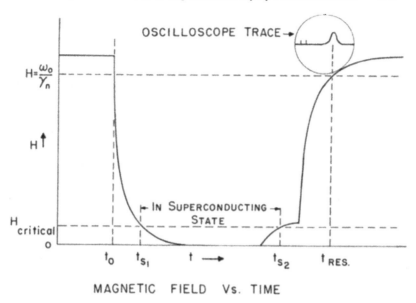

Fig. 3.   The field cycling experiment expressed as magnetic field vs. time. The time dependence is shaped to satisfy the conditions of adiabaticity

the lattice temperature via the spin-lattice relaxation processes. After some time, t_s, in the superconducting state, the magnet is switched on again to H_0, an adiabatic remagnetization. The nuclear resonance apparatus, operating in the steady state mode, is tuned to a resonance frequency somewhat less than $_nH_0$, so that as the magnetic field returns to H_0 the field passes through the resonance condition, producing an NMR signal, $S(t_s)$. The experiment is repeated for varying times t_s. The form of the signal $S(t_s)$ is, to a good approximation,

$$S(t_s) = S_0 \exp(-t_s/T_{1s}) , \qquad (8)$$

where T_{1s} is the spin-lattice relaxation time in the superconducting state, and S_0 a normalization constant. Thus, one can deduce the relaxation time in the superconducting state by measurements of dependence of S on t_s.

For this scheme to work the turn-o and turn-on of the magnetic field must satisfy the two conditions of adiabaticity. The first requirement is that the process be slow enough to be reversible. This requirement is equivalent to saying that at all times the spin system must be describable by a spin temperature $_s$, so that the internal parts of the system must be able to exchange energy su ciently e ectively that internal temperature gradients do not arise. In practice this means that at those parts of the magnetization cycle where the Zeeman and dipolar heat capacities are comparable, those two energy reservoirs must be able to exchange energy rapidly (A discussion of the general situation, distinguishing between slow and sudden turn-o s can be found in the textbook Principles of Magnetic Resonance [21]). The second requirement is that heat transfer occurs only when the system is in the superconducting state. The time to turn the magnet o must be substantially shorter than the spin-lattice relaxation time in either the normal or superconducting states. This condition ruled out many materials with high values of T_c.

For the experiments of Hebel and Slichter, the second condition led them to choose as their superconductor the metal aluminum, which has a long relaxation time of about 1 second in the liquid helium range. Even with aluminum, it was necessary to make a special magnet to achieve a fast enough switching time. They constructed the magnet using laminated iron that could be switched from 500 gauss to zero in a millisecond. The major di culty of aluminum was that its transition temperature was so low. The normal way of cooling was to use liquid He, pumping on it to obtain temperatures below 4.2 K. It is very hard owing to the Rollin film to get below 1   .2 K by this method. By use of a triple Dewar system, to combat the Rollin film, Hebel and Slichter succeeded in reaching 0 .94 K. They observed the resonance at a field of 360 gauss. By the fall of 1956, they had data which are shown in Fig. 4.

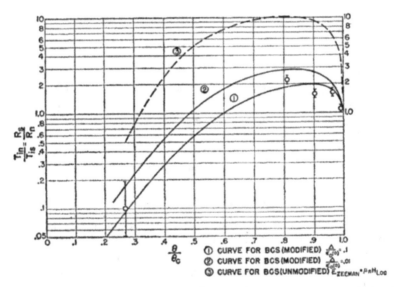

Fig. 4.   Relaxation rate in a superconductor, R_s, relative to the zero-field value extrapolated from the normal state, $R_n(0)$, versus reduced temperature $/$ $_c$. Three theoretical curves from the BCS theory are also shown. Reproduced with permission from [20]. Copyright 1959 by the American Physical Society

These data show that the transition to the superconducting state causes the nuclear relaxation to speed up. This result was a great surprise. One of the important models of the superconducting state was the two-fluid model of Gorter and Casimir [22, 23]. The electrons were divided into a superfluid component and normal component. As one cooled below T_c the fraction of superfluid component grew from zero to 100%. One would expect that only the "normal" component could relax nuclei, hence T_1 should lengthen as one cooled. If one viewed the problem in terms of the existence of an energy gap in the density of electron states, one would suppose that at low temperatures there would be few excitations to states above the gap, leading to a strong inhibition of the relaxation process. In examining their data, Hebel and Slichter thought of one possible explanation. Equation (5) shows that 1 $/T_1$ involves the square of the density of states. If the creation of a gap caused the density of states to pile up on the edges of the gap, this e ect might enhance W as long as the gap was not too large. They showed their data to Bardeen together with this possible explanation but of course lacking a true theory of the superconducting state, one could not judge the validity of this explanation. As we see below, this is part of the explanation of the enhanced relaxation rate, but the other part, dealing with the scattering matrix element in the superconducting state, is more important since it is the crucial manifestation of the pairing of electrons, the central feature of the later successful theory by Bardeen, Cooper, and Schrie er.

3.2 The Bardeen, Cooper, Schrie er Theory

In the spring of 1956, Leon Cooper, who succeeded David Pines as Bardeen's post doctoral associate, made an important discovery. In thinking about metals, he considered a pair of electrons with energies above a filled Fermi sea. He found that if they had an attractive interaction, they could form a bound state. These so-called Cooper pairs demonstrated the essential role of degeneracy and its resolution. He published these results in the fall of 1956, just as Hebel and Slichter were getting their first experimental results. Bardeen, Cooper, and Bardeen's graduate student Bob Schrie er were deeply immersed in an e ort to generalize Cooper's result to include all the electrons, not just a pair above the Fermi sea. In his Nobel Prize lecture, Schrie er describes how the solution to the zero temperature wave function came to him while   away on a trip (actually while riding the New York City subway). It was wave a function made up entirely of Cooper pairs in which each pair state (k, spin up; $-k$, spin down) was occupied fractionally.

Utilizing this wave function, Bardeen, Cooper, and Schrie er [5, 6] solved the Hamiltonian at absolute zero and then rapidly extended the analysis to temperatures up to the transition temperature. They found that there was an energy gap, 2   (T), that varied with temperature, which signaled the transition to the normal state at the temperature at which it vanished. (Note that the quantity   is defined by BCS as the di erence in energy from the gap edge to the Fermi energy. Thus the di erence in energy between the upper and lower gap edges is 2   ). The relationship between T_c and the gap is

$$T_c \quad \frac{2 \ (0)}{3.52} .$$
(9)

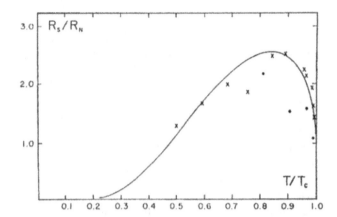

Fig. 5.   Nuclear spin-lattice relaxation rate in the superconducting state of Al.
The circles represent data of Hebel and Slichter, the crosses data of Redfield and
Anderson, and the   solid line   the BCS theory. Reproduced with permission from [25].
Copyright 1965 by the American Physical Society

They were able to calculate the results of many experiments with excellent
agreement, all with only one parameter, the size of the energy gap at absolute
zero. BCS showed Hebel and Slichter their theory, and the latter used it to
analyze their data [19, 20]. (Details of their calculation of T_1 in the normal
and superconducting states can be found in the appendix of Reference [20]).
They found that theory accounts for the increase in 1   /T_1 just below T_c and
predicts a fall-o at lower temperatures where it says

$$\frac{1}{T_1} \quad e^{-(T)/k_B T} . \tag{10}$$

Redfield independently thought of the same experiment as Hebel and
Slichter, but started work somewhat after they began. He likewise studied
Al, but to cool his sample he obtained some ^{3}He, thus avoiding the di cul-
ties of the Rollin film which arose with ^{4}He [24]. Figure 5 shows the data of
Redfield and Anderson combined with that of Hebel and Slichter (used in the
Nobel Prize Lecture of Leon Cooper) [25].

Figure 6 shows data of Morse and Bohm [26] on the absorption rate of
ultrasound, likewise from Cooper's Nobel Prize Lecture. The temperature
dependence is strikingly di erent from that of the NMR relaxation rate. For
ultrasound, the rate drops precipitously just below T_c. How can this be? The
absorption of sound waves is much like nuclear spin-lattice relaxation. Both
are low energy scattering events. For NMR, the nucleus flips, flipping the
electron and scattering its wave vector. In sound absorption, a sound quantum
(phonon) is absorbed, scattering the electron's wave vector (the electron spin
is not flipped). For both processes, the energy exchange is much smaller than
$k_B T$. One would expect that in a one-electron theory of metals, both rate

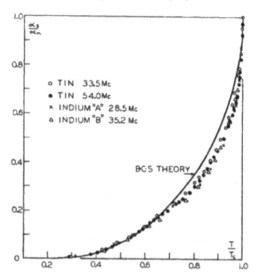

Fig. 6. Comparison of observed ultrasonic attenuation with the ideal theory. Reproduced with permission from [26]. Copyright 1959 by the American Physical Society

processes would be described by an equation such as (4), so they should have the same temperature dependence in the superconducting state.

One of the major triumphs of the BCS theory is that it explains this di erence. Indeed, the explanation is considered to be one of the best proofs of the correctness of the essential idea of the BCS theory, the pairing concept.

One can in fact see the origin of the di erence in detail by going back to (5),

$$W = \frac{2}{\hbar} \quad |\ i|\hat{V}|f\ |^2 F(E_i)(1 - F(E_i))\ (E_i)^2 dE_i\ . \tag{11}$$

As we have remarked, nuclear relaxation and sound absorption both involve electron scattering, but in the former case the electron spin is flipped whereas it is not flipped in the latter case.

The BCS wave function is made up of electron pairs in which an electron with spin up and wave vector k is paired with another electron of wave vector − k and spin down. As a consequence, there are two matrix elements that join any initial state to the same final state. Denoting these as $\hat{V}_1$ and $\hat{V}_2$, we then replace $\hat{V}$ by $(\hat{V}_1 + \hat{V}_2)$ for NMR and $(\hat{V}_1 - \hat{V}_2)$ for ultrasound, giving

$$W = \frac{2}{\hbar} \quad |\ i|\hat{V}_1 \pm \hat{V}_2|f\ |^2 F(E_i)(1 - F(E_i +\ E))\ (E_i)\ (E_i +\ E) dE_i\ , \tag{12}$$

where the plus sign applies to NMR and the minus sign to sound absorption.

According to the BCS theory, the density of states $_s(E)$ in the superconductor obeys the equation

$$\rho_s(E) = \frac{\rho_n(E)E}{E^2 - \Delta^2}^{1/2},$$ (13)

where $\rho_n(E)$ is the density of states in the normal metal, and where E is measured from the Fermi energy and has a minimum absolute value of $|\Delta|$. Thus, at the edges of the gap where E approaches Δ, $\rho_s(E)$ blows up, as shown at the bottom of Fig. 2. This increase of the density of states close to the edge of the energy gap satisfies the concept that Hebel and Slichter had postulated as to why the relaxation was faster in the superconducting state than in the normal state. But to explain the contrast between the NMR and the sound absorption one also needs the fact that the scattering matrix elements have different energy dependence. For NMR the square of the matrix element between states at E and E' goes as $(EE' + \Delta^2)/EE'$, whereas for sound absorption it goes as $(EE' - \Delta^2)/EE'$. Therefore if one neglects the difference in energy between E and E', the energy dependence of the matrix element cancels the singularity in the square of the density of states for sound waves, but not for NMR. This is directly a consequence of the pair nature of the wave functions. There is a beautiful and detailed explanation of all this in Cooper's Nobel Prize Lecture [27]. [2] The details of the T_1 calculation can also be found in the appendix of [20].

One consequence of the BCS form of the density of states is that the expression for the rate, W, such as that of (5) has a logarithmic infinity. One can then return to (4), including the nuclear energy change, but this still predicts too large a peak. Another approach is to introduce a level broadening function. This was the approach of Hebel and Slichter [20, 29]. One can also consider anisotropy of the gap arising from lattice effects. In a beautiful set of experiments, Masuda and Redfield [30] perfected the field cycling technique and the ^{3}He cooling apparatus. Their data fit the BCS theory at both high and low values of T/T_c with a gap equal to $3.2 k_B T_c$, slightly less than the BCS value, and a 10% broadening of the temperature dependent energy gap.

So, together the NMR T_1 and the ultrasonic absorption were among the first evidence that gave strong confirmation of the central concept of the BCS theory, correlated electron pairs.

3.3 The General Pairing Conditions of the BCS Theory

The original BCS paper explained superconductivity as resulting from the effect of the lattice degrees of freedom on the interactions of electrons. [3] There is an energy gap "parameter" Δ_k that is the solution of an integral equation

---

[2] These lectures are also in the appendix of the book by J.R. Schrieffer [28]. We caution the reader that there appears to be a typographical error in Cooper's formula for the matrix element for NMR. The expression $(1 + \Delta E^2/\Delta^2)$ should be $(1 + \Delta^2/E^2)$.

[3] Excellent accounts of the BCS theory can be found in the books by Schrieffer [28] and Tinkham [31].

called the gap equation. As basis states, we use free electron Bloch states with energy measured with respect to the Fermi energy denoted as ϵ_k. In the superconducting state, the interaction between the electrons $V_{kk'}$ changes the energy to E_k where

$$E_k = (\epsilon_k^2 + \Delta_k^2)^{1/2} \tag{14}$$

and Δ_k is the solution of the integral equation (we give the equation at absolute zero for simplicity)

$$E_k = -\sum_{k'} \frac{V_{kk'} \Delta_k}{2E_k} . \tag{15}$$

BCS approximated Δ_k as a single number, Δ_0, for electrons within a given cutoff energy of the Fermi energy, and zero outside this range. The energy gap parameter was independent of the direction in k-space, depending only on energy. One then gets that

$$E_k = (\epsilon_k^2 + \Delta_0^2)^{1/2} \tag{16}$$

for electrons within the cutoff range from the Fermi energy.

In fact, the theory is more general. Depending on the details of the interaction, one may get other pairings. In the state of zero current flow, all pairs have zero momentum. However, the pairs may have orbital angular momentum about their mass centers with quantum numbers, L, of 0, 1, 2, 3, etc. Owing to the exclusion principle, the spin, S, of the pair will be 0 if L is an even number and 1 if L is an odd number. The standard BCS state is ($L = 0, S = 0$). But other possibilities are ($L = 1, S = 1$) (so called p-wave superconductors); ($L = 2, S = 0$) (d-wave superconductors) and so on. Such pairings are often called "unconventional".

An example of unconventional pairing is liquid ^{3}He. The superfluid state of ^{4}He was a challenging problem which appeared to have some similarity to superconductivity, but the ^{4}He atoms obey Bose-Einstein statistics, making the parallel unlikely. However, ^{3}He atoms obey Fermi-Dirac statistics, which lead theorists to speculate in the 1960's that there should be a superfluid state described in terms of a BCS-like theory. Early estimates of the transition temperature were overly optimistic. Experimenters looked unsuccessfully, pushing the possible temperature lower and lower. Because at short distances the interaction between ^{3}He atoms is repulsive from the overlap of the electron clouds, it had been suspected that the most favorable pairing state might not be L = 0. Then, in 1972, D. D. Osheroff, R. C. Richardson, and D. M. Lee [4, 32] found the transition at 2.7 mK using NMR to study ^{3}He. These data were explained by A. J. Leggett [33] as arising from a BCS-like phase in which ($L = 1$, $S = 1, S_z = \pm 1$). Osheroff, Richardson, and Lee shared the Nobel Prize in 1996 for their work, and Leggett received the Nobel Prize in 2003 for his.

In solving the Hamiltonian for interactions that give other pairings, one finds that Δ_k varies with direction in k-space as well as with the energy. The conventional BCS state has an isotropic gap so the gap function is simply a single number. More generally, it is a function of wave vector, and may even change sign as one moves around in k-space. In fact, one must think of Δ_k as being a function that has an algebraic sign. The sign has significance and can indeed be directly measured in certain tunnelling experiments [34].

Superconductivity was discovered in elemental metals (Hg and Pb). At first it was thought to be a highly unusual phenomenon, however, in subsequent times it has been found to have very widespread occurrence. Materials such as Nb$_3$Sn and Nb$_3$Ge had much higher transition temperatures than Hg, Pb, or Al, getting above 20 K for Nb$_3$Ge. Some of the so-called heavy-fermion metals also exhibited superconductivity. More recently the cuprates, the alkali fullerides, and several classes of organic crystals have been shown to be superconductors. A major issue has become what mechanism explains the gigantic leap in T_c represented by the cuprates.

The discovery of the cuprates as superconductors by George Bednorz and Alex Müller resulted from careful and imaginative thought about methods of enhancing the electron-phonon interaction, thinking about electron-lattice coupling from a background that benefited from deep experience with topics such as the role of the Jahn-Teller effect, especially by Alex Müller. This approach led to a revolutionary result, long sought after by experimenters and theorists alike. However, other mechanisms have been proposed, such as mechanisms involving the electron spins.

As we discuss later, any given mechanism tends to favor a particular pairing state. In a given theory, only certain pairings can result, so one can rule out such a theory if one finds a different pairing state. Thus, much effort has gone into trying to determine the pairing states for the various materials that are superconductors. We now turn to several examples and show what NMR can say about the pairing state.

4 Type I and II Superconductors

One of the earliest theories of superconductivity was that of the London Brothers [35]. A natural outcome of their theory was the concept of the penetration depth, λ, which described the penetration of a magnetic field into a superconductor. Their theory expressed the fact that the diamagnetic shielding described by Meissner took place over a non-zero distance, λ, and gave a formula for it,

$$\lambda^2 = \frac{mc^2}{4\pi n_s e^2}, \tag{17}$$

where m is the electron mass, e its charge, and n_s the number density of superconducting electrons. At absolute zero, n_s is expected to be of the order of the number density of electrons.

In 1953, A. B. Pippard [36] discovered that an additional length was needed to describe experiments on microwave electrical conductivity. This length, which he called the coherence length,   , describes the fact that sometimes one must think of the properties of a material in terms of non-local equations. Thus, when the microwave p enetration depth gets less than the electron mean free path, Ohm's Law is no longer accurate at a point in space since the electric field varies so much between electron collisions. He argued that a similar thing should apply in superconductors, and estimated the length by an uncertainty principle argument as

$$= \frac{a\ v_F}{k_B\, T_c}\, ,$$ (18)

where a is a constant of order unity, v_F the velocity of electrons at the Fermi energy, and T_c the superconducting transition temperature.

A similar concept had arisen earlier in the theory of V. L. Ginzburg and L. D. Landau [37]. One may think of   as describing a sti ness of the supercon-ducting wave function: if one subjects a superconductor to a spatially varying disturbance, it is very costly in energy for the superconducting wave function to respond over lengths shorter than   . In the conventional superconductors such as Al, the penetration depth is much shorter than the coherence length:

$$-\quad 1\,.$$ (19)

In 1957, A. A. Abrikosov published a paper [38] in which he considered what would happen to the Ginzburg-Landau theory if instead

$$-\quad 1\,.$$ (20)

The result is quite remarkable. Consider an interface between normal and superconducting regions of a metal. In the presence of a magnetic field, (19) corresponds to a positive surface energy (a cost in energy to have such an inter-face), whereas (20) corresponds to a negative surface energy, hence an energy gain by having such an interface. Abrikosov called those materials obeying (19) Type I superconductors, those obeying (20) Type II superconductors.

If an external magnetic field is applied to a Type I superconductor, the material excludes the magnetic field until it reaches a critical value H_c above which the material becomes a normal metal and the magnetic field can pene-trate. If one applies a magnetic field to Type II superconductor, the magnetic field is excluded until a field H_{c1} is reached. Then the magnetic field begins to penetrate, but not uniformly. Rather it consists of flux in tubes, the core of which is much like a normal metal. These tubes, called fluxoids, increase in spatial density as the applied field increases until a field H_{c2} is reached at which the fluxoid separations are of the order of the coherence length. Above that field, the material is no longer superconducting.

For these theoretical contributions, Ginzburg and Abrikosov were awarded the Nobel Prize in 2003. Landau had been awarded the prize in 1962. From the point of view of magnetic resonance, the discovery of these two types of superconductors has great importance since the field penetration in a Type II superconductor makes possible direct observation of NMR in a superconductor in the presence of a magnetic field.

From (17) and (18), one expects that Type II superconductors arise from materials with high transition temperatures and low carrier densities. The alkali fullerides, the cuprates, and many organic superconductors are Type II superconductors.

5 The Alkali Fullerides

5.1 Introduction

The discovery of the C_{60} molecule opened many new avenues of science. ^{13}C NMR made possible interesting studies of solid C_{60} [39]. The addition of alkali atoms, A, to form materials such as $A_x C_{60}$, where x ranges from 0 to 3, added another dimension. The $A_3 C_{60}$ materials are superconductors with remarkably high transition temperatures. For example, some transition temperatures for various A_3's are 19.5 K for K_3, 29.5 K for Rb_3, and 31 K for Rb_2Cs.

All of these materials are ideal for NMR study since all the alkalis give strong NMR signals, and there are strong ^{13}C signals from natural abundance [13]C. Studies have been made for a variety of A_3's including Rb_3, Rb_2K, Rb_2Cs, RbK_2, $RbCs_2$, and K_3 [39, 40].

5.2 NMR in the Normal State

The NMR properties of the alkali fullerides in the normal state are rather similar to conventional metals. Figure 7 shows $log(1/T_1T)$ vs. T for [87]Rb, ^{13}C, and [133]Cs nuclei for Rb_2CsC_{60} from Stenger et al. [40]. The curves at all three sites are uniformly displaced from one another on the log scale, showing that the ratio of the T_1's is independent of temperature. There is, however, a marked deviation from the Heitler-Teller law, $1/T_1T$ = constant. In fact, Stenger reports that there is an 80% increase from 30 K to 292 K in Rb_2CsC_{60}. For K_3C_{60} the increase is 45% and for Cs_2RbC_{60} it is 140%. Tycko et al. observed a similar phenomenon in Rb_3C_{60} [41]. Figure 8a shows their data. The normal state plots of $1/T_1T$ vs T show that T_1T is nearly independent of T in the normal state, but with a slight upward drift as the temperature rises.

5.3 T_1 in the Superconducting State

Tycko et al. [41] studied [13]C T_1 in K_3C_{60} and Rb_3C_{60} in both the normal and superconducting states. Figures 8 and 9 show their data below T_c. The plot of $log T_1$ vs $1/T$ is a straight line, obeying the equation for the BCS (L = 0, S = 0) pairing

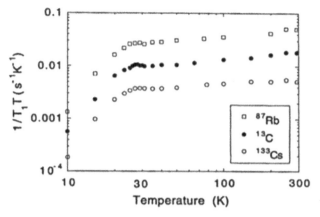

Fig. 7.   Spin-lattice relaxation ($1/T_1T$) versus temperature for ^{87}Rb, ^{13}C, and ^{133}Cs in Rb$_2$CsC$_{60}$ at an applied field of 8.8 T. Uncertainty is estimated as ± 5%. Reproduced with permission from [40]. Copyright 1995 by the American Physical Society

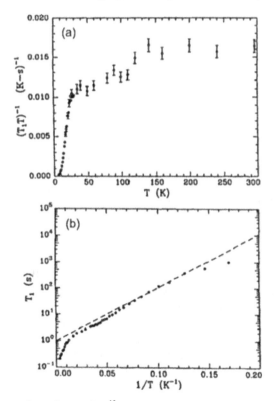

Fig. 8.   Temperature dependence of ^{13}C NMR spin-lattice relaxation in Rb$_3$C$_{60}$: (a) plotted as $(T_1T)^{-1}$ vs. T; (b) plotted as log T_1 vs. T^{-1}. The dashed line is a fit to the BCS ($L = 0, S = 0$) equation from 8 to 12 K. Reproduced with permission from [41]. Copyright 1991 by the American Physical Society

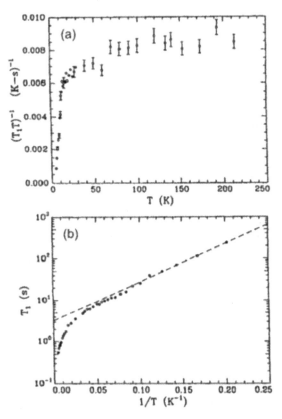

Fig. 9.   Temperature dependence of ^{13}C NMR spin-lattice relaxation in K$_3$C$_{60}$: (a) plotted as $(T_1T)^{-1}$ vs. T; (b) plotted as log T_1 vs. T^{-1}. The dashed line is a fit to the BCS ($L = 0, S = 0$) equation below 9 K. Reproduced with permission from [41]. Copyright 1992 by the American Physical Society

$$\frac{1}{T_1} \propto e^{-\Delta/k_BT} . \tag{21}$$

The fit gives Δ as 21.3 K and 46.8 K for K$_3$C$_{60}$ and Rb$_3$C$_{60}$, respectively. These numbers correspond to $2\Delta/k_BT_c^{NMR}$ of 3.0 and 4.1 respectively, compared to the BCS weak coupling value of 3.5. Thus, these data support the BCS ($L = 0, S = 0$) state found for electron-phonon coupling mechanism of superconductivity. As we shall see, the same is not true for NMR data in the superconducting states of the cuprates or the organics.

There is, however, one difference from the data on conventional superconductors. There is no sign of the coherence peak in $1/T_1$. It is known from the studies of ^{51}V$_3$Sn by Y. Masuda and N. Okuba [42, 43] that a static magnetic field may suppress the coherence peak. Accordingly, Stenger et. al. studied the magnetic field dependence of the ^{13}C T_1 in Rb$_2$CsC$_{60}$. Their results are shown in Fig. 10, which displays the temperature dependence of the ratio of

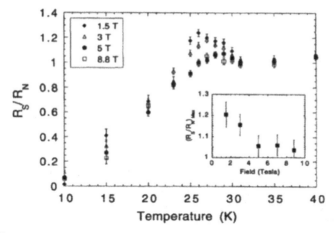

Fig. 10. ^{13} C R_s/R_n vs. T of Rb_2CsC_{60} for several magnetic fields. Inset: Maximum of R_s/R_n vs. applied field. Reproduced with permission from [40]. Copyright 1995 by the American Physical Society

the relaxation rate in the superconductor, R_s, to the rate in the normal state, R_n. We see that a "coherence peak" is present for su ciently low magnetic fields.

5.4 The Knight Shift

Although this article is largely focused on the role of spin-lattice relaxation, useful information can also be obtained from measurements of the Knight shift. We turn to that subject here.

Electron spins in an $S = 0$ state do not couple to an applied magnetic field. This law is quite rigorous. A Zeeman term in the spin Hamiltonian has no non-zero matrix elements to other states. Consequently, the Knight shift in a superconductor with spin singlet pairing should vanish at $T = 0$. In fact, the Knight shift did not appear to vanish in early experiments on colloidal Hg [44, 45] and platelets of Sn [46]. The situation turned out to be rather complicated (e ects of impurities, of small particle sizes, and of changes in the chemical shift arising from narrow metal bands) and is beyond the scope of this article to discuss.

K. Yosida [47] worked out the theory of the Knight shift, K_s, for BCS spin-singlet superconductors. He found that the ratio of K_s to the value K_n in the normal state is

$$\frac{K_s(T)}{K_n(T_c)} = - \frac{\int \,_s(E)\frac{F}{E}dE}{\int \,_n(E)\frac{F}{E}dE}$$

$$= \int \left(\frac{E}{E^2 - \Delta^2} \right)^{1/2} \frac{F}{E} dE , \tag{22}$$

where F is the Fermi function and E is given by (16).

Measurement of the Knight shift involves several complicating factors. They can be seen by studying (23) which gives the expression for the resonance angular frequency ω in an magnetic field B resulting from an applied field B_0:

$$\omega = \omega_n (1 + K^S + K^L) B , \tag{23}$$

where K^S is the Knight shift and K^L is the chemical shift. The former arises from polarizing electron spins, the latter from polarizing electron orbital magnetism. B is the magnetic induction in the sample. In most magnetic resonance $B = B_0$. However, the Meissner effect makes B different from B_0. We write

$$B = B_0 + \Delta B . \tag{24}$$

To a good approximation for Type II superconductors, $\Delta B / B_0$ is small, enabling us to write

$$\omega = \omega_n \left[1 + K^S + K^L + \frac{\Delta B}{B_0} \right] B_0 . \tag{25}$$

Equation (25) is equivalent to saying that there is a shift tensor, K, given by

$$K = K^S + K^L + \frac{\Delta B}{B_0} . \tag{26}$$

Suppose one has more than one nuclear species in the sample. We label the species by a number such as 13 for C and 87 for Rb. Then

$$^{13}K = {}^{13}K^S + {}^{13}K^L + \frac{\Delta B}{B_0} \tag{27}$$

and

$$^{87}K = {}^{87}K^S + {}^{87}K^L + \frac{\Delta B}{B_0} . \tag{28}$$

If we write the electron spin-nuclear spin Hamiltonian as

$$H = \omega_n \hbar_0 I_z + \omega_e \hbar_0 S_z - \omega_n I_z A \omega_e S_z , \tag{29}$$

where we have made explicit the dependence of the hyperfine coupling on the nuclear and electron magnetic moments, then the Knight shift is

$$K_\alpha (T) = \mathcal{A}_\alpha \chi_s (T) , \tag{30}$$

where $\chi_s (T)$ is the electron spin susceptibility and α denotes the nuclear species. The $\mathcal{A}$'s vary with species because they depend on the local wave function but are independent of T.

We can get rid of the unknown quantity $\Delta B/B_0$ by subtracting (27) from (28). This is the procedure used by Stenger et al. [48]. Thus, they get

$$^{87}K(T) - {}^{13}K(T) = {}^{87}A - {}^{13}A \quad \chi_s(T) + \text{constant} \quad . \tag{31}$$

For a BCS superconductor with ($L = 0, S = 0$), $\chi_s(T)$ vanishes at $T = 0$. Unfortunately we do not have an independent method of determining the chemical shifts in these materials, hence we can not prove from the shift measurements that $\chi_s(0)$ vanishes. However, if we assume it does, we can use (31) to test the temperature dependence.

Stenger et al. used (31) to test the relationship between the energy gap and T_c. Their data are shown in Fig. 11. They find a good fit for $2\Delta_0 = 3.52 k_B T_c$, the weak-coupling BCS value.

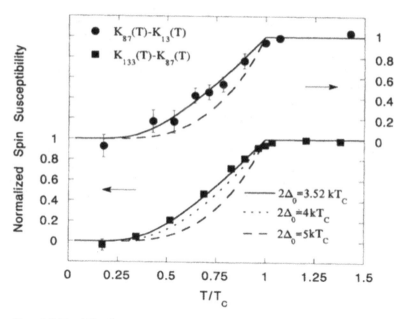

Fig. 11.   NMR shift differences ($K_{87} - K_{13}$) for Rb_3C_{60} (filled circles) and ($K_{133} - K_{87}$) for Rb_2CsC_{60} (filled squares) versus T/T_c. Also shown are BCS spin susceptibility predictions for $2\Delta/k_B T_c = 3.52$ (solid line), 4 (dotted line), and 5 (dashed line). Reproduced with permission from [48]. Copyright 1993 by the American Physical Society

5.5 Conclusions for the Alkali Fullerides

In conclusion, the exponential behavior of T_1 at low temperatures, the existence of a peak in the relaxation rate just below T_c, the temperature dependence of the Knight shift, the fact that the temperature dependence of both

T_1 and Knight shift fit the BCS gap- T_c relationship, all lead to the conclusion that these materials are conventional ($L = 0, S = 0$) BCS superconductors.

6 The Cuprate Superconductors

6.1 Background

The discovery of superconductivity in lanthanum barium copper oxide at a temperature close to 40 K by J. D. Bednorz and K. A. M" uller in 1986 [49] opened a vast array of exciting new physics. Shortly thereafter, they found similar results in lanthanum strontium copper oxide.

La_2CuO_4 is an antiferromagnetic insulator. If one thinks of it as made up of La^{3+}, Cu^{2+}, and O^{2-} ions, one can view this formula as saying that the La atoms give up 3 electrons and the Cu atoms give up 2 electrons in order for the O atoms to form closed p shells, becoming O^{2-} ions. Since the ionic form of Sr is Sr^{2+}, substitution of a Sr atom for a La means that there is one less electron available to contribute to the O atoms. Thus every Sr substituted for a La e ectively adds a hole to the closed shells of the O^{2-} ions. One speaks of the Sr as producing hole doping.

The crystal structure of this material, often abbreviated as LSCO, is shown in Fig. 12.

For low doping, the material remains an insulator, but the antiferromagnetic transition temperature drops with doping. Writing the formula as $La_{2-x}Sr_xCuO_4$, one can make a phase diagram, shown in Fig. 13. Above x   0.05, the material is no longer an antiferromagnet but becomes both a conductor and a superconductor.

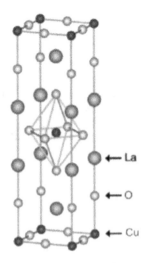

Fig. 12.   Crystal structure of La $_2CuO_4$

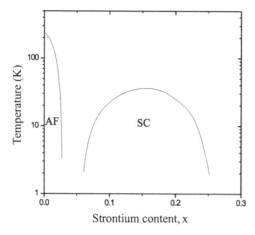

Fig. 13. Phase diagram of La$_{2-x}$Sr$_x$CuO$_4$

It is customary to say that La$_2$CuO$_4$ is the parent compound of this family of materials.

Not long after the discovery of the systems based on La$_2$CuO$_4$, a material with superconducting transition temperature near 94 K, YBa$_2$Cu$_3$O$_7$, was discovered by C.W. Chu, M.K. Wu and collaborators [50]. Their material has a parent compound, YBa$_2$Cu$_3$O$_6$, which is also an antiferromagnetic insulator. If one writes the formula as YBa$_2$Cu$_3$O$_{6+x}$, the phase diagram is that of Fig. 14 [51]. The crystal structure is shown in Fig. 15 [52].

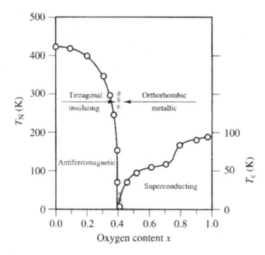

Fig. 14. Phase diagram of YBa$_2$Cu$_3$O$_{6+x}$. Reproduced with permission from [51]. Copyright 1991 by Elsevier

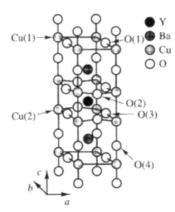

Fig. 15. Crystal structure of YBa $_2$Cu$_3$O$_{7-y}$ for y = 0. Reproduced with permission from [52]. Copyright 1989 by the American Physical Society

According to band theory, both parent compounds should be conductors. It is thought that they are instead Mott insulators, materials in which electron-electron correlations interfere with electron flow because they make it very unfavorable to have two charge carriers simultaneously on the same atom.

A Cu^{2+} ion has one hole in the d-shell, resulting in its having a net electron spin of 1/2. It might be natural to think of the doped materials as having a conduction band formed from holes on the O atoms with Cu spin-1/2 ions imbedded. The system might be much like Cu metal with magnetic impurities, the system which gives rise to the famous Kondo effect. But this is not a correct description and NMR gives some of the best proof. The situation of an isolated hole has been analyzed by Zhang and Rice [53]. They conclude that "Cu-O hybridization strongly binds a hole on each square of O atoms to the central Cu $^{2+}$ ion to form a local singlet. This moves through the lattice in a similar way as a hole in the single band effective Hamiltonian of the strongly interacting Hubbard model." This viewpoint explains why one observes in NMR that both the Cu and the O nuclei experience the transition to superconductivity at the same temperature and are not separable into two independent systems.

From the earliest experiments it became apparent that the cuprate conductors are not ordinary metals. In the normal state of a normal metal, the spin (Knight) shift is independent of temperature, 1$/T_1$ is proportional to temperature, and T_2 is independent of temperature. None of these statements holds true in the electrically conducting cuprates.

For example, we show data from N. J. Curro et al. for the 81 K superconductor YBa $_2$Cu$_4$O$_8$ [54]. Figure 16 shows the normal state Knight shift of the planar ^{63}Cu nuclei. We see it is strongly temperature dependent, reaching a maximum Knight shift at about 500 K. Figure 17 shows 1$/T_1$ versus T and T_1T versus T. At high temperatures T_1 becomes independent of T. In fact,

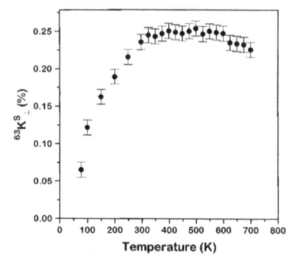

Fig. 16.   The measured ^{63} Cu(2) Knight shift in the perpendicular direction as a function of temperature when the static magnetic field is perpendicular to the crystal c axis. A temperature-independent orbital contribution has been subtracted o . Reproduced with permission from [54]. Copyright 1997 by the American Physical Society

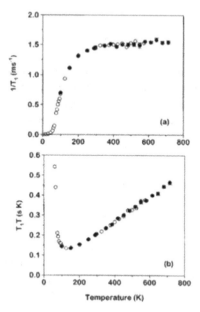

Fig. 17.   (a) ^{63} Cu(2) spin-lattice relaxation rate 1   /T $_1$ as a function of tempera- ture. (b) T$_1$T as a function of temperature. The   open circles are the data [55] of Corey et al. Reproduced with permission from [54] Copyright 1997 by the American Physical Society

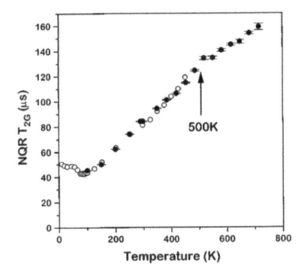

Fig. 18. ^{63} Cu(2) NQR T_{2G} as a function of temperature. The   open circles   are the data of Corey et al. [55]. Reproduced with permission from [54]. Copyright 1997 by the American Physical Society

the plot of $T_1 T$ shows that above about 130 K it is linear in   T. Figure 18 shows the quantity T_{2G}, a measure of the transverse relaxation time, versus T, again showing a strong   T dependence, in contrast to the usual situation in a metal.

The basic problem in understanding these data is that one does not have a theory of the cuprates even for the normal state. The situation is analogous to that which prevailed for the superconducting state of conventional supercon-ductors prior to the BCS theory. At that time, one had quite a good theory of the normal state of metals. Using it one could understand the NMR shifts and relaxation times, things such as the Korringa relation and how it was modified by electron-electron interactions. However, one could not understand exper-imental results in the superconducting state since there was no microscopic theory of superconductivity. It was only after one had the BCS theory that one could explain results such as the di erent temperature dependence of the NMR T_1 and the ultrasonic absorption rate.

For the cuprates, the goal of NMR studies is to help in the development of a theory of the normal state as well as the superconducting state. We give some indication below of what progress has been made. As we have remarked, using NMR we are certainly able to say that these are not conventional metals.

6.2 The Spin Hamiltonian

The cuprates are rich in nuclear moments: ^{63}Cu, ^{65}Cu, ^{89}Y, ^{137}Ba, ^{139}La, and if one does isotopic substitution, ^{17}O. Thus NMR can probe throughout the unit cells of these materials.

In general, the shifts and the magnetic susceptibility vary with doping (i.e. with x in the chemical formulas above) and temperature. Alloul et al. [56] studied the ^{89}Y shifts in YBCO, to use a shorthand notation that covers all dopings, and found that the shifts were a linear function of the magnetic susceptibility as he varied it as a function of temperature and of O dopings. From this he concluded that the ^{89}Y shifts were determined by the Cu electron spin magnetization. Subsequently Takigawa et al. [57] studied the ^{17}O and ^{63}Cu shifts for nuclei in the CuO planes of YBa$_2$Cu$_3$O$_{6.63}$ and found out that not only were both those shifts linear functions of the spin magnetic susceptibility, but in fact were all proportional to one another since at low temperatures in the superconducting state they went to values given by the chemical shifts only. Since one expects that the spin susceptibility arises largely from the spin magnetism of the Cu atoms, the conclusion is that there must be transferred hyperfine coupling between the Cu electron spins and the ^{17}O and ^{89}Y nuclei. It is also found that the isotropic component of the ^{63}Cu spin shift relative to its value at $T = 0$ is positive, showing that it is not due to core polarization, which is always negative. Mila and Rice [58] showed that this could be explained by postulating there was transferred hyperfine coupling from one Cu atom to its four nearest neighbors. They showed theoretically that this mechanism could account for the size of the observations.

These results can be summarized in a spin Hamiltonian for the Cu nuclei:

$$^{63}H = _{63}{}^{63}\hat{I} \cdot 1 + {}^{63}K^L \cdot H_0 + {}^{63}\hat{I} \cdot A \cdot S_0 + B \cdot \sum_n S_n + {}^{63}H_Q , \quad (32)$$

where H_0 is the applied magnetic field, $^{63}K^L$ is the orbital (chemical) shift tensor, A the on-site hyperfine tensor, B the transferred hyperfine tensor, S_0 the on-site electron spin, S_n the electron spin of the four nearest neighbor Cu atoms, and $^{63}H_Q$ the electric quadrupole coupling to the on-site Cu nuclear spin.

The Hamiltonian for a ^{17}O nucleus is

$$^{17}H = _{17}{}^{17}\hat{I} \cdot 1 + {}^{17}K^L \cdot H_0 + {}^{17}\hat{I} \cdot C \cdot \sum_n S_n , \quad (33)$$

where $^{17}K^L$ is the oxygen chemical shift, C the transferred hyperfine coupling constant, and S_n the spin of the two nearest neighbor Cu spins.

Note that neither (32) nor (33) makes mention of an electron spin on the O atom. It is assumed, in the spirit of the Zhang-Rice singlet, that the spins are so tightly coupled that all e ects are encompassed in terms of the Cu spins and the hyperfine and transferred hyperfine coupling coe cients.

In calculating the electric quadrupole coupling it is useful to consider small clusters of atoms. Then if the   z axis is perpendicular to the Cu–O planes, and the x and y axes are oriented along Cu– O bonds, to a first approximation Cu^{2+} has a single hole in the $x^2 - y^2$ orbital. The holes are then in the highest energy orbital (since the electrons will be in the lowest energy arrangement), that is, Cu–O antibonding   orbitals.

The presence of the Cu electron spins means that it is useful to think of a spatial Fourier spectrum of electron spins. Thus at a given spatial position, r, we write

$$S(r) = \sum_q S_q e^{iq \cdot r} . \tag{34}$$

The ground state in the absence of an applied static magnetic field is unmagnetized. Application of a static field will cause magnetization. There will also be thermal excitation of magnetization. Using (34) we may think of the wavelength,   , of the excitations or the   q vector. The two are related by

$$q = \frac{2}{\lambda} . \tag{35}$$

It is found experimentally that the T_1's of the O and the Cu nuclei have very different temperature dependences. Shastry [59] and Hammel et al. [60] proposed that the relaxation arose from spin fluctuations, and that the different species responded to different wavelengths. Thus, for long wavelengths (q   0), both species respond, but for fluctuations close to the antiferromagnetic wavelength (q_x, q_y) = (/a, /a), where a is the lattice constant, the Cu responds, but the O does not since the spins fluctuation on its two neighboring Cu neighbors are opposite to each other and thus, by (34), cancel each other.

This point of view can be expressed formally quite elegantly using a formula discovered by Moriya [61] for nuclei that are relaxed by coupling to electron spins,

$$\frac{1}{T_1} = \frac{k_B T}{2\mu_B^2} \sum_q \sum_{=} |A_ (q)|^2 \quad (q, \omega_n)/ \omega_n , \tag{36}$$

where   denotes the direction of the applied static magnetic field,   the two perpendicular directions, $A_ (q)$ the   component of the hyperfine coupling tensor of wave vector q, and (q, ω_n) the imaginary part of the electron spin susceptibility at wave vector   q and angular frequency ω_n, the nuclear Larmor frequency.

Equation (36) has a simple interpretation. A nucleus precessing at angular frequency ω_n exerts an alternating magnetic field on the electron spin system. The strength of the alternating field is given by the transverse hyperfine coupling $A_ $. The electron spin system absorbs energy at a rate given by (q, ω_n).

The quantities $|A_ (q)|^2$ can be obtained from the spin Hamiltonians. For example,

$$^{63}A_x(q) = {}^{63}A_y(q) = A_{xx} + 2B\,[\cos(q_x a) + \cos(q_y a)]$$
$$^{63}A_z(q) = A_{zz} + 2B\,[\cos(q_x a) + \cos(q_y a)], \tag{37}$$

where $A_{xx} = A_{yy}$ and A_{zz} are components of the on-site Cu hyperfine tensor of (32).

Finding a theory of the nuclear T_1's therefore can be considered a search for theoretical expressions for $(q, {}_n)$. Or phrased alternatively, NMR T_1 data measure a weighted average of the imaginary part of the low frequency electron spin susceptibility.

Pennington and Slichter [62] discovered that the T_2 in the cuprates was dominated by an indirect coupling mechanism through the electron spin system so that measurements of T_2 give information about the real part of the electron spin susceptibility. Thus, the interaction of the z components of nuclear spin of a nucleus at position r_1 with a second at position r_2 had the form of a contribution H_{12} in the nuclear spin Hamiltonian given by

$$H_{12} = \hat{I}_{z1} \sum_{r,r} G(r_1, r)\ (r - r)G(r, r_2)\hat{I}_{z2}, \tag{38}$$

where the G's are form factors such as

$$G(0, r) = A_{zz} + B \sum_{r,r+} \tag{39}$$

and $(r - r)$ is the static magnetism produced at position r by a spatial delta function magnetic field of unit amplitude at position r. $(q, 0)$ is the spatial Fourier transform of $(r - r)$.

They calculated the contribution to the second moment, denoting it as $(1/T_{2G})^2$. A more detailed discussion is in the article by Slichter et al. [63].

Thus we see that NMR can provide data about both the real and the imaginary part of the electron spin susceptibility.

We have already seen NMR data for one of the cuprates, $YBa_2Cu_4O_8$. In Fig. 19 we show T_1 data by Imai et al. [64] for $La_{2-x}Sr_xCuO_4$. The striking feature is that at high temperatures the T_1 curves for both the undoped and the doped samples all converge to the same value and become independent of temperature. These data were analyzed by Chubukov and Sachdev [66] using scaling concepts based on the two-dimensional Heisenberg antiferromagnet. They found indeed that at high temperatures the T_1 should become independent of temperature.

Figure 20 shows $1/T_1T$ from Takigawa et al. [67] for $YBa_2Cu_3O_{7-y}$ with $O_{6.63}$ ($y = 0.37$) and O_7 ($y = 0$). Shown are data for both ^{63}Cu and ^{17}O. From (36) it is evident that the two nuclei must see different regions of q-space. The form factors in fact insulate the ^{17}O nuclei from magnetic fluctuations at the antiferromagnetic wave vector, whereas the Cu nucleus sees them fully. We discuss this topic in greater detail below in Sect. 6.4.

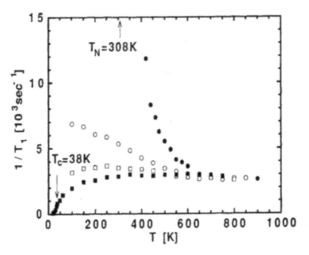

Fig. 19.   Temperature dependence of $1/{}^{63}T_1$ measured by NQR for La$_{2-x}$Sr$_x$CuO$_4$ (•, x = 0; , x = 0.04; , x = 0.075; , x = 0.15). Reproduced with permission from [64]. Copyright 1993 by the American Physical Society

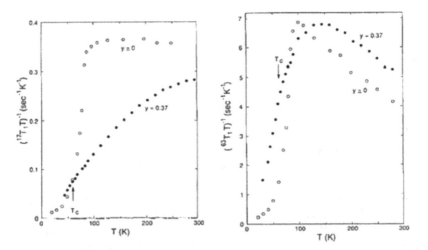

Fig. 20.   Temperature dependence of $1/T_1T$ in YBa$_2$Cu$_3$O$_{7-y}$ with y = 0.37 (open circles) and y = 0 (closed circles). On the left are data for the O(2,3) sites, on the right are data for the Cu(2) site. Reproduced with permission from [67]. Copyright 1991 by the American Physical Society

   Millis, Monien, and Pines (MMP) [68] have proposed an explicit model which has these features. They assume the spin susceptibility is given by the expression

$$(q,) = \frac{2}{1 + (q - Q)^2 \,{}^2 - i\frac{}{{}_{sf}}}, \tag{40}$$

where   is a constant,   a temperature-dependent correlation length, Q the point at which the susceptibility peaks, and $_{sf}$ the inverse of a relaxation time. For Q they pick the antiferromagnetic value ($Q_x, Q_y) = (\pm /a, \pm /a)$. They assume that (40) is accurate near Q, but add a second term, independent of q, to represent the form near the origin.

A series of papers (Sokol and Pines [69], Barzykin and Pines [70], and Zha, Barzykin and Pines [71]) using concepts of scaling as in that of Chubukov and Sachdev led to the realization that at high temperatures $T_1 T$ should be a linear function of T, as is shown in Fig. 17. Figure 21 shows $T_1 T$ versus T for optimally doped LSCO from the paper by Zha et al. [71] showing the linear character at high temperatures.

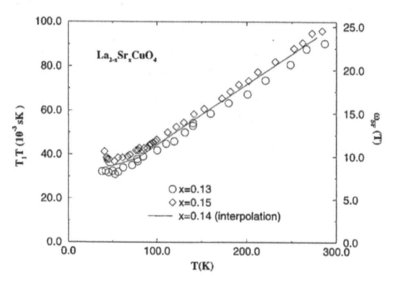

Fig. 21.   The interpolated $^{63}T_{1c}$ for La $_{1.86}$ Sr $_{0.14}$ CuO $_4$ is shown together with the measured values of $^{63}T_{1c}T$ for La $_{1.87}$ Sr $_{0.13}$ CuO $_4$ and La $_{1.85}$ Sr $_{0.15}$ CuO $_4$ of Ohsugi et al. [72]. Shown on the righthand side is the scale for $_{sf}(T)$ $^{63}T_{1c}T$ for La $_{1.86}$ Sr $_{0.14}$ CuO $_4$ inferred from the fit to the neutron-scattering experiments. Reproduced with permission from [71]. Copyright 1996 by the American Physical Society

The upturn in $T_1 T$ at low temperatures is a signature of a phenomenon called the spin gap or spin pseudogap. It is reminiscent of the situation one would have for a nucleus relaxed by a pair of coupled electron spins with a ground state singlet and excited state triplet. The ground state electron spin singlet does not couple to the nuclear spin, hence does not contribute to nuclear relaxation. At high temperatures, the spin-triplet state becomes occupied. Since it is magnetic, it could produce relaxation of the nucleus. But as the temperature drops, the occupancy of the triplet state drops, freezing out the relaxation when $k_B T$ becomes comparable to the singlet-triplet energy splitting.

6.3 The Superconducting State

Relaxation time measurements for YBCO in the superconducting state were first observed by Imai et al. [73] and by Kitaoka et al. [74]. Figure 22 shows data from Imai et al. plotted on a log-log scale. The low temperature behavior seems to fit the power law

$$\frac{1}{T_1} \quad T^3 . \tag{41}$$

Figure 23 displays low field ^{63}Cu and ^{17}O data of Martindale et al. [75] for $YBa_2Cu_3O_7$ showing again the temperature dependence of (41).

Figure 24 shows the data of Martindale et al. in a plot of $\log(\quad ^{63}T_{1c})$ vs. $1/T$, both axes normalized to their values at T_c for $YBa_2Cu_3O_{7-y}$ with y close to 1. The conventional BCS form in such a plot, once one is below the coherence peak, is a straight line with a slope proportional to the energy gap, . The straight line shows the BCS slope using (9) and (10). We note the data have no coherence peak and clearly do not fit an exponential in 1 $/T$.

For an $L = 0$ orbital state, is uniform in k space and grows in size as one goes down in T, levelling o at low T. This clearly does not fit the data.

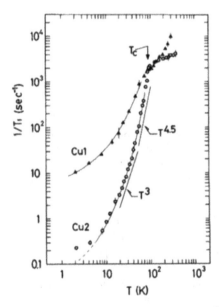

Fig. 22.   Temperature dependence of the 1 $/T_1$ obtained for ^{63}Cu plotted in a log-log scale. Reproduced with permission from [73]. Copyright 1988 by the Physical Society of Japan

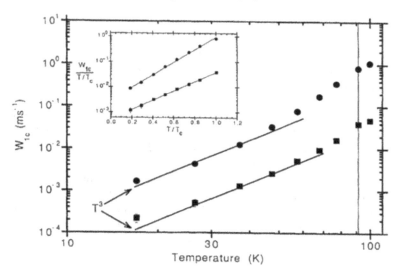

Fig. 23. W_{1c} from sample 2 measured in a 0 .67 T field vs. temperature on a log-log scale. The circles are ^{63} Cu and the squares are ^{17} O. The vertical line is at T_c = 91 .2 K for this field strength and orientation. The lines through the data show the power-law behavior of (41). Reproduced with permission from [75]. Copyright 1993 by the American Physical Society

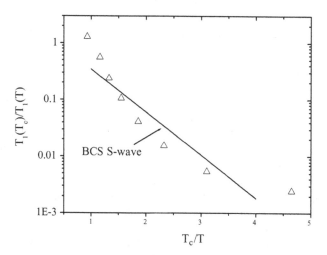

Fig. 24. Data of Martindale et al. [75] for YBa $_2$Cu$_3$O$_{7-y}$ (y 1), contrasting with the straight line appropriate for the BCS (L = 0 , S = 0) state prediction at low temperatures

The slope of the data appears to be continuously changing with temperatures, getting smaller at lower temperatures. As one goes to lower T, the gap appears to be shrinking. Such a situation might arise if the gap function were not uniform in k space. At higher T there are excitations across all parts of the gap, but at low T excitations occur only where the gap is small. Thus the data suggest a non-uniform gap.

Since the Knight shift measurements by Barrett et al. [76] and by Takigawa et al. [67] showed that the electron spins were in the $S = 0$ (spin-singlet) state below T_c, we expect the orbital pairing to be one of $L = 0, 2, 4$, etc. Having ruled out $L = 0$, we turn to $L = 2$.

The fact that the cuprate parent compounds are antiferromagnets suggested that possibly the superconducting mechanism involved electron spins. Several theorists proposed such models [77, 78, 79]. They found that the pairing would be $L = 2$ (d-wave). The gap function would have nodes along the lines $|k_x| = |k_y|$ corresponding to the form

$$(k) \quad k_x^2 - k_y^2 . \tag{42}$$

Moreover, at low temperatures thermal excitations exist only near the nodes and are expected to lead to a T^3 rate for relaxation processes, consistent with the data of Fig. 23.

Thus the NMR data rule out $L = 0$ pairing (unless for some reason crystal asymmetry produces an enormous asymmetry in the $L = 0$ gap function) and strongly suggest d-wave pairing. Historically, NMR gave some of the earliest evidence for d-state pairing, the state now generally accepted as correct.

6.4 T_1 Anisotropy

In the normal state of the cuprates, the T_1 is anisotropic as a result of the anisotropy of the hyperfine coupling ($A_{xx} = A_{yy} = A_{zz}$). However, although the T_1's are temperature dependent the anisotropy is independent of temperature. The temperature dependence comes from the T dependence of . Figure 25 shows data by Barrett et al. [80]. The inset gives the ratio of the rate $1/T_1$ for a static field parallel to that perpendicular to the Cu–O planes.

In the superconducting state they found that the ratio became T-dependent. Martindale et al. [75] explored this further. They discovered that the relaxation rate varied with the strength of the applied static field, and attributed this variation to the presence of the fluxoids. They reasoned that the cores of the fluxoids, which might be expected to be much like the normal material, might be a source of spin-lattice relaxation. The lower the applied field, the lower the fluxoid density, hence the less the fluxoid cores should contribute to relaxation.

The crystal c axis is defined as the normal to the Cu–O planes. We can think of T_{1c}, the relaxation time when the static field is along the c axis, as arising from fluctuating magnetic fields in the a and b directions. Likewise, we

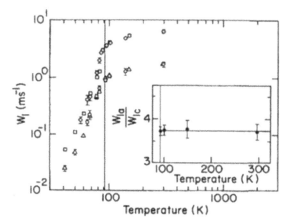

Fig. 25.   The ^{63}Cu(2) spin-lattice relaxation rate $^{63}W_{1a}$ vs temperature for H_0 a
(, sample 1;   , sample 4) and for H_0 c (, sample 1;   , sample 4). The vertical
solid line is at 92 K. Inset: The normal-state ratio W_{1a}/W_{1c} vs. temperature for
sample 4 (•). The horizontal line is at $W_{1a}/W_{1c} = 3.73$, and the vertical line is
at 92 K. Reproduced with permission from [80]. Copyright 1991 by the American
Physical Society

can think of the T_{1a} as arising from fluctuating fields in the   b and c directions.
The c direction is also the axis of quantization for the ^{63}Cu nuclei when the
applied field is zero, as in an NQR experiment. Thus, if one has an applied
field along the   c direction, and then lowers it towards zero, the resulting T_{1c}
should go over smoothly to the NQR value, measuring the effects of the   x and
y components of fluctuating field. There is no way that one can measure the
z components of fluctuating fields from T_1 measurements in zero field. But a
weak applied field transverse to the   c axis makes this possible. Accordingly,
Martindale et al. [75] extended the measurements of Barrett et al. [80] to
low fields as did Takigawa, Smith, and Hults [57]. Their results are shown in
Fig. 26, along with theoretical results from Bulut and Scalapino [81] that we
discuss below.

As we have remarked, the anisotropy of the ^{63}Cu T_1 can be analyzed
in terms of hyperfine coupling terms of (37). These terms depend on the
components of q. It is convenient to define the quantities   F below which are
the square of the terms in (37):

$$^{63}F = {}^{63}A_{xx}(q)^2 = {}^{63}A_{yy}(q)^2 = \frac{1}{2}\left[{}^{63}A_{xx}(q)^2 + {}^{63}A_{yy}(q)^2\right],$$

$$^{63}F = \frac{1}{2}\left[{}^{63}A_{xx}(q)^2 + {}^{63}A_{zz}(q)^2\right],$$

$$^{17}F = 2\left[C^2\left[1 + \frac{1}{2}(\cos(q_x a) + \cos(q_y a))\right]\right]. \tag{43}$$

It is convenient for some calculations to define a quantity $^{63}F^e$ as

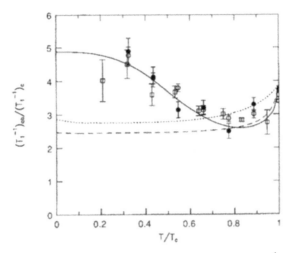

Fig. 26. Temperature dependence of the Cu(2) anisotropy ($T_1^{-1})_{ab} / (T_1^{-1})_c$. The experimental data are from Martindale et al. [82] (squares) and Takigawa, Smith and Hults [57] (solid and open circles). The curves are the theoretical results of Bulut and Scalapino [81]. Reproduced with permission from [81]. Copyright 1992 by the American Physical Society

$$^{63}F^e \;=\; {}^{63}A_{zz}(q)^2 \;. \tag{44}$$

We note that

$$^{63}F^e \;=\; 2\,{}^{63}F \;-\; {}^{63}F \;. \tag{45}$$

^{63}F comes into the calculation of $^{63}T_{1c}$, and ^{63}F comes into the calculation of $^{63}T_{1b} = {}^{63}T_{1a}$. These functions are represented in   q-space in Fig. 27 taken from Thelen, Pines, and Lu [83]. The figure shows the variation along the line from $(0,0)$ to $(/a, 0)$, then along the line from $(/a, 0)$ to $(/a, /a)$, and finally from $(/a, /a)$ back to $(0,0)$. The first Brillouin zone is shown in Fig. 28 from the same paper.

We see that the form factors   F are very large at some parts of the Brillouin zone, very small at others. If   B were zero, the Cu   F's would all have the same shape in q-space (independent of   q), but they would still di er from the oxygen F. The important point is that there are three NMR measurements of T_1 ($^{63}T_{1c}$, $^{63}T_{1a}$, $^{17}T_1$), each of which probes a di erent part of   q-space. Thus NMR not only probes real space, being able to look at the various atomic sites (Cu sites in the planes or chains, O sites in either the planes or chains, Y sites, La sites etc), but also it probes   q-space. This second capability is a result of the existence of transferred hyperfine coupling in addition to on-site hyperfine coupling.

The MMP expression (40) peaks at   q $=$ Q. For YBa$_2$Cu$_3$O$_7$ this is the antiferromagnetic wave vector, $(/a, /a)$. At low temperatures, all excitations are near to the Fermi surface. From Fig. 28 we see how the MMP

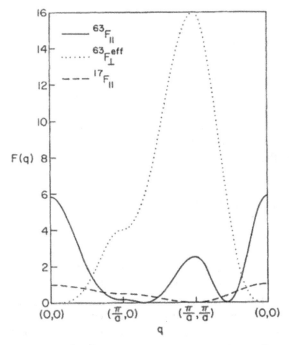

Fig. 27.   Form factors as a function of momentum for oxygen ^{17}F in units of C^2 and for copper sites $^{63}F^e$ and ^{63}F in units of $4 B^2$. Reproduced with permission from [83]. Copyright 1993 by the American Physical Society

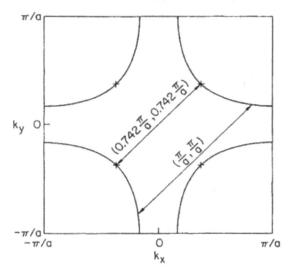

Fig. 28.   Fermi surface for $t = -0.45t$. The + symbols indicate nodes on the Fermi surface for a $d_{s^2-y^2}$ superconductor, connected by momenta $(0   .742 /a,   0.742 /a)$. Reproduced with permission from [83]. Copyright 1993 by the American Physical Society

Q fits the Fermi surface. At low temperatures, the gap nodes are the only places where excitations can occur. These are evidently excited by $q = 0$ or $q = 0.742(\ /a, /a\ )$. We can see from Fig. 27 that this will change the ratios of the various F's and therefore change the anisotropy of the relaxation times. This is in essence the explanation of Bulut and Scalapino and of Thelen, Pines, and Lu for the T-dependent anisotropy of the ^{63}Cu T_1's in the superconducting state. The several theoretical curves show that the anisotropy of T_1 strongly favors d-wave orbital pairing over s-wave, in agreement with the conclusions reached from Fig. 26. Note further that fitting the NMR ratios at low T tells one the location on the Fermi surface of the nodes.

7 The Organic Superconductors

7.1 Introduction

Soon after the discovery of the cuprate superconductors came the discovery in 1988 of $-(\text{BEDT}-\text{TTF})_2\text{Cu(NCS)}_2$ [84], the first in the family of organic superconductors of the form $-(\text{BEDT}-\text{TTF})_2\text{X}$, where BEDT-TTF (further abbreviated ET) represents the organic molecule bis – (ethylenedithiatetrathiafulvalene), X represents one of a number of polymeric anions, and represents the particular packing phase of this family. While superconductivity had first been found in organic compounds in 1980 [85, 86], the $-(\text{ET})_2\text{X}$ family of superconductors distinguished itself by presenting the highest T_c of any organic compound at that date and by demonstrating similarities to the cuprates. Among the characteristics that the $-(\text{ET})_2\text{X}$ family shares with the cuprates are the following

- Highly anisotropic properties, quasi two-dimensional in nature.
- A similar phase diagram if one equates doping in the cuprates with pressure in the $-(\text{ET})_2\text{X}$ family [87]. In particular, antiferromagnetic and superconducting phases are found to be adjacent in the phase diagram, as seen in Fig. 29.
- Unconventional superconductivity, with d-wave being the most likely pairing state.

NMR has played a major role in developing an understanding of the physics of the $-(\text{ET})_2\text{X}$ systems and demonstrating the similarities to the cuprates. After a brief discussion of the structure of the $-(\text{ET})_2\text{X}$ family, we will discuss the contributions of NMR in the normal, superconducting, and antiferromagnetic states of the family.

7.2 Structure

The $-(\text{ET})_2\text{X}$ compounds are composed of conducting layers of ET molecules alternating with insulating layers of polymers, as shown in Fig. 30. We

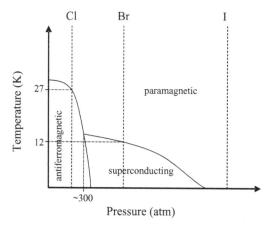

Fig. 29.   Low-temperature phase diagram of the κ−(BEDT−TTF)$_2$Cu[N(CN)$_2$]X family. The vertical lines indicate the ambient pressure characteristics of the partic-ular compounds

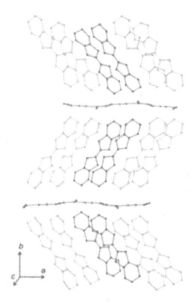

Fig. 30.   Structure of κ−(BEDT−TTF)$_2$Cu[N(CN)$_2$]X. Adjacent dimers have dif-ferent shading. In order to aid viewing, the ^{1}H in the terminal ethylene groups of the molecules have not been shown

will focus on the compounds with the polymer Cu[N(CN)$_2$]Y, where Y = Cl, Br or I, which we will hereafter simply refer to as κ−(ET)$_2$Cl, κ−(ET)$_2$Br and κ−(ET)$_2$I [88, 89, 90, 91]. The unit cell of these compounds is orthorhombic with a and c lying in the plane of the ET layers and   b oriented in the inter-layer direction. The polymers run along   a. Within the conducting layer the

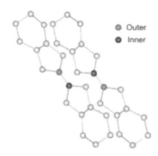

Fig. 31.   An ET dimer with the central   [13] C sites indicated

ET molecules pair up into dimers, as shown in Fig. 31. There are two ET layers in the unit cell with two dimers in each layer, for a total of four dimers per unit cell. The highest occupied molecular orbitals (HOMO's) of the molecule are composed primarily of atomic   p orbitals. In a dimer the HOMO's of the two molecules interfere, producing bonding and antibonding orbitals of the dimer. Each dimer transfers one electron from an antibonding orbital (the higher in energy of the two dimer orbitals) to the polymer layer, resulting in a half-filled conduction band formed of overlapping ET orbitals.

   As seen in the phase diagram of Fig. 29, $-(ET)_2Cl$ is an antiferromagnet at ambient pressure with $T_N = 27\,K$. At pressure of 300 atm this compound becomes a superconductor with $T_C = 13\,K$. $-(ET)_2Cl$ might be said to be analogous to the parent compounds in the cuprates, La $_2CuO_4$ and YBa $_2Cu_3O_6$, in that it is located at the region of lowest pressure (doping) in the phase diagram and is believed to be a Mott insulator. $-(ET)_2Br$ is a superconductor at ambient pressure ($T_C = 12\,K$) and $-(ET)_2I$ is an insulator showing neither magnetism nor superconductivity.

   There are several nuclei in the ET molecule that have been studied by NMR. We will discuss the research involving molecules labelled with   [13] C at the two central carbon sites, as shown in Fig. 31. This particular doping is denoted $-(^{13}C_2-ET)_2X$. As the electronic density is highest at the center of the molecules, the central carbons are ideally located for probing the nature of the electronic interactions. Since the molecules pair up into dimers, the two central carbons on a given molecule exist in slightly different chemical environments, which we refer to as the inner and outer sites, shown in Fig. 31.

7.3 The Normal State

This section concerns the behavior of the materials at temperatures above the superconducting ($-(ET)_2Br$) and antiferromagnetic ($-(ET)_2Cl$) transition temperatures. One of the first tasks of NMR in these systems was to assign the observed resonance frequencies to their corresponding nuclear sites. It was found that for an arbitrary orientation of the applied field, the   [13] C NMR spectrum of $-(^{13}C_2-ET)_2Cl$ will contain 16 resonance lines in a single crystal

sample. A factor of 8 in the number of lines is due to the presence of 8 inequivalent central carbon sites in the compound: a factor 4 for the four inequivalent
dimers in a unit cell and a factor of 2 from the inner/outer site distinction
(Sect. 7.2). There is an additional factor of 2 (bringing the total number of
lines to 16) from the strong nuclear-nuclear dipolar coupling between the two
central ^{13}C nuclei of a given molecule.

From data on the dependence of the room temperature resonance frequencies on the orientation of the external field, De Soto et al. [92] were able to
assign the resonance frequencies to the nuclear sites. An example of the orientation data is shown in Fig. 32. The fit comes from a spin Hamiltonian of
the form

$$H_i = - \quad H_0 \cdot (1 + K_i^L + K_i^S) \cdot \hat{I}_i , \qquad (46)$$

which includes Zeeman (1), orbital (K^L) and spin (K^S) contributions. The
subscript i denotes the particular site of the nucleus, corresponding to one of
the 8 physical sites discussed above. The spin (Knight shift) tensor is found
by subtracting out the orbital contribution, which is known from data on pure
ET molecules [93]. De Soto et al. found that the two Knight shift tensors per
dimer, which must be attributed to the inner and outer sites, had principle
axes that varied slightly from the principle axes of the molecule, one tensor
showing significantly more deviation than the other. Based on the geometry
of the dimer, it is expected that the   p orbitals of the inner sites will have
more overlap with the HOMO of the adjacent molecule than do the orbitals

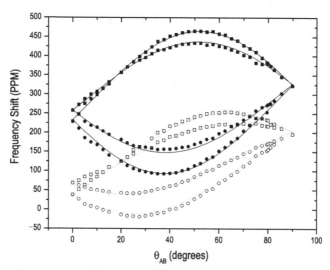

Fig. 32.   An orientation study of the central ^{13}C NMR resonance frequencies (expressed as frequency shift from a reference in parts per million) as function of external field orientation. The data were taken at room temperature with an applied
field of 8.3 T oriented in the   a-b plane. The filled points are from the outer site, the
open points are from the inner site, and the   solid lines are fits using (46)

of the outer sites. Noting that the angle of overlap of the inner site orbitals corresponds to the angle of deviation shown by the tensor that deviated more from the molecular plane, De Soto et al. assigned the tensor with greater deviation to the inner site of the dimer.

The components of the tensors are $K^S_{inner} = (-57, -135, 423) \pm 25$ ppm and $K^L_{outer} = (55, -33, 728) \pm 25$ ppm. Both tensors are dominated by the z component, indicating their p-like nature. Note that the outer site has a larger tensor, suggesting a larger value of the electron spin density at that site. This is somewhat surprising, as it is generally known that the electronic density is larger near the center of the molecules and so one might also expect the electron density to be greatest at the center of the dimer. M.-H. Whangbo and H.-J. Koo have provided an explanation for this result (private communication) by considering the -type HOMO's of the ET molecules in the ET dimer, which has only three electrons to occupy the resulting two orbitals. Their calculations indicate that there is indeed greater charge density at the inner sites for the bonding orbital of the dimer. However, the reverse is true for the antibonding orbital, in which the unpaired electron is located. This explains the higher spin density at the outer site.

The temperature dependence of the Knight shift for $-(^{13}C_2-ET)_2Br$ is shown in Fig. 33(a), along with relaxation data. Above 150 K, T_1T constant while the Knight shift K^S varies, so that the Korringa relation (1) is not followed. The reason for this can be seen in the Korringa factor, shown in Fig. 33(b). As discussed in Sect. 2, this factor is less than one when antiferromagnetic correlations are present, and this is seen to be the case in this temperature range. The presence of the antiferromagnetic fluctuations are seen in T_1T, which decreases with decreasing temperature until 50 K. Below 50 K, T_1T increases and K^S decreases, apparently caused by a freezing out of the spin susceptibility, similar to the pseudogap in the cuprates.

The influence of antiferromagnetic fluctuations on the nuclear relaxation in the normal state can be seen more clearly by comparing the spin-lattice relaxation of $-(ET)_2Br$ to that of $-(ET)_2Cl$, shown in Fig. 34. The spin-lattice relaxation peaks near T_N in $-(ET)_2Cl$, consistent with the slowing down of the antiferromagnetic fluctuations near the magnetic transition. The relaxation in $-(ET)_2Br$ follows that of $-(ET)_2Cl$ down to 50 K, suggesting that antiferromagnetic fluctuations are present in $-(ET)_2Br$. Below 50 K, T_1 drops o in $-(ET)_2Br$, showing the pseudogap-like behavior.

7.4 The Superconducting State

As with the cuprates (Sect. 6.4), the analysis of spin-lattice relaxation in $-(ET)_2Br$ is complicated by the presence of fluxoids. Again, the e ect of fluxoids can be minimized by taking data in smaller fields. Figure 35 shows the superconducting state spin-lattice relaxation data of De Soto et al. As expected, the data in the lower field (0 .6 T) show a lower relaxation rate than the data in the higher field (8 .3 T), a di erence attributable to the extra

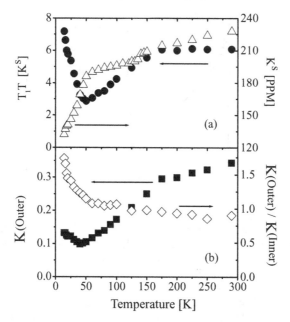

Fig. 33.   (a) The temperature dependence at $H_0 = 8.3\,T$ and H_0 a of: (a) the Korringa product, T_1T (circles) and the Knight shift (triangles), for the outer site (b) the Korringa factor K^{outer} (squares) and the ratio K^{outer} / K^{inner} (diamonds). Reproduced with permission from [92]. Copyright 1995 by the American Physical Society

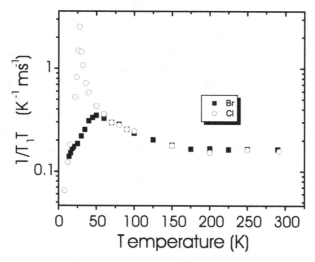

Fig. 34.   The temperature dependence of the ^{13}C NMR spin-lattice relaxation rate in the $-(ET)_2Br$ (squares) and $-(ET)_2Cl$ (circles) compounds

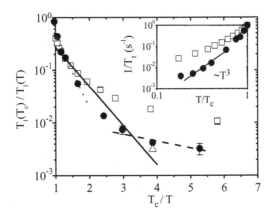

Fig. 35.   The temperature dependence of the spin-lattice relaxation rate in the superconducting state, with H_0 a, and $H_0 = 8.3\,T$ (squares) and $H_0 = 0.6\,T$ (circles). A "zero-field" extrapolation point (triangle) is shown. The   solid lines   indicate fits to the weak-coupling BCS expression (10), T_1^{-1} $\exp(-\ (T)/k_BT)$, treating   as a parameter. Near T_C the dotted line   shows the fit $= 3.0$, and at low T the   dotted line   shows the fit $= 0.3$. The   solid line   shows the BCS result of $(0)/k_BT_C = 1.76$. The inset: the same data shown on a log-log plot, now fit to the d-wave power law, T_1^{-1} T^3. Reproduced with permission from [92]. Copyright 1995 by the American Physical Society

relaxation provided by the fluxoids in the higher field. For this reason, the analysis treats only the data in the 0   .6 T field.

No spin-coherence peak is present in the data of Fig. 35, and the data are clearly not fit by a single exponential, as shown by the di erent values of required to fit the data to the BCS expression (10) at di erent temperatures. Near T_C, $= 3$ is found, while at lower temperatures, $= 0.3$, suggesting an anisotropy of the superconducting gap of at least 3   .0/ 0.3 = 10. The inset shows a reasonable fit of the relaxation data to the T_1^{-1} T^3 power law of (41), consistent with superconductivity mediated by magnetic correlations.

We conclude this section with a direct comparison of the $-(ET)_2Br$ relaxation data to that of the cuprates, provided in Fig. 36. The cuprate data are those of Fig. 24, now combined with the relaxation data of De Soto et al. The similarity between the scaled relaxation rates of the two materials is striking, and it is again demonstrated that neither the cuprates nor the organics have relaxation behavior that is described by the BCS (L = 0 , S = 0) expression (10).

7.5 The Antiferromagnetic State

Just as with the cuprates, theorists have developed theories of superconductivity mediated by antiferromagnetic correlations [94, 95, 96, 97], spurred largely

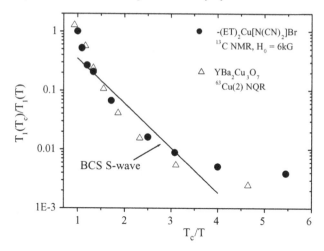

Fig. 36.   A comparison of the temperature dependence of the spin-lattice relaxation rates of κ−(ET)$_2$Br [92] to that of YBa$_2$Cu$_3$O$_7$ [75]. The rates have been scaled to their values at T_C. The solid line indicates the BCS ($L = 0, S = 0$) result. Reproduced with permission from [92]. Copyright 1995 by the American Physical Society

by the proximity of the antiferromagnet state to the superconducting state. Here we review what NMR has revealed about the antiferromagnetic ordering.

The first NMR study of the magnetic state of κ−(ET)$_2$Cl was that of Miyagawa et al. [98]. Their NMR data was taken on the ^{1}H nuclei found in the terminal ethylenes of the ET molecules. The spin-lattice relaxation showed a peak in $(T_1 T)^{-1}$ near 27 K, providing a value for T_N. They also found that the ^{1}H line split into three discrete lines below the magnetic transition, suggesting that the magnetic ordering was commensurate with the crystal lattice, and allowing them to estimate the magnetic moment as (0.4 − 1.0)μ_B/ dimer. The knowledge that the ordering is commensurate strongly suggests that magnetism is a result of a Mott insulator state, rather than a spin-density wave, which would lead to incommensurate ordering. Magnetometry further identified both an easy-axis along b and the presence of weak ferromagnetic canting.

The work of our group [99] on κ−(ET)$_2$Cl began with the study of the central ^{13}C resonance frequencies in the magnetic state. The data for the outer site are shown in Fig. 37, and the data for the inner site are qualitatively similar. The data show the resonance lines developing large orientation-dependent shifts in the antiferromagnetic state. We learned several things from these data. First, the large magnitude of the shifts in the antiferromagnetic state suggests a magnetic moment of approximately 0.5 μ_B per dimer, consistent with the findings of Miyagawa et al. Second, that there is no additional splitting of the lines in the antiferromagnetic state confirms the commensurate ordering and puts constraints on the possible orientation of the electron spins.

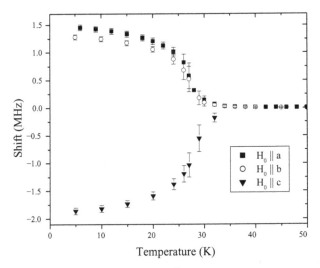

Fig. 37.   The temperature dependence of ^{13} C NMR outer site frequency shift in
$-(ET)_2Cl$, for several orientations of the applied field. Reproduced with permission
from [99]. Copyright 2003 by the American Physical Society

It is quite surprising that for each orientation of applied field the antiferromagnetic ordering gives rise to only one line. We mention two reasons
why one might expect otherwise. The first concerns the potential for domains
of different electron spin orientation. Reversing the orientation of all electron
spins leaves unchanged the energy of a system with isotropic exchange coupling. Therefore, one might expect two configurations of spin ordering (dif-
fering solely by spin reversals) which represent the ground state, and both
configurations would be found in a given sample. If one configuration shifts
the NMR frequency up, the other shifts it down. Thus all NMR lines would
occur as pairs of opposite shift. The fact that we see a single line means
that there cannot be multiple domains of spin orientation in the sample. We
show later that the argument for multiple domains is invalidated because of
the presence of non-isotropic elements of the exchange interaction, specifically
the Dzialoshinskii-Moriya interaction.

A second argument for why more than one line might be expected arises
from the distinct hyperfine tensors of the various dimer sites. As discussed in
Sect. 7.3, there are four hyperfine tensors per unit cell. While in the normal
(paramagnetic) state these four tensors provide degenerate resonance frequencies when the external field is applied along one of the crystal axes (points of
highest symmetry), it is surprising that in the antiferromagnetic state these
distinct hyperfine tensors do not lead to distinct resonance frequencies. To
see this, we consider a basic nuclear spin Hamiltonian with a Zeeman and
hyperfine interaction,

$$H_i = H_{Zeeman,i} + H_{spin,i}$$

$$= - \gamma_n \, \hat{I}_i \cdot H_0 + \hat{I}_i \cdot A_i \cdot \hat{S}_i$$

$$= - \gamma_n \, \hat{I}_i \cdot \left(H_0 - \frac{A_i \cdot \hat{S}_i}{\gamma_n} \right)$$

$$- \gamma_n \, \hat{I}_i \cdot \hat{H}_{e,i}, \tag{47}$$

where $\hat{I}_i$ and $\hat{S}_i$ are respectively the nuclear and electron spin operators at the site i. We define $\hat{H}_{e,i}$ to be the effective field perceived by the nuclear spin at site i.

If the Zeeman term dominates the hyperfine interaction with the field applied along the axis μ, then to first order the nuclear spin Hamiltonian is

$$H_i = - \gamma_n \, \hat{I}_i \left(H_0 - \frac{1}{\gamma_n} (A_i^a \, S_i^a + A_i^b \, S_i^b + A_i^c \, S_i^c) \right)$$

$$- \gamma_n \, \hat{I}_i \, H_{eff,i}, \tag{48}$$

where we have replaced the electron spin operator, $\hat{S}$ with the thermal average expectation value, S.

In the paramagnetic state the electron spin is also polarized along the μ direction, so that the Hamiltonian reduces to

$$H_i = - \gamma_n \, \hat{I}_i \left(H_0 - \frac{1}{\gamma_n} A_i^\mu \, S_i^\mu \right), \tag{49}$$

so that now only a single diagonal component (A_i^μ) of the hyperfine tensor is sampled. Since the diagonal components of the tensors are the same for all four dimer sites, the degeneracy in the paramagnetic state when the field is along a crystal axis is explained.

However, in the magnetic state the electron spins are no longer aligned along the direction of applied field. In fact, a simple antiferromagnet with just the Zeeman and isotropic exchange interactions will have the electrons aligned perpendicular to the field to maximize the Zeeman interaction while keeping the nearest neighbor spins mostly antiparallel (this assumes that the Zeeman interaction is much smaller than the isotropic exchange interaction, a condition true for all of the fields we used in our study). Let's assume that the electron spins align along b when the field is along a. Then if we label the two dimer sites in the A layer as A1 and A2, we find an effective field at A1 of

$$H_{e,A1} = H_0 - \frac{1}{\gamma_n} A_{A1}^{ab} S_{A1}^b, \tag{50}$$

and an effective field at A2 of

$$H_{e,A2} = H_0 - \frac{1}{\gamma_n} A_{A2}^{ab} S_{A2}^b. \tag{51}$$

From symmetry arguments we know that $A_{A2}^{ab} = A_{A1}^{ab}$, and antiferromagnetic ordering of nearest neighbor spins implies $S_{A2}^b = - S_{A1}^B$, which leads to

$$H_{e,A2} = H_0 + \frac{1}{n} A_{A1}^{ab} S_{A1}^{b} .$$ (52)

Thus, for this configuration, $H_{e,A1}$ and $H_{e,A2}$ differ by $\frac{2}{n} A_{A1}^{ab} S_{A1}^{b}$, and two resonance lines result. We can therefore rule out the possibility that the electron spins align along b when the field is along a. From this demonstration we see that the condition imposed by our data that every site see the same effective field does in fact reveal constraints on the electronic ordering. Further constraints result due to the off-diagonal components of the hyperfine tensor changing sign between sites, unlike the diagonal components.

Following through with the analysis, we deduced the electronic spin orderings shown in Fig. 38. These diagrams overlook the slight canting of the spins toward the applied field due to the Zeeman interaction. Note that the type of the interlayer ordering (between A1 and B1, for instance) changes as the direction of the field changes, indicating that the interlayer exchange interaction is not playing a significant role in the ordering. Also, no ordering is shown for $H_0 \parallel c$ because in that case we do not have enough constraints to make an assignment.

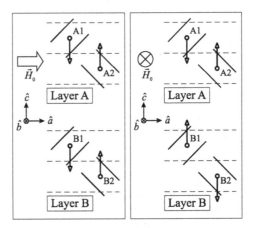

Fig. 38.   Electron spin ordering of the AF state of κ−(ET)$_2$Cl for fields along the a and b directions, as deduced from NMR data. The dashed lines indicate glide planes of symmetry. Reproduced with permission from [99]. Copyright 2003 by the American Physical Society

To explain our NMR results along with the magnetometry data of our group and other groups, we developed an electron spin Hamiltonian that includes an isotropic exchange interaction (to provide for antiferromagnetic alignment of nearest neighbor spins), an antisymmetric or Dzialoshinskii-Moriya (DM) interaction [100, 101] (to provide for the weak ferromagnetism), an anisotropic exchange interaction (to provide for the b easy-axis), and the Zeeman interaction (to couple the spins to the applied field):

$$H_{spin} = H_{iso} + H_{DM} + H_{anis} + H_{Zeeman}$$

$$= \frac{|J|}{2} \sum_{n,m} \hat{S}_n \cdot \hat{S}_m + \frac{1}{2} \sum_{n,m} D_{nm} \cdot (\hat{S}_n \times \hat{S}_m)$$

$$+ \frac{J}{2} \sum_{n,m} \hat{S}_n^b \hat{S}_m^b + g\mu_B H \cdot \sum_n \hat{S}_n . \tag{53}$$

Here n and m represent nearest neighbor (intralayer) dimer sites.

The DM interaction leads to weak ferromagnetism through the cross product in the DM expression. Consider two neighboring spins, S_{A1} and S_{A2}, which are antiferromagnetically aligned along the c axis, as shown in Fig. 39. If the DM vector between the spins ($D_{A1,A2}$) is along b, and the field is along a, then as the spins cant toward a to minimize the Zeeman interaction, they will develop a cross product that points along -b to minimize the DM interaction. Note that if one switches the spins S_{A1} and S_{A2} the cross product points in the opposite direction, and the DM energy is not minimized. Thus, the presence of the DM interaction explains why the argument for multiple domains of spin orientation does not hold up.

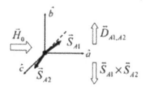

Fig. 39.   A diagram showing the e ect of the DM interaction on two neighboring spins. The spins are drawn with the same origin to aid viewing

To determine the electron spins orientations that minimize the DM interaction in -(ET)$_2$Cl for particular orientation of the applied field, it was necessary to know something about the orientation of the DM vectors (one for each layer). We were able to identify several constraints on the DM vectors through a symmetry analysis in the spirit of the analysis carried out by Coffey et al. [102] for La$_2$CuO$_4$. With the information gained from the symmetry analysis, we learned that the electronic ordering implied by the DM vectors was in fact the ordering determined through analysis of our NMR data. Furthermore, additional magnetometry measurements, when interpreted with our model, enabled us to determine values for both the isotropic exchange and DM interaction.

8 Conclusion

From the classic BCS superconductors to the more exotic cuprate and organic superconductors, NMR has been an important part of the extensive body of research on these intriguing materials. NMR work on aluminum was central to the confirmation of the BCS theory of superconductors. Later relaxation studies of both the normal and superconducting states of other superconductors have informed theoretical work on the pairing state, leading us closer to an understanding of the fundamental mechanisms of superconductivity.

This work was supported by the U.S. Department of Energy, Division of Materials Sciences under Award No. DEFG02-91ER45439, through the Frederick Seitz Materials Research Laboratory at the University of Illinois at Urbana-Champaign.

References

1. W. Heitler, E. Teller: Proc. Roy. Soc. A   155 , 629 (1936)
2. E.M. Purcell, H.C. Torrey, R.V. Pound: Phys. Rev.   69 , 37 (1946)
3. N. Bloembergen, E.M. Purcell, R.V. Pound: Phys. Rev.   73 , 679 (1948)
4. D.D. Oshero , R.C. Richardson, D.M. Lee: Phys. Rev. Lett.   28 885 (1972)
5. J. Bardeen, L.N. Cooper, J.R. Schrie er: Phys. Rev.   106 , 162 (1957)
6. J. Bardeen, L.N. Cooper, J.R. Schrie er: Phys. Rev.   108 , 1175 (1957)
7. W.D. Knight: Phys. Rev.   76 , 1259 (1949)
8. J. Korringa: Physica   16 , 601 (1950)
9. R.E. Norberg, C.P. Slichter: Phys. Rev.   83 , 1074 (1951)
10. D.F. Holcomb, C.P. Slichter: Phys. Rev.   98 , 1074 (1955)
11. D. Pines: Solid State Physics , Vol. 1, Chapter Electron Interaction in Metals (Academic, New York 1955) p 367
12. J.J. Spokas, C.P. Slichter: Phys. Rev.   113 , 1462 (1959)
13. A.G. Anderson, A.G. Redfield: Phys. Rev.   116 , 583 (1954)
14. H. Kamerlingh Onnes: Comm. Phys. Lab. Univ. Leiden, (119, 120, 122) (1911)
15. E. Maxwell: Phys. Rev.   78 , 477 (1950)
16. C.A. Reynolds, B. Serin, W.H. Wright, L.B. Nesbitt: Phys. Rev.   78 , 487 (1950)
17. J. Bardeen, D. Pines: Phys. Rev.   99 , 1140 (1955)
18. J. Bardeen: Phys. Rev.   97 , 1724 (1955)
19. L.C. Hebel, C.P. Slichter: Phys. Rev.   107 , 901 (1957)
20. L.C. Hebel, C.P. Slichter: Phys. Rev.   113 , 1504 (1959)
21. C.P. Slichter: Principles of Magnetic Resonance   (Springer-Verlag, Berlin 1996) p 223–231
22. C.J. Gorter, H.G.B. Casimir: Phys. Z.   35 , 963 (1934)
23. C.J. Gorter, H.G.B. Casimir: Z. Tech. Phys.   15 , 539 (1934)
24. A.G. Redfield: Phys. Rev. Lett.   3 , 85 (1959)
25. M. Fibich: Phys. Rev. Lett.   14 , 561 (1965)
26. R.W. Morse, H.V. Bohm: Phys. Rev.   108 , 1094 (1959)
27. S. Lundquist, editor: Nobel Lectures, 1971– 1980, World Scientific (1992)
28. J.R. Schrie er: Theory of Superconductivity   (Addison-Wesley, 1988)
29. L.C. Hebel: Phys. Rev.   116 , 79 (1959)
30. Y. Masuda, A.G. Redfield: Phys. Rev.   125 , 159 (1962)
31. M. Tinkham: Introduction to Superconductivity   (McGraw-Hill, 1996)
32. D.D. Oshero , W.J. Gully, R.C. Richardson, D.M. Lee: Phys. Rev. Lett.   29 , 920 (1972)

33. A.J. Leggett: Phys. Rev. Lett.   29, 1227 (1972)
34. D.A. Wolman, D.J. Van Harlingen, W.C. Lee, D.M. Ginsberg, A.J. Leggett: Phys. Rev. Lett.   71, 2134 (1993)
35. F. London, H. London: Proc. Roy. Soc. A   149, 71 (1935)
36. A.B. Pippard: Proc. Roy. Soc. A   216, 547 (1953)
37. V.L. Ginzburg, L.D. Landau: Zh. Eksperim. I Teor. Fiz.   20, 1064 (1950)
38. A.A. Abrikosov: Zh. Eksperim. i Teor. Fiz.   32, 1442 (1957)
39. R. Tycko, G. Dabbagh, R.M. Fleming, R.C. Haddon, A.V. Makhija, S.M. Zuharak: Phys. Rev. Lett.   67, 1886 (1991)
40. V.A. Stenger, C.H. Pennington, D.R. Bu nger, R.P. Ziebarth: Phys. Rev. Lett.   74, 1649 (1995)
41. R. Tycko, G. Dabbagh, M.J. Rosseinsky, D.W. Murphy, A.P. Ramirez, R.M. Flemming: Phys. Rev. Lett.   68, 1912 (1992)
42. Y. Masuda, N. Okuba: Phys. Rev. Lett.   20, 1475 (1968)
43. Y. Masuda, N. Okuba: J. Phys. Soc. Jpn.   26, 309 (1969)
44. F. Reif: Phys. Rev.   102, 1417 (1956)
45. F. Reif: Phys. Rev.   106, 208 (1957)
46. G.M. Androes, W.D. Knight: Phys. Rev.   121, 779 (1961)
47. K. Yosida: Phys. Rev.   110, 769 (1958)
48. V.A. Stenger, C. Recchia, J. Vance, C.H. Pennington, D.R. Bu nger, R.P. Ziebarth: Phys. Rev. B   48, R9942 (1993)
49. J.G. Bednorz, K.A. Muller: Z. Phys. B   64, 189 (1986)
50. M.K. Wu, J.R. Ashburn, C.J. Torng, P.H. Hor, R.L. Meng, L. Gao, Z.J. Huang, Y.Q. Wang, C.W. Chu: Phys. Rev. Lett.   58, 908 (1987)
51. J. Rossat-Mignod, L.P. Renault, C. Vettier, P. Burlet, J.Y. Henry, G. Lapertot: Physica B   196, 58 (1991)
52. J.B. Boyce, F. Bridges, T. Claeson, M. Nygren: Phys. Rev. B   39, 6555 (1989)
53. F.C. Zhang, T.M. Rice: Phys. Rev. B   37, 3759 (1988)
54. N.J. Curro, T. Imai, C.P. Slichter, B. Dabrowski: Phys. Rev. B   56, 877 (1997)
55. R.L. Corey, N.J. Curro, K. O'Hara, T. Imai, C.P. Slichter, K. Yoshimura, M. Katoh, K. Kosuge: Phys. Rev. B   53, 5907 (1996)
56. H. Alloul, T. Ohno, P. Mendels: Phys. Rev. Lett.   63, 1700 (1989)
57. M. Takigawa, J.L. Smith, W.L. Hults: Phys. Rev. B   44, 7764 (1991)
58. F. Mila, T.M. Rice: Physica C   157, 561 (1989)
59. B.S. Shastry: Phys. Rev. Lett.   63, 1228 (1988)
60. P.C. Hammel, M. Takigawa, R.H. He ner, Z. Fisk, K.C. Ott: Phys. Rev. Lett.   63, 1992 (1989)
61. T. Moriya: J. Phys. Soc. Jpn.   18, 516 (1963)
62. C.H. Pennington, C.P. Slichter: Phys. Rev. Lett.   66, 381 (1991)
63. C.P. Slichter, R. Corey, N.J. Curro, J. Haase, C. Milling, D. Mohr, J. Schmalian, R. Stern: Molec. Phys.   95, 897 (1998)
64. T. Imai, C.P. Slichter, K. Yoshimura, K. Kosuge: Phys. Rev. Lett.   70, 1002 (1993)
65. K. Yoshimura, Y. Nishizawa, Y. Ueda, K. Kosuge: J. Phys. Soc. Jpn.   59, 3073 (1990)
66. A.V. Chubukov, S. Sachdev: Phys. Rev. Lett.   71, 169 (1993)
67. M. Takigawa, A.P. Reyes, P.C. Hammel, J.D. Thompson, R.F. He ner, Z. Fisk, K.C. Ott: Phys. Rev. B   43, 247 (1991)
68. A.J. Millis, H. Monien, D. Pines: Phys. Rev. B   42, 167 (1990)

69. A. Sokol, D. Pines: Phys. Rev. Lett.   71, 2813 (1993)
70. V. Barzykin, D. Pines: Phys. Rev. B   52, 13585 (1995)
71. Y. Zha, V. Barzykin, D. Pines: Phys. Rev. B   54, 7561 (1996)
72. S. Ohsugi, Y. Kitaoka, K. Ishida, G.-Q. Zheng, K. Asayama: J. Phys. Soc. Jpn. 63, 700 (1994)
73. T. Imai, T. Shimizu, H. Yasuoka, Y. Ueda, K. Kosuge: Phys. Soc. Jpn.   57, 2280 (1988)
74. Y. Kitaoka, S. Hiramatsu, T. Kondo, K. Asayama: J. Phys. Soc. Jpn.   57, 30 (1988)
75. J.A. Martindale, S.E. Barrett, K.E. O'Hara, C.P. Slichter, W.C. Lee, D.M. Ginsberg: Phys. Rev. B   47, 9155 (1993)
76. S.E. Barrett, D.J. Durand, C.H. Pennington, C.P. Slichter, T.A. Friedmann, J.P. Rice, D.M. Ginsberg: Phys. Rev. B   41, 6283 (1990)
77. N.E. Bickers, D.J. Scalapino, R.T. Scalettar: Int. J. Mod. Phys. B   1, 687 (1987)
78. D. Pines: Physica B   163, 78 (1990)
79. T. Moriya, Y. Takahashi, K. Ueda: J. Phys. Soc. Jpn.   59, 2905 (1990)
80. S.E. Barrett, J.A. Martindale, D.J. Durand, C.H. Pennington, C.P. Slichter, T.A. Friedman, J.P. Rice, D.M. Ginsberg: Phys. Rev. Lett.   66, 108 (1991)
81. N. Bulut, D.J. Scalapino: Phys. Rev. Lett.   68, 706 (1992)
82. J.A. Martindale, S.E. Barrett, C.A. Klug, K.E. O'Hara, S.M. DeSoto, C.P. Slichter, T.M. Friedmann, D.M. Ginsberg: Phys. Rev. Lett.   68, 702 (1992)
83. D. Thelen, D. Pines, J.P. Lu: Phys. Rev. B   47, 9151 (1993)
84. H. Urayama, H. Yamochi, G. Saito, K. Nozawa, T. Sugano, M. Kinoshita, S. Sato, K. Oshima, A. Kawamoto, J. Tanaka: Chem. Lett.   1, 55 (1988)
85. K. Beckgaard, C.S. Jacobsen, K. Mortensen, H.J. Pedersen, N. Thorup: Solid State Commun.   33, 1119–1125 (1980)
86. D. Jerome, A. Mazaud, M. Ribault, K. Bechgaard: J. Phys. Lett.   41, L95 (1980)
87. R.H. McKenzie: Science   278, 820–821 (1997)
88. A.M. Kini, U. Geiser, H.H. Wang, K.D. Carlson, J.M. Williams, W.K. Kwok, K.G. Vandervoort, J.E. Thompson, D.L. Stupka, D. Jung, M.-H. Whangbo: Inorg. Chem.   29, 2555 (1990)
89. J.M. Williams, A.M. Kini, H.H. Wang, K.D. Carlson, U. Geiser, L.K. Montgomery, G.J. Pyrka, D.M. Watkins, J.M. Kommers, S.J. Boryschuk, A.V. Strieby Crouch, W.K. Kwok, J.E. Schirber, D.L. Overmyer, D. Jung, M.-H. Whangbo: Inorg. Chem.   29, 3272 (1990)
90. H.H. Wang, A.M. Kini, L.K. Montgomery, U. Geiser, K.D. Carlson, J.M. Williams, J.E. Thompson, D.M. Watkins, W.K. Kwok, U. Welp, K.G. Vandervoort: Chem. Mater.   2, 482 (1990)
91. H.H. Wang, K.D. Carlson, U. Geiser, A.M. Kini, A.J. Schultz, J.M. Williams, L.K. Montgomery, W.K. Kwok, U. Welp, K.G. Vandervoort, S.J. Boryschuk, A.V. Strieby Crouch, J.M. Kommers, D.M. Watkins: Synth. Met.   42, 1983 (1991)
92. S.M. De Soto, C.P. Slichter, A.M. Kini, H.H. Wang, U. Geiser, J.M. Williams: Phys. Rev. B   52, 10364–10368 (1995)
93. T. Klutz, I. Hennig, U. Haeberlen, D. Schweitzer: Appl. Mag. Res.   2, 441–463 (1991)
94. J. Schmalian: Phys. Rev. Lett.   81, 4232 (1998)
95. H. Kino, H. Kontani: J. Phys. Soc. Jpn.   67, 3691 (1998)

96. H. Kondo, T. Moriya: J. Phys. Soc. Jpn.   **67**, 3695 (1998)
97. K. Kuroki, T. Kimura, R. Arita, Y. Tanaka, Y. Matsuda: Phys. Rev. B   **65**, 100516 (1998)
98. K. Miyagawa, A. Kawamoto, Y. Nakazawa, K. Kanoda: Phys. Rev. Lett.   **75**, 1174–1177 (1995)
99. D.F. Smith, S.M. De Soto, C.P. Slichter, J.A. Schlueter, A.M. Kini, R.G. Daugherty: Phys. Rev. B   **68**, 024512 (2003)
100. I.E. Dzialoshinskii: Sov. Phys. JETP   **5**, 1259–1272 (1957)
101. T. Moriya: Phys. Rev.   **120**, 91–98 (1960)
102. D. Co ey, K.S. Bedell, S.A. Trugman: Phys. Rev. B   **42**, 6509–6513 (1990)

NMR in Magnetic Molecular Rings and Clusters

F. Borsa [1,2], A. Lascialfari [1] and Y. Furukawa [3]

[1] Dipartimento di Fisica "A.Volta" e Unita' INFM, Universita' di Pavia, 27100 Pavia, Italy
lascialfari@fisicavolta.unipv.it

[2] Department of Physics and Astronomy and Ames Laboratory, Iowa State University, Ames, IA 50011
borsa@ameslab.gov

[3] Division of Physics, Graduate School of Science, Hokkaido University, Sapporo 060-0810, Japan

Abstract. Molecular nanomagnets (MNM) are magnetic molecular clusters containing a limited number of transition ions in a highly symmetric configuration and coupled by strong exchange interaction (either ferromagnetic (FM) or more often antiferromagnetic (AFM)). The magnetic intermolecular interaction is very weak and thus the clusters behave as single nanosize units. NMR has proved to be an excellent probe to investigate the static magnetic properties and the spin dynamics of this new fascinating class of magnetic materials. The chapter contains a comprehensive review of the work performed in the last few years by the present authors with only a brief reference to work performed by other researchers. Most of the NMR measurements were performed on protons but important results were obtained also using other nuclei like ^{55}Mn, ^{57}Fe, ^{7}Li, ^{23}Na, ^{63}Cu, ^{19}F. In some cases the NMR was observed at low temperature in zero external field. Some novel NMR phenomena specific of the systems investigated were discovered and explained. For example in the anisotropic ferrimagnetic clusters Mn12 and Fe8, the ground state is a high total spin $S = 10$ state whereby the crystal field anisotropy generates an energy barrier typical of superparamagnets. It is shown how NMR and relaxation measurements can detect the microscopic local spin configuration in the ground state and the dynamics of quantum tunnelling of the magnetization (QMT). Another example is the case of the AFM rings, Fe10, Fe6 and Cr8, in which the ground state is a singlet, $S = 0$, separated from the first triplet excited state by an energy gap of about 5– 10 K. By applying a magnetic field one can observe level crossing effects. These effects were studied by proton NMR and relaxation measurements vs field at low temperature (1.5–3 K). Finally, the nuclear relaxation rate as a function of temperature in the above mentioned AFM rings displays a field dependent peak at a temperature of the order of the exchange constant J, which can be fitted with a general scaling law. From these data, the lifetime broadening of the energy levels can be determined.

F. Borsa et al.: NMR in Magnetic Molecular Rings and Clusters , Lect. Notes Phys. **684** , 297–349 (2006)
www.springerlink.com

1 Introduction

In the last years there has been a great interest in magnetic systems formed by a cluster of transition metal ions covalently bonded via superexchange bridges, embedded in a large organic molecule [1, 2, 3, 4, 5]. Following the synthesis and the structural and magnetic characterization of these magnetic molecules by chemists, the physicists realized the great interest of these systems as a practical realization of zero-dimensional model magnetic systems. In fact the magnetic molecules can be synthesized in crystalline form whereby each molecule is magnetically independent since the intramolecular exchange interaction among the transition metal ions is dominant over the weak intermolecular, usually dipolar, magnetic interaction.

Magnetic molecules (see Figs. 1 (a)–(d) for some example systems) can be prepared nowadays with an unmatched variety of parameters: (i) the size of the magnetic spin can be varied, spanning from high "classical" spins to low "quantum" spins by using different transition metal ions i.e., Fe^{3+}, Mn^{2+} ($s = 5/2$); Mn^{3+} ($s = 2$); Cr^{3+}, Mn^{4+} ($s = 3/2$); Cu^{2+}, V^{4+} ($s = 1/2$); (ii) the exchange interaction can go from antiferromagnetic (AFM) to ferromagnetic (FM) with values of the exchange constant J ranging from a few K to more than 1000 K; (iii) the geometrical arrangement of the magnetic core of the molecule can vary from simple coplanar regular ring of magnetic ions as found in many Fe and Cr rings to totally asymmetric three dimensional clusters such as $[Fe_8(N_3C_6H_{15})_6O_2(OH)_{12}]^{8+} \times [Br_8 \cdot 9H_2O]^{8-}$ (in short Fe8); (iv) the symmetry of the magnetic Hamiltonian can go from isotropic Heisenberg type as in most cases to easy axis or easy plane. Choosing from this variety of model systems one can investigate fundamental problems in magnetism taking advantage of the fascinating simplicity of zero-dimensional systems. Examples of issues of interest are the transition from classical to quantum behavior, the effect of geometrical frustration, the form of the spectral density of the magnetic fluctuations, the spectrum of the low-lying excitations with the connected problem of quantum spin dynamics and tunneling.

NMR has proved to be a powerful tool to investigate both static and dynamic properties of magnetic systems. In particular, it has been very successful in addressing some special features in low dimensional magnetic systems. For example, in one dimensional magnetic Heisenberg chains the long-time persistence of spin correlation has dramatic consequences on the field dependence of the nuclear spin-lattice relaxation rate T_1^{-1} which directly probes the low-frequency spectral weight of spin fluctuations. From T_1 and T_2 measurements one can detect the crossover of spin dimensionality from Heisenberg to XY to Ising as a function of temperature in both one- and two-dimensional systems. One final example among the many is the study of the gap in energy in Haldane $s = 1$ spin chains.

With the above scenario in mind we have undertaken a systematic NMR investigation of molecular nanomagnets since back in 1996. The present review tries to give an account of the main results obtained so far and of the many

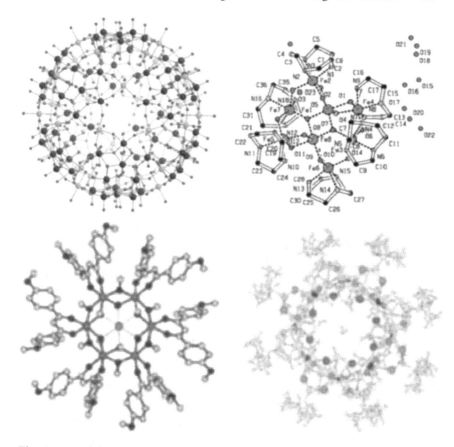

Fig. 1.   top left : Structure of Fe30 [2]. Light gray: iron, dark gray: molybdenum, small gray spheres: oxygen. Hydrogen atoms are not shown for simplicity.
top right : Structure of Fe8 [1]. The hydrogen atoms are not shown for simplicity.
bottom left : Structure of Fe6(X) (X = alkaline ion). Dark gray: iron, black: oxygen, light gray: carbon, central sphere: alkaline atom (Li, Na). Hydrogen atoms omitted for clarity.
bottom right : Structure of Cr8. Main atoms: black: chromium, dark gray: fluorine, light gray: oxygen

exciting projects that still lie ahead. The work was done through a continuous very fruitful collaboration among three NMR laboratories: at the University of Pavia, Italy, at Iowa State University and Ames Laboratory, Ames, IA, USA, and at Hokkaido University, Sapporo, Japan with occasional very useful collaborations with the high field NMR laboratory in Grenoble, France. None of the work could have been done without the precious collaboration and help of our colleagues in chemistry at the University of Florence and of Modena, Italy, and at Ames Laboratory in USA who synthesized and characterized the samples used in the NMR work.

In Sect. 2 we will mention some of the problems encountered in doing NMR in molecular nanomagnets. These are related for example to the presence of many non-equivalent nuclei, to very broad and structured resonance lines, to very short relaxation rates, to vanishingly weak signals. In the three following Sects. (3, 4, 5) we present and discuss the experimental data. To organize the considerable amount of data we chose somewhat arbitrarily to divide them according to the temperature range. In fact, the physical issues regarding the magnetic properties and the spin dynamics of the molecular nanomagnets depend on the relative ratio of the thermal energy $k_B T$ and the magnetic exchange energy J. At high T the individual magnetic ions in the molecule behave as weakly correlated paramagnetic ions; at very low T the individual spins are locked into a collective quantum state of total spin S; at intermediate T the interacting spins develop strong correlations in a way similar to what happens in magnetic phase transitions in three dimensional systems. In the illustration of the physical issues encountered in the different temperature ranges we utilize the most representative results for different kind of molecules. The magnetic molecules which were investigated by NMR but whose results are not mentioned in Sects. 3, 4 and 5 are reviewed separately in Sect. 6.

2 Challenges of NMR in Molecular Nanomagnets

Molecular nanomagnets offer a wide variety of nuclei which can be used to probe the magnetic properties and the spin dynamics. Most of the measurements were done on proton NMR. In this case the signal is very strong but the width of the spectrum and the presence of many inequivalent protons in the molecule require some special attention in the analysis of the results. Due to the above reasons the recovery of the nuclear magnetization was found in many cases to be strongly non-exponential. There are two sources for non-exponential recovery. The first is due to incomplete saturation of the broad NMR spectrum which leads to an initial fast recovery of the nuclear magnetization due to spectral diffusion. If the spectrum is not too wide (at most twice the spectral width of the rf pulse) one can still saturate the whole line by using a sequence of rf pulses provided that the condition of T_1 much longer and T_2 much shorter than the pulse spacing can be met. In nanomagnets with magnetic ground state like [Mn $_{12}$O$_{12}$(CH$_3$COO)$_{16}$(H$_2$O)$_4$] (in short Mn12) and Fe8 [1] the proton spectrum can be as wide as 4 MHz and structured at low temperature. In this case the spectrum has to be acquired by sweeping the field and/or the frequency and plotting the amplitude of the echo signal after proper correction for changes of irradiation conditions. In this case the relaxation can only be measured at given points of the spectrum and the spectral diffusion effect cannot be avoided. One can try to establish the percentage of the fast initial recovery which is affected by spectral diffusion and exclude that from the measurement but a large systematic error can still be unavoidable. The second source of non-exponential recovery is due to the presence in the

molecule of protons having a di erent environment of magnetic ions and thus
having a di erent relaxation rate. If T_2 is fast compared to T_1 then a common
spin temperature is achieved during the relaxation process and the recovery is
exponential with a single T_1^{-1} which is a weighted average of the rates of the
inequivalent protons in the molecule. In the opposite limit encountered when
T_1 is comparable to T_2, each nucleus or group of nuclei relaxes independently
with its own spin temperature and the recovery of the nuclear magnetization
results in the sum of exponentials:

$$n(t) = \frac{M(\ )-M(t)}{M(\ )} = \sum_i p_i e^{-t/T_{1i}} . \tag{1}$$

If there is a continuous distribution of T_1's the recovery follows a stretched
exponential function $\exp(\ -(t/T_1)\ )$ where $ < 1$ is the smaller the wider is
the distribution and T_1 is a relaxation parameter related to the distribution of
T_1's in a non-trivial manner. When the recovery is non-exponential it is best
to measure the T_1 parameter from the recovery of the nuclear magnetization
at short times. In fact the slope at $t \quad 0$ of the semilog plot of $n(t)$ vs t yields
an average relaxation rate $T_1^{-1} = \sum_i p_i \exp(-t/T_{1i})$. Unfortunately, in most
cases the situation is intermediate between the two above limiting cases. In
this circumstance there is no simple way that one can define a spin-lattice
relaxation parameter. Since in many instances one is interested in the relative
changes vs T and H, one can simply define an e ective relaxation parameter
R by taking the time at which the recovery curve $n(t)$ reduces to $1/e$ of the
initial value.

Other isotopes which have been utilized for NMR studies of magnetic
molecules include ^2H, ^{13}C, ^7Li, ^{23}Na, $^{63,65}Cu$. The disadvantage of a weaker
signal in ^{13}C is in part compensated by the advantage of having a nucleus
with strong hyperfine coupling to the magnetic ions and with less number of
inequivalent sites with respect to protons. For the remaining quadrupole nuclei there is the additional information obtained by the quadrupole coupling
with the electric field gradient. When the quadrupole interaction is su ciently
strong to remove the satellite transitions from the central line the non exponential decay of the nuclear magnetization becomes very di cult to analyze
because besides the non-equivalent sites one has to take into account the intrinsic non-exponential decay due to unequal separation of the Zeeman levels.
The $^{63,65}Cu$ case is the only one where, to our knowledge, a pure NQR experiment has been performed in molecular clusters (i.e., [Cu $_8$(dmpz) $_8$(OH) $_8$] $\times$
$2C_6H_5NO_2$, in short Cu8). The NQR spectrum was found to contain several
lines in the frequency range 16–21 MHz.

Very useful information was obtained from the ^{55}Mn and ^{57}Fe NMR in
Mn12 and Fe8 clusters, respectively. The NMR of the above nuclei can be
observed only at low temperature ($T < 4K$) since with increasing temperature the relaxation times T_1 and T_2 become too short. The ^{55}Mn and ^{57}Fe
(in isotopically enriched sample) NMR was detected both in zero field and

in an externally applied field. The zero field ^{55}Mn NMR spectrum in Mn12 consists of three quadrupole broadened lines (i.e., each several MHz wide) in the frequency range 230–370 MHz while the ^{57}Fe NMR spectrum in Fe8 is made of eight di erent rather narrow lines (i.e., 100 kHz) in the frequency range 63–73 MHz. Both Mn12 and Fe8 are ferrimagnetic molecules at low temperature. However, since there are no domain walls and the anisotropy is very high no signal enhancement due to the rf enhancement in domain walls and/or domains is present contrary to normal ferro- or ferri-magnetic long range ordered systems. As a consequence, the NMR signal intensity in zero external field is small (particularly in Fe8) even at low temperature since the frequency range of the overall spectrum is quite broad.

3 NMR at High Temperature ($k_B T$ J)

Most of the magnetic molecular clusters investigated are characterized by exchange constants J which are well below the room temperature energy value $k_B T$. Exceptions to this are the Cu8 ring and to a certain extent also Mn12 and Fe8 clusters. If $k_B T$ J , the magnetic moments in the cluster are weakly correlated and the system behaves like a paramagnet at high temperature. In this case the nuclear spin lattice relaxation due to the coupling to the paramagnetic ions should be field independent as indeed found in paramagnets such as MnF$_2$. On the other hand, the T_1 in molecular nanomagnets is strongly field dependent as shown for a number of systems in Figs. 2 (a)– (f). All data reported here refer to proton NMR. The field dependence of T_1 is a characteristic feature of the zero dimensionality of the magnetic system. A similar field dependence is well known to occur in one dimensional magnetic chains and, to a lesser extent, in two dimensional paramagnets. The fundamental reason for this is that in Heisenberg isotropic paramagnets the time dependence of the spin correlation function has a long time persistence in low dimensions. We will review briefly this result in the following.

In the weak-collision approach T_1^{-1} can be expressed as [6]:

$$T_1^{-1} \quad \sum_{ij} \,_{ij} J_{\pm}^{ij} (\,_e) + \quad \sum_{ij} \,_{ij} J_z^{ij} (\,_L), \tag{2}$$

where i,j number the electronic spins, $_e$ and $_L$ are the Larmor frequencies of the electron and of the nucleus, respectively, $_{ij}$ and $_{ij}$ are geometrical factors and $J_{\pm,z}^{ij}$ are the transverse and longitudinal spectral densities of the spin fluctuations. In (2), $J_{\pm,z}^{ij}(\,)$ can be expressed by the Fourier transform (FT) of the spin correlation function (CF):

$$J_{\pm,z}^{ij}(\,) = \quad G_{ij}(r,t) e^{i\,t} \, dt . \tag{3}$$

An approximate expression for the correlation function can be obtained for an infinite Heisenberg classical chain at high temperature by matching

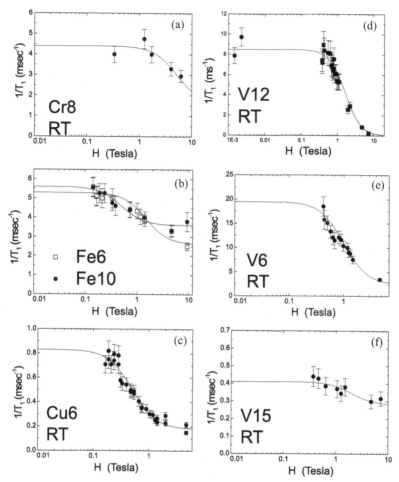

Fig. 2. The external field dependence of ^{1}H-$1/T_1$ at room temperature. (a) Cr8; (b) Fe6 and Fe10; (c) Cu6; (d) V12; (e) V6; (f) V15. The solid lines in the figure are calculated results using (7) with a set of parameters listed in Table 1. All samples are in powder form. The two points at very low field in (d) for V12 refer to spin-lattice relaxation rates in the rotating frame, T_1 , measured at 4 .7 T

the short time expansion to the long time di usive behavior due to the con-servation of the total spin and of its component in the direction of the ap-plied field [7]. For temperatures   T J/k_B the conservation property can be incorporated for spins on a ring by means of a discretized di usion equation to which cyclic boundary conditions are applied. For this model it is found [8, 10] that the auto-correlation function (CF) decays rapidly at short times until it reaches a constant value which depends on the number of spins in the cluster. The plateau in the CF is reached after a time of the order of 10 $_D^{-1}$ where $_D$ is the exchange frequency given at the simplest level of approximation by [6]:

$$D = \frac{2J}{h}[S(S+1)]^{\frac{1}{2}}, \tag{4}$$

with J the exchange constant between nearest neighbor spins   S. The same result is obtained for the CF by using a one-dimensional hopping model on a closed loop [9] or by calculating the spin correlation function with a mode-coupling approach [16]. The leveling-o of the time dependence of the CF at a value approximately given by 1 /N with N the number of spins in the cluster is the result of the conservation of the total spin component for an isotropic spin-spin interaction. In practice the anisotropic terms in the spin Hamiltonian will produce a decay of the CF via energy exchange with the "lattice".

A sketch of the time decay of the CF and of the corresponding spectral density is shown in Fig. 3. The decay at long time of the CF has a cut-o at a time $_A^{-1}$ due to the anisotropic terms in the spin hamiltonian [11, 16]. In the following we will discuss the magnetic field dependence of the nuclear relaxation rate at room temperature in terms of a simplified model which incorporates the theoretical understanding of the spin dynamics in clusters as described above.

On the basis of the time dependence of the CF discussed above and sketched in Fig. 3, we model the spectral function in (2) as the sum of two components [12]:

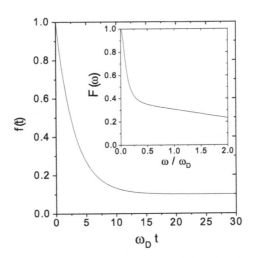

Fig. 3.   Sketch of the decay in time of the autocorrelation function of the spins on magnetic ions. The initial fast decay is characterized by the constant $_D^{-1}$ while the slow decay at long time is characterized by the constant $_A^{-1}$. In the inset the behavior of the Fourier transform i.e. the spectral density is shown

$$J^{\pm}(\ ) = J^{z}(\ ) = J'(\ ) + \frac{A}{2 + \frac{2}{A}} = \frac{D}{2 + \frac{2}{D}} + \frac{A}{2 + \frac{2}{A}}, \tag{5}$$

where we assume the same CF for the decay of the transverse ($\pm$) and longitudinal (z) components of the spins. The first term in (5) represents the Fourier transform (FT) of the initial fast decay of the CF while the second term represents the FT of the decay at long time of the CF due to anisotropic terms in the spin Hamiltonian and we model this second part with a Lorentzian function of width $_A$.

From (4) one can estimate that the exchange frequency $_D$ is of the order of 10^{13} Hz for typical values of J/k_B (10–20 K) and spin values S (1/2–5/2). The spectral function $J'(\ )$ in (5) reaches a plateau and becomes almost frequency independent for $< _D/10$. For the magnetic field strength used in the experiment (see Fig. 3) both $_n$ and $_L$ are smaller than $_D/10$. Thus we will assume $J'(_n) = J'(_e) = _D^{-1}$ in (5) where the characteristic frequency $_D$ is of the same order of magnitude as $_D/10$. Finally, by assuming $_L$ $_A$ in (5), (2) can be rewritten as:

$$\frac{1}{T_1} = K\ \frac{1}{2}A^{\pm}\ \frac{A}{2 + \frac{2}{e}} + \frac{1}{2}\frac{A^{\pm}}{D} + A^{z}\ \frac{1}{D} + \frac{1}{A}, \tag{6}$$

where the constants $A^{\pm}$ and A^{z} are averages over all protons in the molecule of the products of the hyperfine dipolar tensor components $_{ij}$ and $_{ij}$, respectively (see (2)). The constant K which has been factored out from the dipolar tensor coefficients is given by $K = \frac{(h\ _n\ _e)^2}{4} = 1.94 \times 10^{-32}$ (s^{-2} cm^6). The width $_A$ of the narrow component in the spectral function represents the frequency which characterizes the exponential time decay of the spin CF in the cluster due to anisotropic terms in the spin Hamiltonian.

The experimental data in Figs. 2 (a)– (f) were fitted by using an expression of the form

$$\frac{1}{T_1} = \frac{A}{1 + (H/B)^2} + C\ (\text{ms}^{-1}), \tag{7}$$

where the magnetic field H is expressed in Tesla and $B = _A/_e$ (Tesla). The fitting parameters for the different rings and clusters are summarized in Table 1.

The most significant parameter in Table 1 is B which measures the cut-off frequency $_A$ of the electronic spin-spin correlation function. Except for the Cr8 case (complete formula: [Cr$_8$F$_8$Piv$_{16}$], HPiv=pivalic acid) B is around 1 T corresponding to $_A$ 10^{11} rad s^{-1} or h $_A/k_B$ 1 K. The cut-off effect is provided, in principle, by any magnetic interaction which does not conserve the total spin components. In practice, such small terms stem from a variety of mechanisms including intracluster dipolar and anisotropic exchange interaction, single ion anisotropies, inter-ring dipolar or exchange interactions etc. [11]. A detailed calculation for Fe6 based on intraring dipolar interaction yielded $_A = 1.5 \times 10^{11}$ s^{-1} [16]. A similar estimate for Cu6 based on known

Table 1.   The fitting parameters in (7) for the di erent molecular rings and clusters (see text for complete chemical formula), in powders

Single Molecular Magnet	$A\,(\mathrm{ms}^{-1})$	$B\,(\mathrm{T})$	$C\,(\mathrm{ms}^{-1})$
Cr8 (AFM ring – s = 3 / 2)	2.7	5	1.7
Fe6 (AFM ring – s = 5 / 2)	2.7	1.5	2.6
Fe10 (AFM ring – s = 5 / 2)	2	0.5	3.6
Cu6 (FM ring – s = 1 / 2)	0.65	0.5	0.18
V12 (AFM square – s = 1 / 2)	8.5	1.6	0
V6 (AFM triangle – s = 1 / 2)	17	1	2.5
V15 (AFM ring – s = 1 / 2)	0.13	2	0.28

anisotropic nearest neighbor (exchange and dipolar) contributions to nearest neighbor interactions yielded 1 .4× 10^{11} s^{-1} [16]. Both these results can account very well for the experimental findings in Table 1. From the comparison of (6) and (7) one has A = KA$^{\pm}$/2 $_A$ and C KA^z/ $_A$ (since $_D$ $_A$). Thus the order of magnitude of the hyperfine constants is A$^{\pm}$ A^z 1÷10× 10^{46} cm^{-6}. Since A$^{\pm}$, A^z are the product of two dipolar interaction tensor components they are of order of r^{-6} where r is the distance between a ^{1}H nucleus and a transition metal local moment. For most of the rings the value of the hyperfine constants is consistent with a purely nuclear-electron dipolar interaction.

For V12 (complete formula: (NHEt) $_3$[V$_8^{IV}$ V$_4^V$ As$_8$O$_{40}$(H$_2$O)]× H$_2$O) and V6 (complete formula of one variant: Na $_6$[H$_4$(V$_3$L)$_2$P$_4$O$_4$]× 18H$_2$O), the A$^{\pm}$ (A^z) hyperfine constant is one order of magnitude higher indicating the presence of an additional contribution probably due to a contact interaction due to the admixing of the hydrogen s wave function with the d wave function of the Vanadium ions. An alternative way to explain the anomalous values (A C) for V6 and particularly for V12 in Table 1 is to go back to (7) and assume that B = $_A$/ $_N$ instead of B = $_A$/ $_e$. This implies that the cut-o frequency $_A$ is much less than in other clusters, namely of order of the nuclear Larmor frequency in (6). In this case the value of the constant C in (7) is close to zero in agreement with the experiments as can be seen easily by modifying in the appropriate way the approximate expression (6). It is, however, di cult to justify such small value for the cut-o frequency in V12 [17].

4 NMR at Intermediate Temperatures ($k_B T$ J)

As the temperature is lowered and it becomes comparable to the magnetic exchange interaction J strong correlations in the fluctuations of the magnetic moments of the molecule start building up. The situation is analogous to macroscopic three-dimensional magnetic systems when the temperature approaches the critical temperature for the transition to long range magnetic order. In molecular magnets, as a result of the finite size of the system the low lying magnetic states are well separated among themselves. Therefore the correlation of the magnetic moments at low temperature has to be viewed as the result of the progressive population of the collective low lying quantum total spin states of the magnetic molecule without any phase transition. In this intermediate temperature range two very interesting situations arise which can be investigated by NMR and relaxation. On one hand, one can follow the evolution of the electronic spin correlation function as the system crosses over from an uncorrelated finite size paramagnet to a total spin   S collective quantum state. On the other hand, one can investigate the nature of the fluctuations of the local electronic spin while the system is in its ground state but at a temperature for which excitations to higher states are important. In the following we will consider examples of the two situations. For the evolution of the spin correlation function we will refer to simple antiferromagnetic rings having a total spin   S = 0 ground state. For the thermal fluctuations in the ground state we will refer to ferromagnetic clusters having a total high spin S = 10 ground state.

In order to describe the nuclear spin-lattice relaxation in a magnetic system in presence of a correlated spin dynamics it is more convenient to express the nuclear T_1 in terms of the q components of the electronic spins [6, 18, 19]:

$$\frac{1}{T_1} = \frac{(h_{\ n}\ _e)^2}{4} \quad dt\,\cos(\ _n t) \quad dq\ 1/4 A^{\pm}(q)\ S_q^{\pm}(t) S_{-q}^{\pm}(t)$$
$$+ A^z(q)\ S_q^z(t) S_{-q}^z(t) \qquad (8)$$

or in terms of the response functions by using the fluctuation-dissipation theorem [20]:

$$\frac{1}{T_1} = \frac{(h_{\ n}\ _e)^2}{4\,g^{\,2}\mu_B^2} k_B T \quad 1/4 \sum_q A^{\pm}(q)\ ^{\pm}(q) f_q^{\pm}(\ _e)$$
$$+ \sum_q A^z(q)\ ^z(q) f_q^z(\ _n)\ , \qquad (9)$$

where $_n$ and $_e$ are the gyromagnetic ratios of the nucleus and of the free electron, respectively, g is Lande's factor, μ_B is the Bohr magneton, k_B is the Boltzmann constant. The coecients $A^{\pm}(q)$ and $A^z(q)$ are the Fourier transforms of the spherical components of the product of two dipole-interaction tensors [describing the hyperfine coupling of a given proton to the magnetic moments] whereby the symbols ± and z refer to the transverse and longitudinal components of the electron spins with respect to the quantization direction which is here the external magnetic field. The collective   q-dependent spin correlation function is written as the product of the static response function and the normalized relaxation function $f_q^{\pm,z}(\ )$.

At high temperature ($k_B T$ J) one can neglect in (9) the q-dependence of the generalized susceptibility   (q) and of the spectral density function

f_q ( ). If one assumes an isotropic response function $1/2^\pm(q) = {}^z(q) = (q = 0)$ and one takes a q-independent average value for the dipolar hyperfine interaction of the protons with the local moment of the electronic spins, i.e., $A^\pm(q) = A$, $A^z(q) = A^z$ in units of cm^{-6}, then (9) reduces in this high temperature limit to:

$$\frac{1}{T_1} = \frac{(h_{n\ e})^2}{4 g^2 \mu_B^2} k_B T \ (q = 0) \ 1/2 A^\pm J^\pm(\ _e) + A^z J^z(\ _n) \ . \quad (10)$$

If one further assumes for the spectral density of the spin correlation the expressions (5), then (10) reduces to (6) used in the previous paragraph to analyze the field dependence of $1/T_1$ at room temperature. By decreasing the temperature to values such that $k_B T$ becomes comparable to J one expects that the nuclear spin-lattice relaxation rate displays a characteristic temperature dependence related to the correlated spin dynamics according to (8) and (9).

4.1 Thermal Fluctuations in AFM Rings

Measurements of proton T_1 as a function of temperature in a number of antiferromagnetic molecular rings has shown a surprisingly large enhancement of the relaxation rate at low temperatures, resulting in a field dependent peak of T_1^{-1} centered at a temperature of the order of the magnetic exchange constant J/k_B (see e.g., [12], [21] and [22]). The results are shown for three di erent rings in Fig. 4 (a), in Fig. 4 (b), and in Fig. 4 (c). The systems investigated are: Cr8 ($s = 5/2$, J 17.2 K) [15]; Fe6(Na) ($s = 5/2$, J 28.2 K) [5]; Fe10 ($s = 5/2$, J 13.8 K) [5, 21] (for the last two compounds the respective complete formulas are: [NaFe $_6(\mu_2$-OMe)$_{12}$(dbm)$_6$]Cl and [Fe(OMe)$_2$(O$_2$CCH$_2$Cl)]$_{10}$). In all three samples the ground state is nonmagnetic with total spin $S_{Total} = 0$ and the energies of the lowest lying exchange multiplets can be described to first approximation by Lande's interval rule $E_S = 2 J S (S + 1)/N$, where N is the number of spins in the ring [5]. The main feature in the temperature dependence of T_1^{-1} is the strong

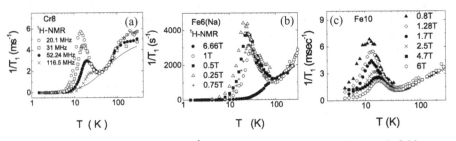

Fig. 4. Temperature dependence of ^{1}H-$1/T_1$ under various external magnetic fields. (a) Cr8, (b) Fe6(Na), and (c) Fe10. The solid line in (a) shows the temperature dependence of T in arbitrary unit

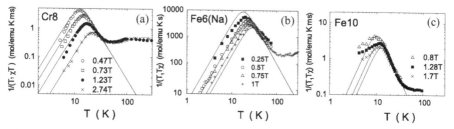

Fig. 5. $1/(T_1 T)$ as a function of temperature in log-log plot. (a) Cr8; (b) Fe6(Na); (c) Fe10. The solid lines in the figures are theoretical calculation (see text). All samples are in powders

enhancement at low T and the presence of a maximum at a temperature T_0 for each of the samples investigated. For $T < T_0$, T_1^{-1} decreases approaching at low T, an exponential drop due to the "condensation" into the $S_{total} = 0$ singlet ground state as discussed later on. It should be noted that the behavior of the relaxation rate is di erent than the behavior of the uniform magnetic susceptibility. The latter, when plotted as T, shows a continuous decrease with an exponential drop at very low temperature, consistent with what is expected for an AFM system with a singlet ground state. When the T_1^{-1} is plotted together with T as shown in Fig. 4 (a) for Cr8, one can see that the two quantities are approximately proportional in the whole temperature range except for the region where the peak in T_1^{-1} occurs. In order to emphasize the critical enhancement it is more e cient to plot the relaxation rate divided by T as shown in Fig. 5 (a), Fig. 5 (b) and Fig. 5 (c) for the cases of Cr8, Fe6 and Fe10, respectively. As shown in these same figures, the peak in the relaxation rate is depressed by the application of an external magnetic field and the position of its maximum moves to higher temperature on increasing the external field. One should also note the field dependence of the intensity of the maximum which decreases as the field is increased. In Fig. 6 we plot $(T_1^{-1})/(T)$ vs. temperature for Fe6(Li), obtained by using two di erent nuclei, i.e., ^1H and ^7Li. The qualitative temperature behavior for temperatures below the peak is very similar for the two nuclei. Finally, in Fig. 7 we plot the renormalized $(T_1^{-1})/(T)$ data as a function of reduced temperature $t = T/T_0$ where T_0 is the temperature of the maximum for di erent samples and di erent fields. The sets of data overlap suggesting that the occurrence of a maximum is a universal e ect of antiferromagnetically coupled rings with $S_{Total} = 0$ ground state.

Recently, it has been shown [23] that the relaxation rate data around the peak for Cr8 can be fitted very well by the simple expression for T_1 (7) used in the high temperature approximation, provided that the constant B is assumed to depend on temperature as $B = DT^n$ with the exponent n close to the value of three. The same kind of fit appears to work reasonably well also for Fe6(Na) and Fe10. This simple result is very surprising because it implies that the expression (10) which is obtained in the limit of high temperature

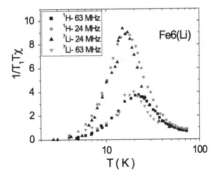

Fig. 6.   Plot of renormalized 1 /T₁ T as a function of T for Fe6(Li), in two different magnetic fields for two different nuclei

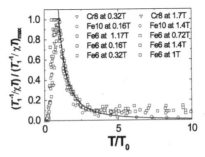

Fig. 7.   Plot of renormalized 1 /T₁ T as a function of T/T₀ for different magnetic fields and systems. All samples are in powders

remains valid even at low temperature provided that the spectral density of the fluctuations in (5) is allowed to narrow down and to become peaked at very low frequency. Thus from (10) and (5) one can write down an expression for the nuclear relaxation of the form:

$$\frac{1}{T_1 T} = A \frac{B}{1 + (H/B)^2} + C \ (\mathrm{ms}^{-1}) , \qquad (11)$$

which can be used to fit the data around the temperature of the peak. The constant C in (11) groups together the terms in (10) which are weakly   T-dependent, particularly in the region of the peak. The quality of the fit obtained using (11) is exceptionally good for Cr8 as shown in Fig. 5 (a) and moderately good for Fe6(Na) and Fe10 as shown in Figs. 5 (b) and 5 (c) (Note that only the first term in (11) in shown in Fig. 5). The constant   B in (11) can be identified either with $_D$/ $_e$ or with $_A$/ $_n$. In the first case the peak would be due to the slowing down of $_D$ as T^3 from a value at room temperature much higher than $_e$ to a value of the order of $_e$ in the region of the peak. In the second case the peak would be due to the slowing down of $_A$ from a value of the order of $_e$ at room temperature (see previous paragraph) to a value of the order of $_n$ in the region of the peak. Measurements of proton T_1 alone cannot distinguish between these situations. On the other hand, measurements of relaxation on two nuclear species in the same molecular cluster can give us the answer. These measurements have been performed in the AFM ring Fe6(Li), in all identical to Fe6(Na) except for the replacement of Na with Li which induces a change of J from J = 28 K for Fe6(Na) to J = 23 K for Fe6(Li). The results are shown in Fig. 6. It is quite clear that the maxima in T_1^{-1} overlap when the resonance frequency is the same and not the magnetic field. This is a direct proof that the constant   B in (11) has to be identified with $_A/_n$, namely that the width of the Lorenzian in (5) becomes of order of $_N$ at the temperature of the peak. Thus it appears that the peak arises from the spectral density of the longitudinal fluctuations $J^z(_n)$ in (10).

As a direct consequence of the fitting formula (11) (where   C can be neglected in the region of the peak) one finds that the renormalized plot of $1/(T_1T)$ vs t = T/T_0 shown in Fig. 7 has the simple form

$$\frac{1/(T_1T)}{(1/(T_1T))_{max}} = \frac{2t^n}{1 + t^{2n}} .$$
(12)

In Fig. 8, we show the renormalized plot compared to the function in (12). The fit is remarkably good with   n = 3 and no adjustable parameters [23].

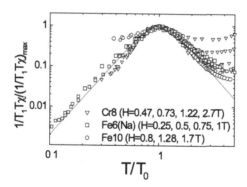

Fig. 8.   Log-log plot of renormalized 1 $/(T_1 T)$ as a function of T/T_0. The solid line is the theoretical curve (12) with   n = 3. The samples are in form of powders

At present there is no theoretical description which explains the critical enhancement manifested in the peak of T_1^{-1} for AFM rings. One possible scenario is that the e ect arises from the increase of the antiferromagnetic correlation which generates an enhancement and a slowing down of the spin fluctuations at the staggered wave vector   q = . To see this, we rewrite (9) in an approximate form where we neglect any di erence between the transverse (±) and longitudinal (z) terms and we divide the response functions in a non-critical part which is described by the   q = 0 term and in a critical part described by the q =   term:

$$\frac{1}{T_1} = \frac{(h\,_n\,_e)^2}{4\,g\,^2\mu_B^2}\,k_B\,T\quad(0)\quad A(0)\,f_0(\,_e)\quad 1 + \frac{A(\,)}{A(0)}\,\frac{f\,(\,_e)}{f_0(\,_e)}\,\frac{(\,)}{(0)}$$

$$+\ A(0)f_0(\,_n)\quad 1 + \frac{A(\,)}{A(0)}\,\frac{f\,(\,_n)}{f_0(\,_n)}\,\frac{(\,)}{(0)}\ .\qquad\qquad(13)$$

By comparing (13) with (10) which fits so well the experimental data one would deduce that the critical term $f\,(\,_e)/f_0(\,_e)$, times $(\,)/\,(0)$, has a simple Lorenzian form with a correlation frequency $_D = _e B$ which displays a critical behavior as T^n. This result is di cult to justify on the basis of dynamical scaling arguments similar to the ones used in phase transitions [20]. First principles theoretical calculations of the spin correlation function could give the answer.

An alternative scenario is one in which the magnetic critical slowing down plays no relevant role in these finite size systems. In this case the peak in the nuclear relaxation rate could be simply related to a decrease of the cut-o frequency $_A$ which reflects the anisotropy terms in the magnetic Hamiltonian which do not commute with the Heisenberg Haniltonian. These terms determines the electronic spin lattice relaxation via spin-phonon coupling.

An important clue for the understanding of the problem is the dependence of the position in temperature of the peak of 1 /T_1 from the exchange interaction J in the AFM ring. The fits of the experimental data with (11) yields B = DT^3 with a value of D di erent for the three AFM rings. The values of the constant D are plotted in Fig. 9 (a) vs the exchange constant J of the three AFM rings in a log-log plot. In Fig. 9 (b) the same values are plotted vs the energy gap separating the ground state $S_T = 0$ from the $S_T = 1$ excited state according to Lande's rule = 4 J/N . In both cases there appears to be a negative power dependence i.e., D J^- and D $^-$ with = 4 ± 0.5 and = 2 .3 ± 0.5.

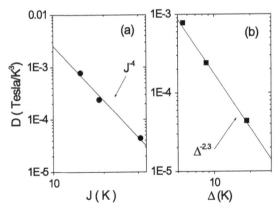

Fig. 9. (a) A relation between D and J. (b) A relation between D and . The data refer to Cr8, Fe6(Na) and Fe10

4.2 Thermal Fluctuations of the Magnetization in Nanomagnets: Mn12, Fe8

The cluster containing twelve Mn ions, Mn12, has a high total spin ($S = 10$) ground state in spite of the large antiferromagnetic coupling between the local Mn moments [14]. This circumstance together with the large magneto crystalline anisotropy generates an easy axis nanoferrimagnet with spectacular superparamagnetic effects [24]. Fe8 is similar to Mn12 with regards to the high spin ground state ($S = 10$) but it has a more complex anisotropy and a lower barrier for the reorientation of the magnetization (about 27 K against 67 K for Mn12) [25].The interest in molecular magnets is largely related to the possibility of observing quantum resonant tunneling of the magnetization (QTM) [26]. For the proper description of the quantum dynamical effects in the high spin ground state one has to take into account the environmental effects represented by spin-phonon coupling, intermolecular magnetic interactions and hyperfine interactions with the nuclei. NMR is a very suitable local microscopic probe to investigate the above-mentioned environmental effects. The NMR spectrum at low temperature shows a multiplicity of spectral lines, which yield directly the local hyperfine field at the nuclear site due to the coupling with the local moment of the magnetic ions in the high total spin ground state. Furthermore, the nuclear spin-lattice relaxation is driven by the fluctuations of the orientation of the total magnetization of the molecular cluster and thus yields information about spin-phonon coupling which limits the lifetime of the m components (i.e., along the easy axis) of the total spin S.

At very low temperature ($4 K$) both Mn12 and Fe8 are in their magnetic ground state and the magnetization of the molecule is frozen in the time scale of an NMR experiment. Thus the nuclei experience a large static internal field. This allows to detect ^{55}Mn NMR in zero external field in Mn12 and ^{57}Fe NMR in zero external field in Fe8. For protons the internal field is small, being generated by the proton-Mn (Fe) moment dipolar interaction and thus the NMR in zero field is weak and can be observed only at low frequency over a broad frequency interval (2–4 MHz) due to the presence of many inequivalent proton sites in the molecule. On the other hand, the proton NMR in an external magnetic field shows a broad structured spectrum with a field independent shift of the lines of order of the internal field. For the analysis of the hyperfine field at the proton site and at the deuteron site in deuterated Fe8, as well as of the temperature dependence of the spectrum, we refer to [27]. At intermediate temperatures (4–30 K) the magnetization of the Mn12 and Fe8 molecules are subject to large and fast fluctuations due to the thermal populations of the higher energy magnetic quantum states $m = \pm 9$, $\pm 8, \pm 7...$ separated by crystal field anisotropy within the total spin ground state $S = 10$. As a result the zero field NMR lines progressively disappear on increasing the temperature and the spin-lattice relaxation becomes very fast.

We have utilized this circumstance to investigate the thermal fluctuations of the magnetization by measuring the temperature and field dependence of T_1^{-1}. The study was possible for proton NMR [27] and Muon Spin Rotation (μSR) [28] in Fe8 over the whole temperature range. In Mn12 the proton relaxation becomes too fast in the temperature range (4–20 K) and we had to measure the muon longitudinal relaxation by   μSR technique [29]. For ^{55}Mn NMR in Mn12 the signal can be detected only up to about 4 K and for ^{57}Fe NMR the signal is lost at about 1.8 K because of a very short T_2. The proton T_1 vs temperature for Fe8 is shown in Fig. 10. The proton T_1 in Mn12 is shown in Fig. 11 vs magnetic field at low temperature below 4.2 K. Regarding the temperature dependence in the range 4–30 K we refer to the   μSR relaxation rate since the proton NMR cannot be detected due to the short T_1. The results for the longitudinal muon relaxation rate (a parameter analogous to $1/T_1$) are shown in Fig. 14 for different applied longitudinal fields in Mn12 powders. The ^{55}Mn T_1 as a function of temperature is shown in Fig. 12 in the narrow temperature range in which the signal is observable in zero external field. Finally, the field dependence of the ^{55}Mn T_1 is shown in Fig. 13.

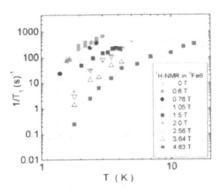

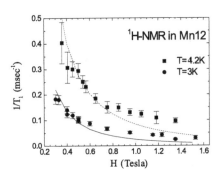

Fig. 10.   Temperature dependence of ^{1}H-1/T_1 in Fe8 "oriented" powders at different magnetic fields, parallel to the easy-axis. The lines are theoretical estimation calculated by (18) with a set of parameters described in the text

Fig. 11.   The external field (parallel to the easy-axis) dependence of ^{1}H-1/T_1 in Mn12 "oriented" powders measured at T = 4.2 and 3 K. The curves are fitting results obtained by (18)

All the relaxation rate data shown here can be interpreted in terms of a simple model of nuclear spin-lattice relaxation via a direct process driven by the fluctuations of the molecular magnetization. Since the fluctuations of the magnetization are a consequence of the finite lifetime of the   m magnetic substates due to spin-phonon interactions, one can obtain from the fit of the NMR data the spin-phonon coupling constant, a quantity not easily derived by other techniques. We give in the following a brief account of the model.

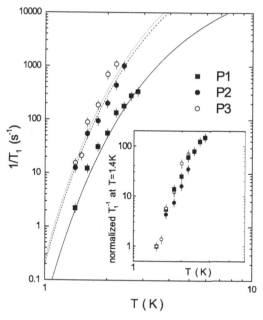

Fig. 12. Temperature dependence of ^{55}Mn-$1/T_1$ for each Mn site in Mn12 oriented powders.   Solid curves are fitting curves according to (18). The inset shows the temperature dependence of $1/T_1$ for the three peaks renormalized at the same value at T = 1.4 K

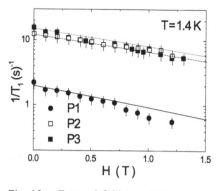

Fig. 13.   External field (parallel to the easy-axis) dependence of ^{55}Mn-$1/T_1$ for each Mn site in Mn12 oriented powders measured at T = 1.4 K. The solid lines are calculated results by (18)

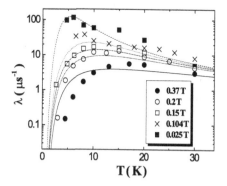

Fig. 14.   Muon longitudinal relaxation rate for different applied longitudinal fields in Mn12 powders, as a function of temperature. The lines are fits to the data following (18)

We start from a semi-classical approach by expressing the nuclear spin-lattice relaxation rate (NSLR) in terms of the correlation function of the transverse component $h_\pm(t)$ of the time dependent transverse hyperfine field at the proton site:

$$\frac{1}{T_1} = \frac{1}{2} \gamma_N^2 \int \overline{h_\pm(t) h_\pm(0)} \, e^{i\omega_L t} \, dt, \tag{14}$$

with γ_N the nuclear gyromagnetic ratio and ω_L the Larmor frequency. The fundamental assumption in this model is that, although the hyperfine field $h_\pm(t)$ is due to the interaction of the protons with the local moments of the Fe^{3+} (Mn) ions, its time dependence is the same as the time dependence of the orientation of the total magnetization of the molecular cluster. Thus one can use for the correlation function of $h_\pm(t)$ an exponential function with the correlation time τ_m determined by the lifetime broadening of the m sublevels:

$$\overline{h_\pm(t) h_\pm(0)} = \sum_{m=+10}^{-10} \overline{h_\pm^2} \, e^{-\frac{t}{\tau_m}} \frac{e^{-\frac{E_m}{k_B T}}}{Z}, \tag{15}$$

where Z is the partition function. The term $\overline{h_\pm^2}$ is the average square of the change of the hyperfine field when the magnetization of the molecule changes orientation (i.e., $\Delta m = \pm 1, \pm 2$). For sake of simplicity we assume an average value independent of the Δm transition considered. The assumption should be acceptable since in the temperature range investigated most of the fluctuations occur between few m states close to the $m = 10$ ground state. The lifetime τ_m for each individual m state is determined by the probability of a transition from m to $m \pm 1$, $W_{m,m \pm 1}$, plus the probability for a transition with $\Delta m = \pm 2$, $W_{m,m \pm 2}$, i.e,

$$\frac{1}{\tau_m} = W_{m,m+1} + W_{m,m-1} + W_{m,m+2} + W_{m,m-2}. \tag{16}$$

The transition probabilities are due to spin-phonon interaction and can be expressed in terms of the energy level differences as [30, 31]:

$$W_{m,m\pm 1} = W_{\pm 1} = C \, s_{\pm 1} \frac{(E_{m\pm 1} - E_m)^3}{e^{\beta(E_{m\pm 1} - E_m)} - 1},$$

$$\tag{17}$$

$$W_{m,m\pm 2} = W_{\pm 2} = 1.06 \, C \, s_{\pm 2} \frac{(E_{m\pm 2} - E_m)^3}{e^{\beta(E_{m\pm 2} - E_m)} - 1},$$

where

$$s_{\pm 1} = (s \mp m)(s \pm m + 1)(2 m \pm 1)^2,$$
$$s_{\pm 2} = (s \mp m)(s \pm m + 1)(s \mp m - 1)(s \pm m + 2).$$

The spin-phonon parameter C in (17) is given by $C = D'^2/(12 \, v^5 h^4)$ with the mass density and v the sound velocity and D' a constant related to the crystal field anisotropy. Finally we can write for the NSLR:

$$\frac{1}{T_1} = \frac{A}{Z} \sum_{m=+10}^{-10} \frac{\tau_m \, e^{-\frac{E_m}{k_B T}}}{1 + \tau_L^2 \tau_m^2}, \tag{18}$$

where $A = \frac{2}{N} h_\pm^2$.

The energy levels E_m in the above equations can be obtained from the Hamiltonian of the molecules expressed in terms of the total spin S:

$$H = -DS_z^2 - BS_z^4 + E(S_x^2 - S_y^2) + g\mu_B S_z H . \tag{19}$$

For Mn12 one has $D = 0.55\,K$, $B = 1.2 \times 10^{-3}\,K$, and $E = 0$ while for Fe8 one has $D = 0.27\,K$, $B = 0$ and $E = 0.046\,K$.

The experimental data both as a function of temperature at different fields (Figs. 10, 14 and 12) and as a function of field at a fixed temperature (Figs. 11 and 13) were fitted to (18) by using (17) and (19) and treating A and C as adjustable parameters. From the fit of the proton relaxation in Fe8 one obtains the parameters: $C = 31\,Hz\,K^{-3}$ and $A = 1.02 \times 10^{12}$ $(rad^2 s^{-2})$. The fit of the proton relaxation data in Mn12 (as well as the μSR relaxation data, see Fig. 10) was obtained with a somewhat larger value of C and a smaller value of $A = 0.45 \times 10^{12}$ $(rad^2 s^{-2})$. The coupling constant A represents the average hyperfine interaction squared between protons and transition metal magnetic moments. The value found for Fe8 is larger than the value obtained in Mn12 indicating that in Fe8 the protons are subject to non-negligible hyperfine interaction due to contact terms in addition to the dipolar interaction. From the knowledge of the spin-phonon coupling parameter C one can estimate the lifetime of the m sublevels by using (16) and (17). For a detailed discussion of the fitting parameters we refer to the original papers. Finally, the ^{55}Mn relaxation data were fitted with reasonable values of C and A, although for this case the two parameters cannot be determined independently. It should be mentioned that ^{55}Mn relaxation data have also been reported by Goto et al. [32] and analyzed with a model based on a two-state pulse fluctuation corresponding to the hyperfine fields in the two lowest energy levels $m = \pm 10, \pm 9$. This model is indeed a better model for the very low temperature region ($< 4.2\,K$) while our model has the advantage to be applicable even at higher temperatures where the higher energy levels become populated. A different approach has been proposed by Yamamoto and collaborator [33], based on a spin-wave approximation for the description of the energy levels and a Raman two-magnon scattering mechanism to describe the ^{55}Mn T_1^{-1} data. Since both the phenomenological models based on a direct relaxation process [49, 32] and the spin-wave model [33] appear to fit the low temperature ^{55}Mn relaxation data, a connection between the two approaches should be established.

5 NMR at Low Temperatures ($k_B T \ll J$)

When the temperature is much lower than the exchange interaction among magnetic moments in the molecule, i.e., $T \ll J/k_B$, the system is mostly in its collective quantum ground state characterized by a total spin S. We have to distinguish the two cases of singlet ground state $S = 0$ and of high spin ground state $S > 0$. In the first case which pertains to AFM rings the residual weak magnetism of the molecule at low temperature is due to the thermal population of the first excited state which is normally a triplet $S = 1$ state. In the second case the molecule at low temperature behaves like a nanomagnet with a spontaneous magnetization proportional to the value of the ground state spin S. If there is no anisotropy the molecule acts like a soft nanomagnet with a magnetization which can be aligned by an external magnetic field with no hysteresis in the magnetization cycle. This is the case of Cu6 FM ring which is discussed in Sect. 6. In presence of an anisotropy the molecule behaves as a hard nanomagnet with hysteresis in the magnetization cycle. However, since there is no long range order each molecule acts as a superparamagnetic particle. At temperatures much lower than the anisotropy barrier the relaxation of the magnetization can be dominated by quantum tunneling. NMR can give interesting information in this low temperature regime. We will review the main results treating separately the case of a non-magnetic ground state (AFM rings) and the case of a magnetic ground state (Mn12 and Fe8).

5.1 Energy Gap of AFM Rings in the Magnetic Ground State

AFM rings such as Fe10, Fe6, Cr8 already discussed in Sect. 3, are characterized by a single nearest neighbor exchange interaction J which generates a singlet ground state of total spin $S = 0$ separated by an energy gap Δ from the first excited triplet state $S = 1$. From simple Lande's interval rule one has $E(S) = 2 J/N S (S + 1)$. Thus in absence of crystal field anisotropy the gap is $\Delta = 4 J/N$ where N is the number of magnetic moments in the ring. In presence of crystal field anisotropy with axial symmetry characterized by the parameter D (see (19)) the gap between $S = 0$ and $S = 1$, $M = \pm 1$ is $4J/N + D/3$ for the case of positive axial anisotropy. In Table 2 we summarize the magnetic parameters for the above mentioned three rings, for the Cu8 ring and for the cluster V12 which can be assimilated to a square of V^{4+} magnetic ions.

From the inspection of the gap value in Table 2 it appears that below liquid helium temperature the molecular magnets will be mostly in the non-magnetic ground state. In Cu8 the gap is so big that the ring is in its singlet ground state even up to room temperature. As a result the magnetic susceptibility goes to zero and the nuclear relaxation rate becomes also very small. Measurements of T_1^{-1} in this low temperature range can yield interesting information about the energy gap Δ and about the quantum fluctuations in the $S = 0$ ground state.

Table 2.   Comparison of single ion spin values, exchange coupling constant (J) and energy gap Δ (K) between the ground state and first excited state at $H = 0$ for different AFM rings

AFM Ring	Magnetic Ion Spin	J (K)	Δ (K) – No Anisotropy
Fe10	5/2	13.8	5.5
Fe6(Na) Trigonal	5/2	28	18.7
Cr8	3/2	17	8.5
Cu8	1/2	1000	500
V12	1/2	17.2	17.2

The electronic spin correlation function entering the expression of T_1^{-1} (see (2) and (3)) is defined as:

$$G_{ij}^{\alpha\beta}(r,t) = \sum_{n}\sum_{l} \langle n|S_i^{\alpha}|l\rangle \, e^{-E_n\beta+iE_nt/\hbar \; -iE_lt/\hbar} \langle l|S_j^{\beta}|n\rangle , \qquad (20)$$

where n,l number the eigenstates, E_n, E_l are the energy eigenvalues, $\beta = 1/k_BT$, $S_{i(j)}^{\alpha}$ are the spin operators of the ith (j^{th}) spin and $\alpha\beta = x,y,z$. For a finite system the energy difference between eigenstates is very large compared to the Larmor Zeeman energy. Therefore for a direct process (see (3)) only the matrix elements with $n = l$ in (20) need to be considered and a broadening of the energy levels has to be introduced in order to fulfill energy conservation (i.e., in order to have some spectral density of the fluctuations at ω_e and ω_L in (2)). It should be noted that an alternative approach is to describe the nuclear relaxation in terms of a Raman process [33]. Even in this case one needs to have a broadening of the levels or a spin wave band. We have not explored this possibility since the direct relaxation process appears to be able to explain the experimental data.

As a consequence of the presence of the Boltzmann factors in (20) the NSLR at very low temperature will be simply proportional to the population of the excited states. For temperatures less than the energy gap Δ one has approximately:

$$T_1^{-1} = A \frac{e^{-\Delta/k_BT}}{1+3e^{-\Delta/k_BT}} , \qquad (21)$$

where A is a fitting constant which contains the hyperfine coupling constants and the matrix elements in (20). The gap Δ depends on the applied magnetic field as a result of the Zeeman splitting of the excited triplet state of the AFM rings. The validity of the simple prediction of (21) was tested in the AFM rings described in Table 2. The dependences of T_1^{-1} vs T in Cr8, Fe6 and V12 are shown in Figs. 15(a), 15(b) and 15(c), respectively. For these systems the condition $T < \Delta/k_B$ can be met in the temperature range (1.4–4.2 K). For Fe10 one has to perform the measurements at lower temperature since the gap is smaller [34].

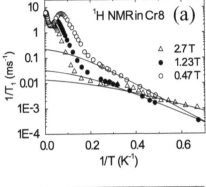

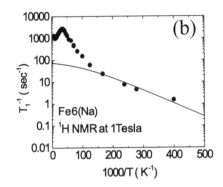

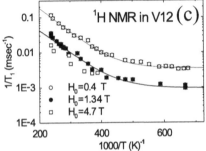

Fig. 15. Semilog plot of ^{1}H-1/T_1 as a function of 1 /T (Cr8, (a)) and 1000 /T (Fe6(Na), (b), and V12, (c)). The solid lines are calculated results using (21). All samples are in the form of powders

As seen in the figures the temperature dependence of T_1^{-1} appears to be fitted well by (21) in the low   T limit for Cr8 and Fe6 with values of the gap in qualitative agreement with the values of the gap in Table 2. Also the decrease of the measured gap with increasing field in Cr8 is in agreement with the notion that the application of an external field should close the gap. It should be noted that in presence of crystal field anisotropy the value of the gap is different from the one in Table 2 and, in presence of a magnetic field, the gap depends on the angle between   H and the symmetry axis of the molecule. Thus measurements in a powder sample cannot be expected to give better quantitative agreement than found here. The case of V12 is particularly interesting. In fact the low temperature proton relaxation data (see Fig. 15(c)) can be fitted by the sum of a term described by (18) with a gap value in agreement with Table 2 and of a constant term. In other words, it appears that the relaxation rate saturates at low   T instead of decreasing as expected for the "condensation" of the AFM ring in the singlet ground state. There are several reasons to exclude the possibility that the constant T_1^{-1} at low T in V12 is due to paramagnetic impurities [35]. Then one can speculate that the low T contribution to T_1^{-1} is due to quantum fluctuations in the singlet ground state for the quantum   s = 1 / 2 system (V12), fluctuations not evident for the classical s = 3 / 2 (Cr8) and s = 5 / 2 (Fe6) rings.

The case of Cu8 appears to be the ideal case for the study of the energy gap and the singlet ground state since the gap is larger than room temperature. The Cu8 ring has been investigated both by proton NMR and [63,65] Cu NMR and NQR [36]. The [63,65] Cu spectrum is complicated by the presence of four non equivalent Cu sites. The [63,65] Cu NQR spectra are composed of four separate lines for each isotope plus an additional line (probably due to impurities), spanning over the frequency range 16–22 MHz. The [63,65] Cu NMR spectrum in high field displays the powder pattern of a central line transition broadened by second order quadrupole e ects (di erent for each of the four inequivalent sites) and anisotropic paramagnetic shift. On the other hand, the proton NMR line is narrow (40 KHz) and field independent. The nuclear spin-lattice relaxation rate (NSLR) decreases exponentially on decreasing temperature for all nuclei investigated as shown in Fig. 16 and Fig. 17. In both figures we have drawn the curves corresponding to (21) with the gap value given in Table 2 and derived from susceptibility measurements. One can see that both for ^{1}H NMR and   [63,65] Cu NQR-NMR the experimental NSLR deviates from the exponential behavior at relatively high temperature in a manner similar to the case of V12 discussed above. However, in the case of Cu8 it appears more di cult to rule out the e ect of paramagnetic defects and thus it is still to be proved that the NSLR in rings of spin 1/2 in the singlet ground state is driven by quantum fluctuations.

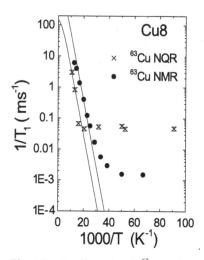

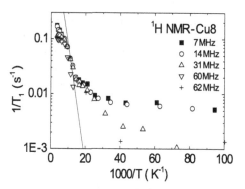

Fig. 16.   Semilog plot of ^{63}Cu-1/T_1 as a function of 1000 /T   in Cu8 powders

Fig. 17.   Semilog plot of ^{1}H-1/T_1 as a function of 1000 /T   in Cu8 powders

5.2 Level Crossing in AFM Rings

In AFM ring of the kind discussed above with $S = 0$ singlet ground state and $S = 1$ triplet excited state, an external magnetic field H removes the residual Kramers degeneracy of the triplet state and induces multiple level-crossings (LCs) at specific magnetic field values H_{ci}, whereupon the ground state of the molecule changes from $S = 0$ to $S = 1$ (H_{c1}), from $S = 1$ to $S = 2$ (H_{c2}), and so on. Because of magnetic anisotropy, the values of H_{ci} depend on the angle between H and the molecular axis z [15, 37]. The situation of near-degeneracy of the magnetic levels near the LC fields, and particularly the situation of level repulsion or anticrossing (LAC), is favorable to the observation of quantum tunneling and quantum coherence [38]. A situation of LC is also found between singlet and triplet states in 1D gapped quantum magnets [39], but the physical context and the continuum of excited states make the situation not comparable to that in finite-size magnets. For the spin dynamics near LC, a crucial issue is the role played by the coupling between magnetic molecular levels and the environment such as phonons and/or nuclear spins [40]. Essential information on this problem is accessed through measurements of the nuclear spin-lattice relaxation rate $1/T_1$ since the nuclei probe the fluctuations of the local field induced at the nuclear site by the magnetic moments of the transition ions. The spin dynamics at LC was studied by means of ^{1}H NMR in Fe10 powder sample [41], Fe6:Li(BPh4) single crystal [42] and Cr8 single crystal [43] close to H_{c1} and H_{c2}. We will review in the following the main results and conclusions derived in the above mentioned studies.

One crucial issue in the study of LC in AFM rings is the structure of the magnetic levels at the critical field where the $S = 0$ state becomes degenerate with the $S = 1$, $M = 1$ state (first level crossing), and similarly for other LC's. If the magnetic Hamiltonian does not contain terms that admix the two degenerate levels, one can have in principle a true LC. If on the other hand the Hamiltonian contains terms which strongly admix the levels, one expects a gap at the critical field resulting in what we call level anticrossing (LAC). The spin Hamiltonian describing an N-membered ring-like spin topology can be written as:

$$H = J \sum_i s_i \cdot s_{i+1} + \sum_i U(s_i) + \sum_{i>j} U_{i,j}(s_i, s_j) + g\mu_B H \sum_i s_i, \qquad (22)$$

where i and j run from 1 to N. The first term is the nearest-neighbor Heisenberg exchange interaction, the second and third terms represent the crystal-field anisotropies and the anisotropic spin-spin interactions (including eventually a Dzyaloshinski-Moriya (DM) term $C_{ij} s_i \times s_j$), respectively, and the last one is the Zeeman term. In (22), $s_{N+1} = s_1$ and the exchange interactions are supposed to be limited to nearest neighbors. Magnetic torque and specific heat measurements have been performed as a function of magnetic field in molecular rings [15, 37, 42]. The former is a powerful tool to locate level crossings, which lead to abrupt variations of the torque signal at low temperature, and to determine the ZFS parameters (D_S) of excited spin states. Specific heat measurements, when performed at sufficiently low temperature, are in principle able to distinguish a true LC from a LAC [42]. Particularly interesting is the possibility to extract the value of the energy gap at the anticrossing field, which is directly proportional to the matrix element connecting the two states involved.

Measurements of proton spin-lattice relaxation as a function of the magnetic field at fixed T are shown in Figs. 18, 19 and 20 for the three AFM rings Fe10, Fe6(Li) and Cr8. In all three cases a strong enhancement of T_1^{-1} is observed in correspondence to critical field values for LC (or LAC). The enhancement is clearly related either to cross relaxation effects or to magnetization fluctuations due to the almost degeneracy of the crossing magnetic levels as will be discussed further on. In Fe10 the peaks of T_1^{-1} are at 4.7, 9.6 and 14 T. The field dependence of the gap in Fe10 leading to the first LC for H parallel to the z axis should be $(H) = (0) + D_1/3 - g\mu_B H$, where the anisotropy parameter $D_1 = 3.2\,K$ and $(0) = 5.5\,K$ from Table 2. The gap closes for $H = 6.3\,K/1.33\,K/T \quad 4.7\,T$ in excellent agreement with the NMR data. Since the measurements in Fe10 were performed in a powder sample

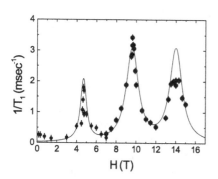

Fig. 18.   Magnetic field dependence of ^1H-$1/T_1$ in Fe10 powders measured at low temperature $T \quad 1.5\,K$

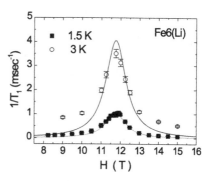

Fig. 19.   Magnetic field dependence of ^1H-$1/T_1$ in Fe6(Li) single crystal measured at $T = 1.5$ and $3\,K$, for $\quad 20$ ($= $ angle between H and the molecular axis)

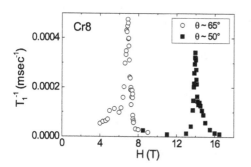

Fig. 20.   Magnetic field dependence of ^1H-$1/T_1$ in Cr8 single crystal measured at $T = 1.5\,K$, for two slightly different values of the angle between H and the molecular axis

there is a distribution of LC fields which generate a broadening of the T_1^{-1} vs H curve. In Fe6(Li) single crystal the maximum of T_1^{-1} can be observed at 11.7 T in excellent agreement with torque measurements for the same angle 25 between the z axis and H. In Cr8 two peaks in T_1^{-1} could be observed at 6.85 T and 13.95 T. The measurements were performed in single crystal but the orientation of the crystal is not defined. From the knowledge of $D_1 = 2.3$ K and (0) = 8.5 in Table 2 one can deduce that the NMR experiments were performed at 65 (1st crossing) and 50 (2nd crossing).

Having established that the peaks in the proton T_1^{-1} vs H curve are indeed evidence for LC (or LAC) it remains to be seen what kind of information one can obtain about the spin dynamics at level crossing from NMR measurements. The processes involved in nuclear spin-lattice relaxation close to LC or LAC are currently not well understood. Ultimately, the challenge is to be able to distinguish between thermal and quantum spin fluctuations in the molecule. In the presence of true LC or very small LAC, there are two magnetic field values for which the level separation $_1 = g\mu_B |H_{ci} - H|$ matches the ^{1}H Zeeman energy h $_L$. At these two field values, one expects a large enhancement of $1/T_1$ due to a direct exchange of energy between electronic and nuclear reservoirs, i.e., a cross-relaxation [18, 41]. The cross-relaxation depends on the relative values of the nuclear and electron spin-lattice relaxation rates T_1 and T_1^{el} respectively, which limit the energy transfer to the "lattice", and on the rate of internal energy exchange between the nuclear Zeeman reservoir and the molecular magnet reservoir. If T_1^{el} is much shorter than the nuclear T_1 and the exchange time between reservoirs is very short, one expects a large T-independent double-peak of T_1^{-1} close to LC. On the other hand, if T_1^{el} T_1, the cross-relaxation may not be e ective. However, one should keep in mind that in practice this situation can only occur in the case of ultra small LACs (as small as the nuclear Zeeman energy, i.e., of the order of mK) and that the separation between the two peaks of 1 /T$_1$(H) would also be as small as the nuclear Zeeman energy. Furthermore, to be experimentally observable as two separate peaks with finite width, this mechanism would require some broadening of the levels. If the LAC is large (with respect to nuclear energies), the nuclear relaxation should have dominant contributions from indirect processes (e.g., Raman-like, Orbach-like [18, 40]). The width of the peak and its T-dependence may be influenced by many parameters, and should certainly be di erent from the above-cited case of cross-relaxation. To obtain a phenomenological expression for 1 /T$_1$ near the LC fields, we suppose that the fluctuations of the magnetization between two adjacent magnetic states will drive the nuclear relaxation, that can be written as:

$$\frac{1}{T_1} = A^2 J(_L),$$ (23)

where J($_L$) is the spectral density of spin fluctuations at the nuclear Larmor frequency $_L$ and A is an average hyperfine coupling constant. In a phenomenological model where the spin-spin correlation function is supposed to decay exponentially with time, we assume

$$\frac{1}{T_1(T,H)} = \frac{A^2}{^2 + (h_L - (H))^2},\tag{24}$$

where   is a temperature-dependent damping factor associated with level broadening. In the case of nuclear relaxation induced by a purely quantum process, one does not expect any   T dependence of T_1. The relaxation can still be described by an expression similar to (24), with a   T-independent   , now having a di erent meaning. Near the LC fields we can write

$$(H) = [g\mu_B (H_{ci} - H)]^2 + {}^2_i \Big]^{1/2},\tag{25}$$

where $_i$ is the temperature independent "gap" at the anticrossing. The use of (24) near the LC condition is justified by the fact that the next excited state is at least several Kelvin higher in energy. Three di erent cases regarding the peak of T_1^{-1} vs H can be distinguished:

a) for $_i$ (T), the width of the peak is determined by $_i$ while the intensity depends mainly on   (T). Qualitatively, for   H = H_{ci} one has $1/T_1(T)$ (T). This situation corresponds to a well-identified LAC.
b) for $_i$ (T), both the width and the intensity of the peak are determined by   (T), and for   H = H_{ci} one has $1/T_1(T)$ 1/   (T). In this situation, it is not possible to identify a LAC, either because there is a true LC or because $_i$ is too small to influence NMR relaxation.
c) for $_i$ (T), the width of the peak is due to both $_i$ and   (T).

Note that in general a dependence $T_1^{-1}(H = H_{ci})$ vs. T shows a maximum if   (T) = | $_i$ - h $_L$|.

Let us now focus on the experimental results. As previously noted, in Fe10 three broad peaks were observed in the powder sample at the first three LC fields (see Fig. 18). In the range 1   .5 < T < 4.2 K, it was shown that $1/T_1(T) =$ const. , thus suggesting that   is temperature independent. Out of the crossing field values, 1 /T_1 has an exponential behavior [41] (as also shown by the authors of [34]). Since Fe10 data can be fitted by (24)) assuming $_1 = 0$, the physical situation could correspond to the case discussed above at point b), i.e., a true LC or small LAC. First principles T_1^{-1} calculations confirmed this scenario [44].

In the case of Fe6(Li) the peak of the dependence 1   /T_1 vs H (see Fig. 19) is quite broad in spite of the fact that the measurements are done in single crystal. The NMR data appear to confirm the specific heat results which are suggestive of a LAC at $H_{c1} = 11.8$ T with an anticrossing gap $_1/k_B = 0.86$ K [42]. In fact, the proton relaxation data can be fitted well near the maximum by using (24) with the above values for the parameters and with   A = 8.5 × 10^7 rad s^{-1}. The two curves in Fig. 19 at 1.5 K and 3 K are well reproduced by assuming a damping factor varying quadratically with temperature T^2

[42]. We note that the   T-dependent broadening   of the magnetic levels is
small (= 0.26 T at 3 K) and thus it a ects only the magnitude of 1   /T$_1$,
while the width of the 1 /T$_1$ peak is determined by the anticrossing gap $_1$.
This corresponds to the situation of the above-cited case (a). The observation
of a large LAC has raised the problem of explaining its occurrence in terms
of antisymmetric interaction in (22) [42, 45, 46].

In Cr8, two LCs are observed in the NMR (see Fig. 20). Both peaks
show a high field shoulder whose origin is not yet understood and will be
disregarded in the discussion which follows. The peaks of 1   /T$_1$ are narrower
than in Fe6:Li(BPh4) by a factor   2.5 for the first crossing and by a factor
   4 for the second crossing. Actually, the dipolar or hyperfine interaction extracted from ^{1}H NMR spectra were estimated to yield a broadening of about
0.1 ÷ 0.2 T, not far from the measured width of the peak of 1   /T$_1$ on the second
crossing. This suggests that LAC e ects are smaller in Cr8 than in Fe6, or
that the damping factor in Cr8 is comparable to or greater than the energy
gap at the anticrossing. This experimental observation seems to be in qualitative agreement with specific heat data [47] that indicate anticrossing gap
values much smaller than in Fe6(Li). However, from   C(H) data it was not
possible to obtain more than an upper limit for $_1$ (0.2 K). As a consequence, one has to determine all three parameters   A, $_1$ and   in (24) from
T$_1$ data alone. This requires extensive   T$_1$ measurements as a function of   H
and T, currently not available. We note that the above discussion in terms of
the phenomenological expression (24), is strictly valid only in the very vicinity
of the LC critical fields. The field dependence at a fixed very low temperature
value over a wide range of H on both sides of H$_{ci}$'s follows a behavior of
the type 1 /T$_1$ = A(H, T) exp(−   (H)/k$_B$ T). The exponential dependence of
1/T$_1$ with a field-dependent gap parameter   (H) reflects the same nuclear
spin-lattice direct relaxation process described by (21) in the previous section.

5.3 Local Spin Configuration in the Ground State
of Nanomagnets Mn12, Fe8

As mentioned before, at very low temperature (4 K) both Mn12 and Fe8
are in their magnetic ground state and the magnetization of the molecule is
frozen in the time scale of an NMR experiment leading to the possibility of
observing zero field NMR. The ^{55}Mn NMR spectrum in zero external field in
Mn12 was first observed by Goto et al. [48]. A detailed study was later reported by us [49]. The ^{55}Mn NMR spectrum is shown in Fig. 21. The narrow
low frequency line originates from the Mn $^{4+}$ ions while the two broader lines
are from the Mn $^{3+}$ ions whereby the broadening of the latter two lines is of
quadrupole origin [49, 50]. The ^{57}Fe NMR signal in zero external field was
detected in isotopically enriched Fe8 [51, 52] and the complete spectrum is
shown in Fig. 22. The eight narrow lines correspond to the eight non equivalent Fe sites in the molecule. The identification of the eight lines with the
corresponding Fe sites was possible from measurements as a function of the

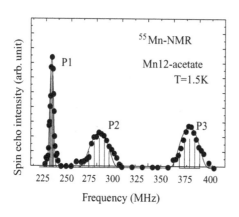

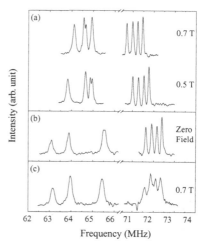

Fig. 21. ^{55} Mn-NMR spectra in Mn12 oriented powders measured at T = 1.5 K under zero magnetic field

Fig. 22. ^{57} Fe-NMR spectra on Fe8 powders measured at T = 1.5 K. (a) H = 0.5 and 0.7 T parallel to the easy axis; (b) H = 0; (c) H = 0.7 T perpendicular to the easy-axis

applied magnetic field [51, 52]. For protons the internal field is small and the NMR in zero field is weak and can be observed only at low frequency over a broad frequency interval (2–4 MHz) due to the presence of many inequivalent proton sites in the molecule. We have observed the signal in zero field but the complete spectrum has never been published. On the other hand the proton NMR in an external magnetic field can be observed easily. It shows a broad structured spectrum with a field independent shift of the lines of order of the internal field. The proton spectrum in Fe8 at 1.5 K is shown in Fig. 23 [27, 53]. The proton spectrum in Mn12 at low temperature is shown in Fig. 24 [60]. Further details about the zero field NMR and about the analysis of the hyperfine field at the different nuclear sites are given in [49] and [50].

One important issue in molecular nanomagnets is the knowledge of the local spin configuration corresponding to the collective quantum state described by the total spin i.e., S = 10 for both Mn12 and Fe8. This issue has been addressed very successfully by NMR and we will summarize the results for both Mn12 and Fe8 in the following.

Mn12

Figure 25 shows the external magnetic field dependence of the ^{55} Mn resonance frequencies of the three signals in the spectrum, with the magnetic field applied along the easy-axis in Mn12. With increasing parallel field, P1 shifts to higher frequency while the other two peaks P2 and P3 (pertaining to Mn^{3+} ions) shift to lower frequency. Since the resonance frequency is proportional to the

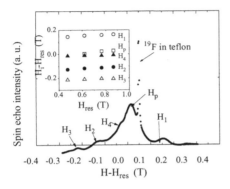

Fig. 23. ^{1}H-NMR spectrum of Fe8 oriented powders at 36 .8 MHz (H_{res} = 0.864 T) and T = 1 .4 K. In the inset the shifts of the lines at di erent fields $H_{1...4,p}$ as a function of H_{res} are shown to be constant. Here H is parallel to the easy-axis

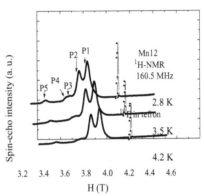

Fig. 24. ^{1}H-NMR spectra in Mn12 oriented powders measured at 160.5 MHz for three di erent temperatures. Here H is parallel to the easy-axis

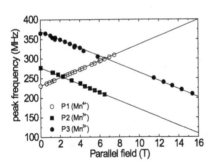

Fig. 25. Parallel field dependence of resonance frequency for each ^{55}Mn-NMR peak at T = 1 .5 K

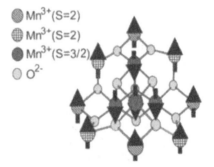

Fig. 26. Schematic structure of the magnetic core of Mn12 and orientation of the Mn moments in the ground state

vector sum of the internal field (H_{int}) and the external field (H_{ext}) i.e., $R = N|H_{int} + H_{ext}|$, this result indicates that the direction of the internal field at the Mn sites for Mn $^{3+}$ ions is opposite to that for Mn $^{4+}$ ions. Since H_{int} originates mainly from the core-polarization [49], H_{int} is negative and the direction of the internal fields at nuclear sites is opposite to that of the Mn spin moment. Thus one can conclude that spin direction of Mn $^{4+}$ ions is antiparallel to the external field, while that of Mn $^{3+}$ ions is parallel to the external field, corresponding to the standard spin structure of magnetic core of Mn12 cluster (see Fig. 26) [54].

In the case when the external magnetic field is applied perpendicular to the easy axis (which is the common axis of the oriented powder), the field dependence of the resonance frequencies is the one shown in Fig. 27. As described

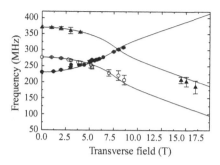

Fig. 27.   Transverse field dependence of resonance frequency for each   [55] Mn-NMR peak at T = 1 .5 K

above, the resonance frequency is proportional to the e ective internal field at the nuclear site, which is the vector sum of H_{int} due to spin moments and H_{ext} due to the external field i.e., $|H_{eff}| = |H_{int} + H_{ext}|$. Thus the opposite field dependence of $|H_{eff}|$ for Mn^{4+} and Mn^{3+} ions indicates that the direction of Mn^{4+} spin moments is antiparallel to that of Mn $^{3+}$ spin moments. This leads to the conclusion that the individual spin moments of both Mn^{4+} and Mn^{3+} ions do not cant independently along the direction of the transverse field but rather rotate rigidly maintaining the same relative spin configuration. For a detailed quantitative analysis of the results we refer to the original paper [55].

Fe8

A similar study has been performed in Fe8 at 1.5 K where the nanomagnet is in its ground state. The field dependence of the eight   [57] Fe resonance frequencies as a function of a magnetic field applied along the main easy axis is shown in Fig. 28. An analysis analogous to the one performed in Mn12 leads to the conclusion that the internal spin configuration of the ground state is as

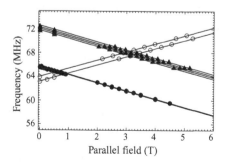

Fig. 28.   Parallel field dependence of resonance frequency for each   [57] Fe-NMR peak at T = 1 .5 K

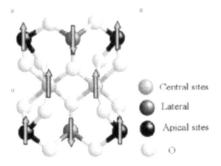

Central sites

Lateral

Apical sites

○

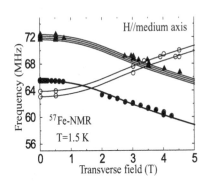

Fig. 29. Structure of Fe8 and orientation of the Fe moments in the ground
state

Fig. 30. Transverse field dependence
of resonance frequency for each [57] Fe-
NMR peak at T = 1 .5 K

depicted in Fig. 29 [56]. When the magnetic field is applied perpendicular to
the main easy axis and parallel to the medium axis in the   xy hard plane,
the field dependence of the resonance frequencies is quite different, as shown
in Fig. 30. Again, a quantitative analysis of the results leads to the same
conclusion as for Mn12 i.e., that the local spin configuration remains intact
up to at least 4 T transverse field whereby the effect of the field is the one of
rotating rigidly the eight Fe [3+] moments. For details see [52].

5.4 Spin Dynamics and Quantum Tunneling of the Magnetization
in Nanomagnets: Mn12, Fe8

At very low temperature, when the nanomagnet occupies mostly the magnetic
ground state, the thermal fluctuations of the magnetization become vanishingly small. In this temperature range the spin dynamics is dominated by
quantum fluctuations. In this paragraph we will concentrate on the issue of
what information NMR can give about the phenomenon of quantum tunneling of the magnetization (QTM). Let us first briefly summarize the QTM
phenomenon. The ground state of both Mn12 and Fe8 clusters is a high total
spin ground state i.e., $S = 10$. The $S = 10$ ground state is split into eleven
sublevels by a strong easy axis anisotropy [13, 14, 57]. The remaining Kramers
degeneracy is removed by an external magnetic field directed along the z easy
axis. The energy levels for H // z are obtained from Hamiltonian (19) as:

$$E_m = -Dm^2 - Bm^4 + g\mu_B Hm . \tag{26}$$

Assuming $g = 2$ one has $g\mu_B H = 1.33H$ (K) for H in Tesla. For Mn12 one
has $D = 0.55$ K, $B = 1.2 \times 10^{-3}$ K; for Fe8 one has $D = 0.27$ K, $B = 0$ and
in the total Hamiltonian $H = -DS_z^2 - BS_z^4 + gBH \cdot S$, the rhombic term
$E(S_x^2 - S_y^2)$, with $E = 0.046$ K, must be added. This term modifies the (26)
of the energy levels in a non-simple way.

Below liquid helium temperature the clusters occupy mostly the $m = \pm 10$
states and the reorientation of the magnetization between these two states becomes extremely long (about one day for Mn12 at 2.4 K) due to the anisotropy
barrier giving rise to a pronounced superparamagnetic behavior [13, 14, 57].
When the relaxation rate of the magnetization is measured in response to a
varying magnetic field H_z along the easy axis, peaks are observed which have
been interpreted as a manifestation of resonant tunneling of the magnetization [26, 57]. The qualitative explanation is that the relaxation rate of the
magnetization is maximum at zero field and at field values where the total
spin states become pairwise degenerate again. The longitudinal field at which
this occurs can be easily calculated from (26) with the parameters given for
Mn12 and Fe8, respectively. It is this degeneracy which increases the tunneling probability and thus shortens the relaxation time. The size of the e ect
depends on terms not shown in the Hamiltonian (19), i.e., terms which couple
the pairwise degenerate states. In particular, a transverse magnetic field component can greatly enhance the tunneling splitting of the degenerate levels
and thus the QTM. On the other hand, the QTM is reduced by the smearing out of the energy levels of the spin states due to spin-phonon coupling,
intermolecular interactions, and/or hyperfine interactions with the nuclei [58].

NMR can detect the QTM in two quite di erent ways which will be briefly
described in the following.

Measurement of the Relaxation Rate of the Magnetization by Monitoring the NMR Signal Intensity in O -Equilibrium State

The idea is quite simple. We have seen above that in the NMR spectrum at
low temperature in Mn12 and Fe8 the position of the resonance lines depend
upon the internal field due to the magnetization of the molecule. Thus if an
external field is also applied the position of the line depends on the vector sum
of the external field and the internal field, the latter being directed along the
magnetization of the molecule. If the direction of the external field is suddenly
reversed (or the sample is flipped by 180   ) the position of the NMR line
changes in the new o -equilibrium situation. The intensity of that particular
line starts from zero and grows back to the full intensity as the magnetization
of the molecule relaxes back to equilibrium along the applied field. The method
was first described for proton NMR in Mn12 by looking at the echo intensity at
the Larmor frequency when the external field is turned on adiabatically at low
temperature [59]. A more straightforward implementation was later applied
to the signal intensity of shifted proton lines in Mn12 as described below. The
detailed results for proton NMR in Mn12 are discussed in [60].

The magnetization of the Mn12 clusters is initially prepared in equilibrium
conditions with the magnetic field along the easy c-crystal axis. By inverting
the magnetic field, or better, by flipping the oriented powder by 180   one creates an o -equilibrium condition whereby the magnetization of each molecule
wants to realign along the external field ($m = -10$ to $m = +10$ transition). At

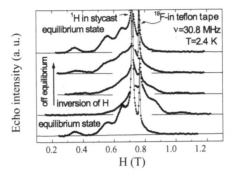

Fig. 31.   Time evolution of ^{1}H-NMR spectrum of Mn12 oriented powders measured at 30.801 MHz and T = 2.4 K, for H parallel to the easy-axis. The spectrum at the bottom corresponds to the thermal equilibrium state. The second lowest spectrum corresponds to the o-equilibrium situation following the field inversion. The remaining spectra from bottom up are taken at di erent times after the field inversion

low temperature and in magnetic fields less than 1 T this process is prevented by the crystal field anisotropy and proceeds very slowly via spin tunneling and phonon assisted relaxation [57, 58]. Figure 31 shows the experimental results. The spectrum at the bottom of Fig. 31 corresponds to the thermal equilibrium state before the inversion, where the easy axis of the clusters is along the magnetic field. Just after the inversion of direction of the sample, the observed spectrum changes drastically as shown in the second spectrum from the bottom of Fig. 31. In the figure, the time evolution is from the bottom up. Since the spectra were obtained by sweeping the magnetic field, a process, which takes about 30 minutes for each spectrum, the spectra do not correspond to a precise o-equilibrium state. However, since the overall process of relaxation of the magnetization at this temperature takes a two or three hundred minutes, the di erent spectra give a qualitative idea of the evolution of the NMR spectrum with time. The signals of the shifted peaks with positive hyperfine fields disappear, while new signals can be observed at magnetic fields higher than the Larmor field H_0, where no signal could be detected before the inversion. After a long time (for example, 400 minutes in this case), the spectrum becomes independent of time and recovers the initial shape before the field inversion.

In order to investigate the e ect quantitatively one can sit at fixed field on one of the shifted lines (see Fig. 24) and follow its amplitude as a function of time without need of recording the full spectrum. The signal intensity for each shifted peak (P2 to P5 shown in Fig. 24) in the spectrum at thermal equilibrium corresponds to the total number of clusters occupying the magnetic $m = -10$ ground state. Immediately after the 180 -rotation of the sample the state $m = -10$ becomes $m = +10$. Then the growth of the signal intensity for each peak after the inversion is proportional to the increase of the number of clusters which return to the $m = -10$ new ground state. Therefore, we can measure a relaxation time of magnetization by monitoring the echo intensity as a function of time. Figure 32 shows a typical time dependence of the echo intensity $h(t)$ measured at 0.4212 T (at the position of the P2 peak) and at $T = 2.4$ K. The experimental results can be fitted tentatively by the expression

$$h(t) = a\,(1 - e^{-t/\tau(H)}) + b, \qquad (27)$$

where $\tau(H)$ is a relaxation time and $a + b$ is the echo intensity for the thermal equilibrium state, $h_{T.E.}$. As can be seen in the figure, the growth of the signal intensity is well fitted by a single exponential function (27) except for an initial fast growth, which accounts for about 30% of the signal. From the slope of $1 - h(t)/h_{T.E.}$ on semi log plot, we can estimate $\tau(H)$. In Fig. 33 it is shown the comparison of the field dependence of the relaxation time measured with NMR and the one measured directly by monitoring the magnetization with a SQUID. Although the two sets of data refer to two different ways of extracting the relaxation time from the recovery curves, one can conclude that the results from both methods are in good agreement. In particular, in both cases one sees minima of $\tau(H)$ at the level crossing fields i.e., $H = 0$ (only for the magnetization), $H = 0.45$ T, and $H = 0.9$ T. Except for the minima, indicated by the arrows in Fig. 33, the H-dependence of $\tau(H)$ follows a thermally activated law $\tau(H) = \tau_0 e^{(67 - 13.3H)/k_B T}$ with $\tau_0 \approx 10^{-6}$ s which is consistent with background thermal excitations over the barrier due to the

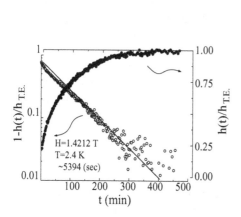

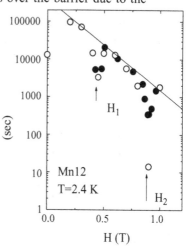

Fig. 32. Time dependence of the echo intensity in Mn12 oriented powders taken at $H = 0.4212$ T and $T = 2.4$ K following the sudden inversion of the orientation of the easy axis of the sample with respect to the applied magnetic field. Solid and open circles represent $h(t)/h_{T.E.}$ and $1 - h(t)/h_{T.E.}$, respectively

Fig. 33. Field dependence of the relaxation time for Mn12 (oriented powders) magnetization measured at $T = 2.4$ K by ^{1}H-NMR (closed circles) and SQUID (open circles). The solid line represents $\tau(H) = \tau_0\, e^{(67 - 13.3H)/k_B T}$ discussed in the text. H is parallel to the easy-axis

anisotropy as modified by the applied magnetic field. The field values corresponding to the minima in the relaxation time agree with critical fields where the magnetic level crossing occurs. It should be pointed out that the NMR method of monitoring the relaxation of the magnetization of the molecule is not entirely equivalent to the thermodynamic method since the NMR signal is a local probe of the magnetization. This subtle difference suggests that in the NMR method there is a microscopic information about the mechanism of reversal of the magnetization. The method has been confirmed [50, 61, 62] by measuring the ^{55}Mn NMR spectrum in Mn12, with results very similar to the ones obtained with proton NMR [60], and it has been extended to very low temperature to observe the avalanche effect of the spin reversal in the magnetization recovery in Mn12 [63]. Also it has been applied to proton NMR in Fe8 at very low temperature to obtain information about Landau-Zener transition as the field is swept through a level crossing condition [64].

Nuclear Spin-Lattice Relaxation Induced by Quantum Tunneling of the Magnetization in Molecular Nanomagnets: Fe8, Mn12

Quantum fluctuations of the magnetization are expected to have a distinctive effect on nuclear spin-lattice relaxation (NSLR). For example, in the case of Fe8 the proton NSLR was observed to decrease very fast down to 400 mK as the result of slowing down of the thermal fluctuations but then to become temperature independent below 300 mK [64]. This leveling-off of proton T_1 was taken as indication of a crossover to a regime of quantum fluctuations of the magnetization. A more direct way to detect the effect of quantum tunneling on NSLR is obtained from the field dependence of T_1 as illustrated in the following.

The occurrence of QTM in zero external magnetic field or in an external field applied along the anisotropy axis is related to the splitting of the pairwise degenerate magnetic levels by an amount Δ_T that is due to off-diagonal terms in the magnetic Hamiltonian arising from anisotropy in the xy plane, intermolecular dipolar interactions, and hyperfine interactions. Normally the tunnel splitting Δ_T in the Mn12 and Fe8 clusters is much smaller than the level broadening so that measurements of Δ_T are difficult. The effect of QTM as a function of the longitudinal magnetic field can be seen as dips in the relaxation time of the magnetization, in correspondence with the critical fields for level crossing as shown in Fig. 33. However, under applied parallel field no clear effect could be observed in the proton T_1. This can be seen in Fig. 34 where we show the longitudinal field dependence of T_1 in both Mn12 and Fe8. The critical fields corresponding to the first level crossing in the ground state (i.e., m = −10 to m = 9) is estimated to be 0.5 T for Mn12 and 0.2 T for Fe8, respectively, from (23). As seen in Fig. 34 the data can be fitted well by the simple model of thermal fluctuations of the magnetization described in Paragraph 4.2 without any marked anomaly at the level crossing fields. Thus we may conclude tentatively that for H // z no effects of quantum tunneling

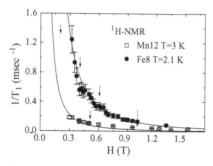

Fig. 34.   Parallel field dependence of ^{1}H-1/T$_1$ in both Mn12 (T = 3 K) and Fe8 (T = 2.1 K), oriented powders. The arrows indicate the critical field values for level crossing across the energy barrier

can be observed on 1 /T$_1$ since the tunneling dynamics is too slow for longitudinal applied fields. As a word of caution we may add that the data in Fig. 34 indicate some enhancement of the NSLR in Mn12 around the level crossing field and in Fe8 the data are incomplete since the first level crossing cannot be reached as the result of the experimental difficulties at low fields thus leaving open the possibility that a small effect of QTM may be observed for longitudinal fields. On the other hand, a recent study of ^{55}Mn NMR in Mn12 at very low temperature (down to 20 mK) has revealed the presence of (temperature independent) tunneling fluctuations [64].

On the other hand, by applying a magnetic field perpendicular to the easy axis (transverse field), one can increase τ of all levels while leaving the symmetry of the double-well potential intact [58, 65]. An increase of the tunneling splitting corresponds to an increase of the tunneling frequency or of the tunneling probability. For   H   1 T, in Fe8, the relaxation (fluctuation) of the magnetization driven by tunneling (coherent and/or incoherent) becomes so fast that it falls within the characteristic frequency domain (MHz) of a NMR experiment. Therefore, when the magnetic field is applied perpendicular to the main easy axis   z (transverse field) a pronounced peak in the spin-lattice relaxation rate, 1 /T$_1$, of protons in a single crystal of Fe8 as a function of external magnetic field can be observed at 1.5 K, as shown in Fig. 35. The effect is well explained by considering that by increasing the transverse field the incoherent tunneling probability becomes sufficiently high as to match the proton Larmor frequency. When the applied field goes through this condition the fluctuation rate of the magnetization is most effective in driving the nuclear relaxation and a maximum appears in 1   /T$_1$. The peak disappears when a parallel field component is introduced in addition to the transverse field, by tilting the single crystal about 5 degrees in   yz plane (see Fig. 35). Since the parallel field component removes the degeneracy of the   ±m magnetic states and consequently the possibility of tunneling, it is clear that the peak in 1 /T$_1$ must be related to a contribution to the nuclear relaxation rate

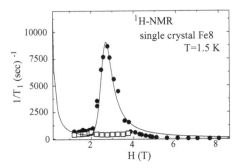

Fig. 35. Transverse field dependence of ^{1}H-1/T_1 at T = 1 .5 K in Fe8 single crystal as a function of the field along the medium axis (closed circles). Open squares are the results obtained when the single crystal is tilted so that the applied field is 5 degrees o the   xy-plane

from the tunneling dynamics. A quantitative interpretation of the e ect and further details can be found in [66].

Measurements of NSLR as a function of transverse field were also per-formed in oriented powder of Mn12 both on proton NMR and on ^{55}Mn NMR [55]. However, in this case we failed to observe a peak of the NSLR such as for Fe8 although the relaxation rate was found to be faster than expected on the basis of purely thermal fluctuations of the magnetization. The results are shown in Fig. 36 where we plot the transverse magnetic field dependence of ^{55}Mn-1/T_1 (P1; Mn $^{4+}$ ions). With increasing the transverse magnetic field, 1/T_1 increases rapidly by about two decades and shows a broad maximum around 5.5 T. We have calculated the expected field dependence obtained from the simple thermal fluctuation model described in Paragraph 4.2, a model which works well to describe the longitudinal field dependence. As seen from the comparison of the dotted line and the experimental points in Fig. 36, there appears to be an additional contribution to NSLR of ^{55}Mn in transverse field. This additional contribution was explained [55] by a phenomenological model which considers the e ect of canting of the magnetization in presence of a transverse field. In fact, the quantization axis of the nuclear spin does not coincide with that of the internal magnetic field due to the electron spins. Therefore the transverse field generates components of the local field perpen-dicular to quantization axis of the nuclear spin, whose fluctuation can be very e ective in producing nuclear relaxation. For the details of the model which generates the fitting curve in the figure we refer to [55].

The transverse field dependence of 1 /T_1 for protons at 1.5 K in Mn12 is also shown in the inset of Fig. 36. Contrary to ^{55}Mn, the proton 1 /T_1 has a rapid initial decrease with increasing field, roughly as H^{-2}, which can be explained very well by (18) in the limit of "slow motion" ($_N$ 1), where T_1^{-1} can be approximated with   A/ ($_N^2$). The key to understand the di erent behavior of 1 /T_1 for ^{55}Mn and for ^{1}H is the fact that the internal magnetic fields at proton sites are very small (at most 0.4 T in comparison with the case of Mn nuclei, 22 T for Mn^{4+} ions) and thus $\nu_N \propto H_{ext}$ for protons while ν_N depends very little on the external field for manganese. At higher fields the proton NSLR levels o to an almost constant value and no peak can be observed contrary to Fe8 (see inset of Fig. 36). The peak due to the matching of the Larmor frequency with the tunneling probability was expected in Mn12 in transverse field around 6– 7 T. We do not have as yet an explanation for the di erent behavior of Mn12 and Fe8 except for the fact that the data in Mn12 were taken in oriented powder since no single crystal large enough for NMR work is available for Mn12. The leveling-o of $1/T_1$ for protons at high fields is due to the same decrease of the lifetime which is the dominant e ect for ^{55}Mn. This can be seen by extracting fom the equation of $T_1^{-1} = A/(\nu_N^2)$ and the experimental points for both ^{55}Mn and ^{1}H. The two sets of data coincide within experimental error as shown in Fig. 37. In the figure, we show also for comparison the transverse field dependence of $^{-1}$ calculated from the spin phonon interaction. The di erence of the experimental points from the solid line represents the contribution to the fluctuation of the magnetization which cannot be related to spin-phonon interaction.

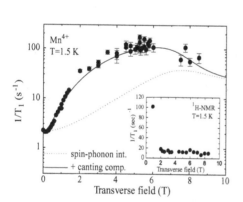

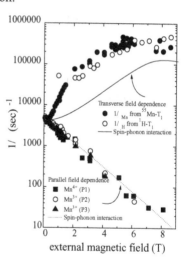

Fig. 36. Transverse field dependence of $1/T_1$ for ^{55}Mn-NMR at the Mn^{4+} site (P1) measured at T = 1 .5 K in Mn12. The inset shows the transverse field dependence of $1/T_1$ of ^{1}H-NMR

Fig. 37. External magnetic field dependence of $1/$ in Mn12 oriented powders, estimated from a relation of $T_1^{-1} = A/(\nu_N^2)$ for both ^{57}Mn and ^{1}H NMR. Solid and dotted lines are transverse and parallel field dependence of $^{-1}$ calculated from the spin-phonon interaction

6 Miscellaneous NMR Studies of Molecular Clusters: Fe2, Fe4, Fe30, Ferritin Core, Cr4, Cu6, V6, V15

In this paragraph we will briefly review the NMR work done in a number of magnetic molecular clusters that were not included in the previous paragraphs because they present specific features which do not fit entirely into the general issues discussed above.

6.1 Iron Clusters: Fe2, Fe4, Fe30 and Ferritin Core

The Iron(III) $S = 5/2$ dimer, [Fe(OMe)(dbm)$_2$]$_2$ (in short Fe2), has a non-magnetic $S = 0$ ground state. The separation between the singlet ground state and the first excited (triplet) state was determined by susceptibility measurements to be about 22 K; proton NMR measurements were performed on Fe2 [67]. The nuclear spin-lattice relaxation rate (NSLR) was studied as a function of temperature at 31 and 67 MHz and as a function of the resonance frequency (10–67 MHz) at $T = 295$ K. The results are shown in Fig. 38. At room temperature the ^{1}H NSLR is independent of frequency (see inset in Fig. 38) contrary to the strong field dependence found in AFM rings as discussed in Paragraph 3. The temperature dependence of the proton NSLR shows a monotonous decrease on lowering the temperature without the peak characteristic of the AFM as described in Paragraph 4.1. 1 $/T_1$ is approximately proportional to T where   is the uniform susceptibility measured at $H = 1$ T.

The cluster of four Iron(III) ions, Fe$_4$(OCH$_3$)$_6$(dpm)$_6$ (in short Fe4), is characterized by a total spin ground state $S_T = 5$ and Ising anisotropy [68]. The cluster behaves at low temperature like a superparamagnet just like Mn12 and Fe8 except that the anisotropy barrier is much lower. The $D = 0.29$ K in (26) corresponds to an energy barrier between the lowest $M_S = \pm 5$ state and

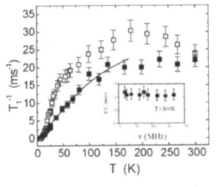

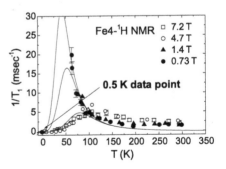

Fig. 38.   Temperature dependence of ^{1}H-1/T_1 in Fe2 powders. The inset shows magnetic field dependence of 1/T_1 at room temperature

Fig. 39.   Temperature dependence of ^{1}H-1/T_1 in Fe4 "oriented" powders

the highest $M_S = 0$ state of only 7.25 K. The Fe4 cluster has been investigated by ^{1}H NMR as a function of temperature (0.5–295 K) and external magnetic field (0.3– 7.2 T) [69]. The results of NSLR are shown in Fig. 39. At very low T (0.5 K) the spectrum becomes very broad indicating the freezing of the Fe^{3+} moments in a superparamagnetic state. The temperature dependence of T_1^{-1} is characterized by a field dependent maximum as shown in Fig. 39. For low magnetic fields the peak of $1/T_1$ becomes so high that the proton NMR signal cannot be detected in the temperature region of the peak. The field dependent maximum as well as the loss of NMR signal in the region of the peak at low magnetic fields (see Fig. 39) is qualitatively similar to the one observed in Mn12 [70]. In Mn12, ^{13}C [71] and ^{2}D NMR [72, 73] were also performed giving similar qualitative experimental results. In [73] an isotope e ect on NMR data was also suggested. It should be noted that the field dependent peaks observed in Fe4 and in Mn12 with maxima located at temperatures of the order of the exchange constant   J have many similarities with the peaks discussed for AFM rings in Sect. 4. A systematic analysis of the nuclear relaxation data in ferrimagnetic molecular clusters is in progress to establish the relaxation mechanism at these intermediate temperatures.

The cluster $\{Mo_{72}Fe_{30}O_{252}(Mo_2O_7(H_2O))_2(Mo_2O_8H_2(H_2O))–(CH_3COO)_{12}(H_2O)_{91}\} \times 150H_2O$ [2], Fe30 in brief, has 30 Fe(III) ions occupying the 30 vertices of an icosidodecahedron. The magnetic properties are characterized by a ground state with total spin state $S_T = 0$ due to O-Mo-O bridges me-diating antiferromagnetic (AF) coupling with the exchange coupling constant $J = 1.57$ K between nearest-neighbor Fe ions. An accurate description of the magnetic properties of the cluster has been based on classical Heisenberg model of spins on the vertices of an icosidodecahedron for arbitrary magnetic fields [2, 74]. ^{1}H nuclear magnetic resonance (NMR) and relaxation mea-surements have been performed in the Keplerate species Mo72Fe30 [75]. The ^{1}H NMR linewidth increases gradually with decreasing   T and it saturates below about 4 K, as expected for a non-magnetic ground state with $S_T = 0$. The results for the magnetic field and temperature dependence of T_1^{-1} are shown in Figs. 40 (a) and (b). The magnetic field dependence of T_1^{-1} at room temperature (see Fig 40 (a)) can be fitted by (7) used to fit the results in AFM rings. The best fit constants $A = 5$ ms^{-1}, $B = 0.1$ T and $C = 2.2$ ms^{-1} compare well with the constants in Table 1. The small value of   B, correspond-ing to $_A = _e B = 1.7 \times 10^{10}$ (rad s^{-1}), is consistent with a highly isotropic Heisenberg system. As shown in Fig. 40 (b) a strong enhancement of T_1^{-1} with a peak around $T = 2$ K is observed, a feature similar to the one observed in antiferromagnetically (AFM) coupled rings (see Paragraph 4.1). It must be mentioned that the first direct experimental confirmation of the transition from $S_T = 0$ to $S_T = 1$ for standard Fe30, was from   µSR experiments where, in correspondence to the first crossing field H_c 0.24 T, the width of the distribution of the local fields at the muon site tends to a plateau [76].

An antiferromagnetic molecular cluster containing a large number of Fe(III) ions (a variable number of the order of 5000) is the biomolecule ferritin.

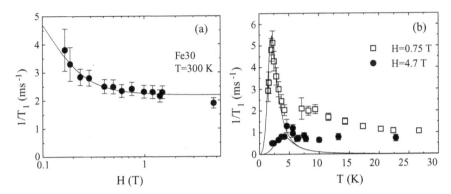

Fig. 40. (a) Magnetic field dependence of ^{1}H-1/T_1 at 300 K in Fe30 powders. (b) Temperature dependence of 1 /T_1 measured at H = 0.75 K (open squares) and H = 4.7 T (closed circles)

It is commonly reported [77] that ferritin becomes antiferromagnetic (AF) although the evidence is largely indirect and the Neel temperature reported varies widely in the range 50 K < T_N < 240 K. Moreover, for T < 30 K, magnetization measurements [78] give evidence of superparamagnetic relaxation and of spin freezing of the total magnetic moment associated with the uncompensated spins at the boundary of the cluster of iron ions. Recently the interest in the magnetic behavior of ferritin has been revived by the confirmation [79, 80] of an early report [81] of the occurrence of macroscopic quantum coherence phenomena at low temperature. Since the naturally occuring ferritin biomolecule contains the largest number of protons in the outer protein shell and thus are not coupled to the iron ions in the core, one has to strip the protein shell in order to use the NMR of the protons inside the core to probe the magnetic behavior of the Fe $^{3+}$ ions.

Proton NMR was performed in ferritin core as a function of temperature from 220 K down to 4.2 K at 4.7 T [82]. The proton spectrum is inhomogeneously broadened and the NSLR depends on the position in the spectrum where the measurements are performed making the analysis of the results difficult and uncertain. The relevant result of the investigation is represented by the maximum in the NSLR observed at a temperature of about 100 K which is a clear indication of the occurrence of an antiferromagnetic transition. The study has to be viewed as preliminary. More experiments are needed to investigate the critical behavior of the NSLR around the ordering temperature and the effect of the superparamagnetic fluctuations and quantum tunneling.

6.2 Ferromagnetic (FM) Ring (Cu6) and Cluster (Cr4)

Both [(PhSiO $_2$)$_6$Cu$_6$(O$_2$SiPh) $_6$] (in short Cu6) and [Cr $_4$S(O$_2$CCH$_3$)$_8$(H$_2$O)$_4$] (NO$_3$)$_2 \times$ H$_2$O (in short Cr4) are molecular nanomagnets with nearest neighbor ferromagnetic exchange interaction. Cu6 is a planar ring of six spin 1 / 2 Cu^{2+}

ions with $J = 61\,K$ and a ground state with $S_T = 3$ separated by an energy gap of $30\,K$ from the first excited state $S_T = 2$ [83, 84]. The ground state is split by an anisotropy $D = 0.435\,K$. Thus in presence of an external magnetic field H, the energy levels of the 7 magnetic substates m in $S_T = 3$ state are given by:

$$\frac{E_m}{k_B} = 0.435\,m^2 - 1.33\,m\,H \quad (Kelvin, with\ H\ in\ Tesla)\ . \quad (28)$$

On the other hand, Cr4 is a cluster formed by four spin 3/2 Cr $^{3+}$ ions on the vertices of almost regular tetrahedron [85]. The exchange constant $J = 28\,K$ and the ground state has $S_T = 6$. The crystal field anisotropy is negligibly small.

Proton NMR measurements have been performed in both systems in the low temperature range where the magnetic clusters are in their total S_T ground state [84, 85]. It is noted that contrary to Mn12 and Fe8 no superparamagnetic e ects are observable since the anisotropy for Cu6 is positive and for Cr4 is negligible. Thus no energy barrier is present for the reorientation of the magnetization even at low temperature. These FM clusters could be thought of as soft nanomagnets in contrast to Mn12 and Fe8 which are hard nanomagnets. Therefore in the temperature range of our measurements (1.4–4.2 K) the thermal fluctuations of the magnetization are still faster than the NMR frequency and no static local field is present at the proton site (i.e., no spin freezing). The systems behave like an assembly of very weakly interacting paramagnetic spins of size $S_T = 3$ and $S_T = 6$ for Cu6 and Cr4, respectively. The results for the proton NSLR are shown in Fig. 41 and Fig. 42 for Cu6 and Cr4, respectively.

The NSLR results can be explained very well by a simple model of spin-lattice relaxation due to the fluctuations of the total magnetization of the cluster. One can assume that the transverse hyperfine field at the proton site due to the interaction with the di erent Cu (Cr) magnetic moments is

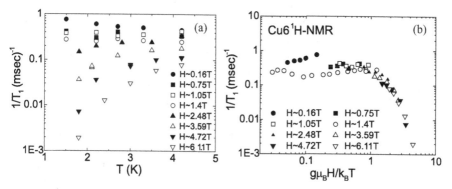

Fig. 41. ^{1}H-1/T_1 in Cu6 powders. (a) 1/T_1 plotted as a function of temperature for assorted values of the magnetic field. (b) 1/T_1 plotted as a function of $g\mu_B H/k_B T$

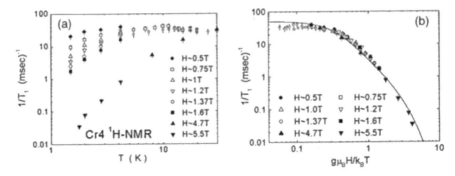

Fig. 42. ^{1}H-1/T$_1$ in Cr4 powders. (a) 1/T$_1$ plotted as a function of temperature for assorted values of the magnetic field. (b) 1/T$_1$ plotted as a function of $g\mu_B$ H/k$_B$ T

proportional to the component of the magnetization of the cluster in the direction z of the applied field i.e., $h_{\pm}(t) = C M_z(T) = C (M_z(T) + M_z(T))$. Then from (14) it follows for the NSLR:

$$\frac{1}{T_1} = \frac{1}{2} {}^2C^2 \quad M(0)^2 e^{-t} e^{-i\omega_L t} dt = A \quad M(0)^2 \frac{}{^2 + \frac{2}{L}}, \quad (29)$$

where we dropped the subscript z for simplicity. The characteristic frequency can be viewed classically as a probability per unit time for the total spin $S_T = 3$ (6) of the Cu6 (Cr4) cluster to change its orientation along the external magnetic field or, quantum mechanically, as the broadening of the corresponding magnetic eigenstate. The average of the square of the fluctuation can be calculated exactly as:

$$M(0)^2 = \frac{k_B T}{(g\mu_b)^2} (H,T) = \sum_m \frac{m^2 e^{-E_m/kT}}{Z} - \left(\sum_m \frac{m e^{-E_m/kT}}{Z} \right)^2, \quad (30)$$

where E_m is given by (28). The data in Figs. 41 (a) and 42 (a) have been replotted in Figs. 41 (b) and 42 (b) as a function of $g\mu_B$ H/k$_B$ T. In Cr4 the NSLR data at different fields and temperatures rescale well with $g\mu_B$ H/k$_B$ T indicating that $_L$ in (29). The data are reproduced very well by a theoretical curve obtained from (29) and (30) assuming to be H- and T-independent and using the energy levels, (28), with negligible anisotropy. On the other hand, the data in Cu6 shown in Fig. 41 (b) do not scale so well vs $g\mu_B$ H/k$_B$ T. This is attributed to the non-negligible anisotropy term in (25) and the H and T dependence of [84]. In conclusion, the above results of proton NSLR demonstrate that in the FM molecular clusters Cu6 and Cr4 the low temperature spin dynamics can be described as a simple thermal fluctuation of the magnetization of the cluster in its collective quantum state of total spin S_T. The simple model for the NSLR leads to the same result as obtained from the first-principles calculation of the equilibrium two-ion time correlation function for the given set of magnetic eigenstates of the molecule whereby the characteristic frequency   for the fluctuations of the total spin of the molecule corresponds to the broadening of the magnetic eigenstates [85].

6.3 Vanadium Clusters: V6 and V15

V15 cluster (complete formula: $K_6[V_{15}As_6O_{42}(H_2O)] \times 8H_2O$) contains 15 paramagnetic vanadium ions V^{4+}, each of which has spin $s = 1/2$. The V15 cluster is a very peculiar Heisenberg spin-triangles system which can be viewed as a triangle formed by three groups of five V^{4+} ions [1]. As a result of frustration the ground state is formed by two $S_T = 1/2$ doubly degenerate states separated by a small gap of about 0.2 K. One expects that the magnetic properties and the spin correlations are dominated by interlayer coupling rather than by intralayer coupling. This situation can be compared with another spin-triangles system, namely V6 [86]. In V6 the ions are arranged in two almost independent trinuclear units, each forming a strongly antiferromagnetic frustrated triangle. The low temperature ground state for each of these triangles can be characterized by a doubly degenerate $S_T = 1/2$, a doubly degenerate first excited state with $S_T = 1/2$ separated by a gap of about 60 K, and a fourfold degenerate second excited state with $S_T = 3/2$ separated by 85 K [86]. ^{1}H NMR and relaxation measurements were performed in both clusters [87, 88], in form of powders. For V6 the T-dependence of proton T_1^{-1} is approximately proportional to   T, where   is the uniform susceptibility. On the other hand, in V15 the T-dependence of T_1^{-1} is quite different from the one of T. This was taken as indication of a critical T-dependence of , a parameter which measures the width of the V^{4+} spin fluctuation spectral density. For details in these two frustrated spin triangles we refer to the [87, 88].

A proton NMR study has been performed also in a frustrated triangular system of Mn spins $[Mn_3O(O_2CCH_3)_6(C_5H_5N)_3] \times C_5H_5N$, in brief Mn3 [89], powders. The Mn system is similar in many respects to the frustrated spin-triangle system V6 although the three Mn ions in the triangle are different, with one Mn^{2+} $(s = 5/2)$ ion and two Mn^{3+} $(s = 2)$ ions. Also the coupling constant J between the two Mn^{3+} spins is three times as large as the coupling J between the Mn^{3+} and the Mn^{2+} spins. The most remarkable NMR result is a strong enhancement of $1/T_1$ with a peak at T_0 J just like in the AFM rings (see Paragraph 4.1).

Another frustrated spin-triangle system is $(NH_4)[Fe_3(\mu_3\text{-}OH)(H_2L)_3(HL)_3]$ (in short Fe3, L = neutral ligand) [90]. Here again one observes a peak in T_1^{-1} at T   J, analogously to Mn3.

7 Summary and Conclusions

In the present review we tried to illustrate in some details a large body of NMR data obtained by our groups starting back in 1996 in magnetic molecular clusters (or molecular nanomagnets or single molecule magnets). We tried to quote briefly also work done by other NMR groups although some reports may have escaped our attention and we apologize for this. As a conclusion of the review we may say that the main characteristics of NMR in these new interesting magnetic systems have been established and the field is now mature for a deeper investigation and a better understanding of some remarkable e ects reported here. In fact, the theoretical analysis of the results in this review has been often rather qualitative and based on simple models to describe the NMR and relaxation e ects observed. We will summarize in the following some of the major issues emerging from our investigation with the focus on what remains to be done.

We have observed a magnetic field dependence at room temperature of proton T_1 in all nanomagnets investigated. Since at room temperature molecular clusters have to be viewed as nanosize paramagnets, the field dependence is a clear signature of the zero dimensionality of the system. From the field dependence of T_1 one can obtain the "cut-o " frequency for the time decay of the electron spin correlation function. The detailed origin of the "cut-o " frequency for the di erent molecules remains to be established.

At intermediate temperatures, namely when the electron spin in the molecular cluster becomes strongly correlated ($k_B T$ J), we have observed a T- and H-behavior of T_1^{-1} which can be explained satisfactorily with simple phenomenological models based on a direct relaxation process. The models are based on the assumption that the relaxation process is a direct process due to the hyperfine field fluctuations related to the lifetime of the magnetic molecular states. In particular, for the cases of Mn12 and Fe8 the low temperature nuclear relaxation can be described in terms of fluctuations of the magnetization in the total spin $S = 10$ ground state manifold due to spin-phonon interaction, and a value for the spin-phonon coupling constant could be derived for these two nanomagnets. The fact that the nuclear T_1 can be explained in such a simple way came as a surprise since normally in a highly correlated spin system such as ferro and antiferromagnetc bulk system close to the phase transition one has to consider the interplay of fluctuations of di erent symmetry i.e., di erent   q-vectors. In nanosize molecular magnets it appears that the discreteness of the magnetic energy levels plays an important role whereby only the lifetime of the total spin quantum state of the molecule is all it matters in determining the nuclear spin-lattice relaxation. One issue which remains to be clarified is the possibility of introducing a spin-wave description of the excited states in these nanomagnets and consequently the relation between a description of the relaxation in terms of a direct process and the description in terms of one- and two-magnons scattering.

At even lower temperature ($k_B T$ J) we have observed dramatic e ects on the nuclear relaxation rate due to level crossing e ects and quantum tunneling of the magnetization. It appears that NMR can be useful to distinguish between level crossing (LC) and level anticrossing (LAC) in AFM rings but a general theory to describe the observed enhancement of T_1^{-1} at the critical field is still lacking. The same is true for quantum tunneling of the magnetization. The phenomenon can be detected by NMR in di erent ways but a detailed description of the coupling between the nuclear Zeeman reservoir and the tunneling reservoir has to be firmly established in order to extract information from NMR measurements.

Finally, it should be mentioned that the NMR spectra at low temperature give important information about hyperfine interactions. In particular, the zero field NMR and the evolution of the spectrum as a funtion of an applied field in single crystals or oriented powders is a unique method which enables one to determine the size and orientation of the local magnetic moments of the molecular cluster in its ground state configuration.

Acknowledgements

We thank all collaborators who made possible this investigation and whose names appear in the many papers quoted in the review. A special thank to our colleagues chemists who introduced us to this field and provided the many samples and helped us to understand their properties. Their names also appear in most of our joint publications.

We remark that all the figures presented by the authors of the current work were drawn using data taken from references listed in the bibliography and referred to along the main text.

References

1. D. Gatteschi et al: Science   265 , 1055 (1994)
2. A. Muller et al: Chem. Phys. Chem.   2, 517 (2001)
3. K. Wieghardt, K. Pohl, H. Jibril, G. Huttner: Angew. Chem. Int. Ed. Engl.   23, 77 (1984)
4. T. Lis: Acta Cryst.   B36 , 2042 (1980)
5. D. Gatteschi, A. Caneschi, R. Sessoli, A. Cornia: Chem. Soc. Rev.   2, 101 (1996); K.L. Taft, C.D. Delfs, G.C. Papaefthymoiu, S. Foner, D. Gatteschi, S.J. Lippard: J. Am. Chem. Soc.   116 , 823 (1994)
6. T. Moriya: Prog. Theor. Phys.   16, 23 (1956); ibidem 28, 371 (1962)
7. F. Borsa, M. Mali: Phys. Rev. B   9, 2215 (1974)
8. J.H. Luscombe, M. Luban: J. Phys.: Condens. Matter   9, 6913 (1997)
9. J. Tang, S.N. Dikshit, J.R. Norris: J. Chem. Phys.   103 , 2873 (1995)
10. J.H. Luscombe, M. Luban, F. Borsa: J. Chem. Phys.   108 , 7266 (1998)
11. J.P. Boucher, M. Ahmed Bakheid, M. Nechschtein, G. Bonera, M. Villa, F. Borsa: Phys. Rev.   13 , 4098 (1976)
12. A. Lascialfari, Z.H. Jang, F. Borsa, D. Gatteschi, A. Cornia: J. Appl. Phys.   83, 6946 (1998)
13. A.L. Barra, P. Debrunner, D. Gatteschi, C.E. Schulz, R. Sessoli: Europhys. Lett. 35, 133 (1996)

14. R. Sessoli, Hui-Lien Tsai, A.R. Schake, Sheyi Wang, J.B. Vincent, K. Folting, D. Gatteschi, G. Christou, D.N. Hendrickson: J. Am. Chem. Soc. 115, 1804 (1993)
15. J. Van Slageren et al: Chem. Eur. J. 8, 277 (2002); S. Carretta, J. van Slageren, T. Guidi, E. Liviotti, C. Mondelli, D. Rovai, A. Cornia, A.L. Dearden, F. Carsughi, M. A ronte, C.D. Frost, R.E.P. Winpenny, D. Gatteschi, G. Amoretti, R. Caciu o: Phys. Rev. B 67, 094405 (2003)
16. A. Lascialfari, D. Gatteschi, A. Cornia, U. Balucani, M.G. Pini, A. Rettori: Phys. Rev. B 57, 1115 (1998)
17. D. Procissi, A. Shastri, I. Rousochatzkis, M. Al Rifai, P. Kogerler, M. Luban, B.J. Suh, F. Borsa: Phys. Rev. B 69, 094436 (2004)
18. A. Abragam: The Principles of Nuclear Magnetism (Clarendon Press, Oxford 1961)
19. T. Moriya: Prog. Theor. Phys. 28, 371 (1962)
20. F. Borsa, A. Rigamonti in: Magnetic Resonance of Phase Transitions (Academic Press, 1979)
21. A. Lascialfari, D. Gatteschi, F. Borsa, A. Cornia: Phys. Rev. B 55, 14341 (1997)
22. D. Procissi, B.J. Suh, E. Micotti, A. Lascialfari, Y. Furukawa, F. Borsa: J. Magn. Magn. Mater. 272–276 (2004) e741–e742
23. S.H. Baek, M. Luban, A. Lascialfari, Y. Furukawa, F. Borsa, J. Van Slageren, A. Cornia: Phys. Rev. B 70, 134434 (2004)
24. B. Barbara, W. Wernsdorfer, L.C. Sampaio, J.G. Park, C. Paulsen, M.A. Novak, R. Ferre, D. Mailly, R. Sessoli, A. Caneschi, K. Hasselbach, A. Benoit, L. Thomas: J. Magn. Magn. Mater. 140–141, 1825 (1995)
25. C. Delfs, D. Gatteschi, L. Pardi, R. Sessoli, K. Wieghardt, D. Hanke: Inorg. Chem. 32, 3099 (1993)
26. For Mn12, J.R. Friedman, M.P. Sarachik, J. Tejada, R. Ziolo: Phys. Rev. Lett. 76, 3830 (1996); L. Thomas, F. Lionti, R. Ballou, D. Gatteschi, R. Sessoli, B. Barbara: Nature (London) 383, 145 (1996); for Fe8, C. Sangregorio, T. Ohm, C. Paulsen, R. Sessoli, D. Gatteschi: Phys. Rev. Lett. 78, 4645 (1997); W. Wernsdorfer, A. Caneschi, R. Sessoli, D. Gatteschi, A. Cornia, V. Villar, C. Paulsen: Phys. Rev. Lett. 84, 2965 (2000)
27. Y. Furukawa, K. Kumagai, A. Lascialfari, S. Aldrovandi, F. Borsa, R. Sessoli, D. Gatteschi: Phys. Rev. B 64, 094439 (2001)
28. A. Lascialfari, P. Carretta, D. Gatteschi, C. Sangregorio, J.S. Lord, C.A. Scott: Physica B 289–290, 110 (2000); D. Gatteschi, P. Carretta, A. Lascialfari: Physica B 289–290, 94 (2000)
29. A. Lascialfari, Z.H. Jang, F. Borsa, P. Carretta, D. Gatteschi: Phys. Rev. Lett. 81, 3773 (1998)
30. M.N. Leuenberger, D. Loss: Phys. Rev. B 61, 1286 (2000)
31. J. Villain, F. Hartmann-Boutron, R. Sessoli, A. Rettori: Europhys. Lett. 27, 537 (1994); F. Hartmann-Boutron, P. Politi, J. Villain: Int. J. Mod. Phys. 10, 2577 (1996)
32. T. Goto, T. Koshiba, T. Kubo, K. Awaga: Phys. Rev. B 67, 104408 (2003)
33. S. Yamamoto, T. Nakanishi: Phys. Rev. Lett. 89, 157603 (2002); H. Hori, S. Yamamoto: Phys.Rev. B 68, 054409 (2003)
34. S. Maegawa, M. Ueda: Physica B 329, 1144 (2003); T. Yamasaki, M. Ueda, S. Maegawa: Physica B 329, 1187 (2003)
35. B.J. Suh, D. Procissi, P. Kogerler, E. Micotti, A. Lascialfari, F. Borsa: J. Mag. Mag. Mater. 272–276 (2004) e759–e761

36. A. Lascialfari, Z.H. Jang, F. Borsa, A. Cornia, D. Rovai, A. Caneschi, P. Carretta: Phys. Rev. B   61, 6839 (2000)
37. A. Cornia et al: Phys. Rev. B   60, 12177 (1999)
38. F. Meier, D. Loss: Phys. Rev. Lett.   23 5373 (1999); O. Waldmann: Europhys. Lett. 60, 302 (2002); A. Chiolero, D. Loss: Phys. Rev. Lett.   80, 169 (1998)
39. For NMR studies, see M. Chiba et al: J. Phys. Soc. Jpn.   57, 3178 (1988); G. Chaboussant et al: Phys. Rev. Lett.   80, 2713 (1998); M. Chiba et al: Physica (Amsterdam)   246B and 247B   , 576 (1998)
40. A. Abragam, B. Bleaney: Electron Paramagnetic Resonance of Transition Ions (Clarendon Press, Oxford 1970)
41. M.H. Julien, Z.H. Jang, A. Lascialfari, F. Borsa, H. Horvatic, A. Caneschi, D. Gatteschi: Phys. Rev. Lett.   83, 227 (1999)
42. M. A ronte, A. Cornia, A. Lascialfari, F. Borsa, D. Gatteschi, J. Hinderer, M. Horvatic, A.G.M. Jansen, M.-H. Julien: Phys. Rev. Lett.   88, 167201 (2002)
43. A. Lascialfari, F. Borsa, M.-H. Julien, E. Micotti, Y. Furukawa, Z.H. Jang, A. Cornia, D. Gatteschi, M. Horvatic, J. Van Slageren: J. Magn. Magn. Mater. 272–276   (2004) 1042–1047
44. A. Cornia, A. Fort, M.G. Pini, A. Rettori: Europhys. Lett.   50, 88 (2000)
45. H. Nakano, S. Miyashita: J. Phys. Soc. Jpn.   71, 2580 (2002)
46. F. Cinti, M. A ronte, A.G.M. Jansen: Eur. Phys. J. B   30, 461 (2002)
47. M. A ronte et al: Phys. Rev. B   68, 104403 (2003)
48. T. Goto, T. Kubo, T. Koshiba, Y. Fuji, A. Oyamada, J. Arai, T. Takeda, K. Awaga: Physica B   284–288   , 1277 (2000)
49. Y. Furukawa, K. Watanabe, K. Kumagai, F. Borsa, D. Gatteschi: Phys. Rev. B 64, 104401 (2001)
50. T. Kubo, T. Goto, T. Koshiba, K. Takeda, K. Agawa: Phys. Rev. B   65, 224425 (2002); T. Koshiba, T. Goto, T. Kubo, K. Awaga: Progr. Theor. Phys. (Suppl.) 145, 394 (2002)
51. S.H. Baek, S. Kawakami, Y. Furukawa, B.J. Suh, F. Borsa, K. Kumagai, A. Cornia: J. Magn. Magn. Mater.   272–276   (2004) e771–e772
52. Y. Furukawa, S. Kawakami, K. Kumagai, S-H. Baek, F. Borsa: Phys. Rev. B 68, 180405(R) (2003)
53. Y. Furukawa, K. Aizawa, K. Kumagai, A. Lascialfari, S. Aldrovandi, F. Borsa, R. Sessoli, D. Gatteschi: Mol. Cryst. Liq. Cryst.   379, 191 (2002)
54. R.A. Robinson, P.J. Brown, D.N. Argyriou, D.N. Hendrickson, S.M.J. Aubin: J. Phys.: Cond. Matter   12, 2805 (2000); M. Hennion, L. Pardi, I. Mirebeau, E. Suard, R. Sessoli, A. Caneschi: Phys. Rev. B   56, 8819 (1997); I. Mirebeau, M. Hennion, H. Casalta, H. Andres, H.U. Gudel, A.V. Irodova, A. Caneschi: Phys. Rev. Lett.   83, 628 (1999); A. Cornia, R. Sessoli, L. Sorace, D. Gatteschi, A.L. Barra, C. Daiguebonne: Phys. Rev. Lett.   89, 257201 (2002)
55. Y. Furukawa, K. Watanabe, K. Kumagai, F. Borsa, T. Sasaki, N. Kobayashi, D. Gatteschi: Phys. Rev. B   67, 064426 (2003); K. Watanabe, Y. Furukawa, K. Kumagai, F. Borsa, D. Gatteschi: Mol. Cryst. Liq. Cryst.   379, 185 (2002)
56. Y. Pontillon, A. Caneschi, D. Gatteschi, R. Sessoli, E. Ressouche, J. Schweizer, E. Lelievre-Berna: J. Am. Chem. Soc.   121, 5342 (1999); R. Caciu o, G. Amoretti, A. Murani, R. Sessoli, A. Caneschi, D. Gatteschi: Phys. Rev. Lett. 81, 4744 (1998)
57. For example, D. Gatteschi, R. Sessoli: Magnetism: Molecules to Materials III edited by J.S. Miller, M. Drillon (Wiley-VHC Verlag GmbH, Weinheim 2002)

58. I. Tupitsyn, B. Barbara in: Magnetism: Molecules to Materials III   edited by J.S. Miller, M. Drillon (Wiley-VHC, Verlag GmbH, Weinheim 2002)
59. Z.H. Jang, A. Lascialfari, F. Borsa, D. Gatteschi: Phys. Rev. Lett.   84, 2977 (2000)
60. Y. Furukawa, W. Watanabe, K. Kumagai, Z.H. Jang, A. Lascialfari, F. Borsa, D. Gatteschi: Phys. Rev. B   62 14246 (2000)
61. Y. Furukawa, K. Watanabe, K. Kumagai, F. Borsa, D. Gatteschi: Physica B 329–333 , 1146 (2003)
62. T. Kubo, H. Doi, B. Imanari, T. Goto, K. Takeda, K. Awaga: Physica B   329– 333 , 1172 (2003)
63. T. Goto, T. Koshiba, A. Oyamada, T. Kubo, Y. Suzuki, K. Awaga, B. Barbara, J.P. Boucher: Physica B   329–333 , 1185 (2003)
64. For Fe8 see: M. Ueda, S. Maegawa, S. Kitagawa: Phys. Rev. B   66, 073309 (2002); for Mn12 see: A. Morello, O.N. Bakharev, H.B. Brom, L.J. de Jongh: Polyhedron 22, 1745 (2003)
65. For example, M. Luis, F.L. Mettes, L. Jos de Jongh in: Magnetism: Molecules to Materials III   edited by J.S. Miller, M. Drillon (Wiley-VHC Verlag GmbH, Weinheim 2002)
66. Y. Furukawa, K. Aizawa, K. Kumagai, A. Lascialfari, F. Borsa: J. Magn. Magn. Mater. 272–276 , 1013 (2004); Y. Furukawa, K. Aizawa, K. Kumagai, R. Ullu, A. Lascialfari, F. Borsa: Phys. Rev. B   69, 01405 (2004); for proton NMR study on oriented powders see also Y. Furukawa, K. Aizawa, K. Kumagai, R. Ullu, A. Lascialfari, F. Borsa: J. Appl. Phys.   93, 7813 (2003)
67. A. Lascialfari, F. Tabak, G.L. Abbati, F. Borsa, M. Corti, D. Gatteschi: J. Appl. Phys. 85, 4539 (1999)
68. A.L. Barra, A. Caneschi, A. Cornia, F.F. de Biani, D. Gatteschi, C. Sangregorio, R. Sessoli, L. Sorace: J. Am. Chem. Soc.   121 , 5302 (1999); G. Amoretti, S. Carretta, R. Caciu o, H. Casalta, A. Cornia, M. A ronte, D. Gatteschi: Phys. Rev. B 6410 , 104403 (2001)
69. D. Procissi, B.J. Suh, A. Lascialfari, F. Borsa, A. Caneschi, A. Cornia: J. Appl. Phys. 91 , 7173 (2002)
70. A. Lascialfari, D. Gatteschi, F. Borsa, A. Shastri, Z.H. Jang, P. Carretta: Phys. Rev. B 57, 514 (1998)
71. R.M. Achey, P.L. Kuhns, A.P. Reyes, W.G. Moulton, N.S. Dalal: Solid State Comm. 121 , 107 (2002); Polyhedron   20, 11 (2001); Phys. Rev. B   64, 064420 (2001)
72. D. Arcon, J. Dolinsek, T. Apih, R. Blinc, N.S. Dalal, R.M. Achey: Phys. Rev. B 58, R2941 (1998)
73. R. Blinc, B. Zalar, A. Gregorovic, D. Arcon, Z. Kutnjak, C. Filipic, A. Levstik, R.M. Achey, N.S. Dalal: Phys. Rev. B   67, 094401 (2003)
74. M. Axenovich, M. Luban: Phys. Rev. B   63, 100407 (2001)
75. J.K. Jung, D. Procissi, R. Vincent, B.J. Suh, F. Borsa, C. Schroder, M. Luban, P. Kogerelr, A. Muller: J. Appl. Phys.   91, 7388 (2002)
76. E. Micotti, D. Procissi, A. Lascialfari, P. Carretta, P. Kogerler, F. Borsa, M. Luban, C. Baines: J. Magn. Magn. Mater.   272–276 , 1099–1101 (2004)
77. Biomineralization: Chemical and Biochemical Perspectives   edited by Stephen Mann, John Webb, Robert J.P. Williams (VCH, Wcinhcim 1989)
78. M-E.Y. Mohie-Eldin, R.B. Frankel, L. Gunter: J. Magn. Magn. Mater.   135 , 65 (1993)

79. J. Tejada, X.X. Zhang, E. del Barco, J.M. Hern´ andez: Phys. Rev. Lett.   79 , 1754 (1997)
80. J.R. Friedman, U. Voskoboynik, M.P. Sarachik: Phys. Rev. B   56 , 10793 (1997)
81. D.D. Awschalom, J.F. Smyth, G. Grinstein, D.P. DiVincenzo, D. Loss: Phys. Rev. Lett. 68 , 3092 (1992)
82. Z.H. Jang, B.J. Suh, A. Lascialfari, R. Sessoli, F. Borsa: J. Appl. Magn. Res. 19 , 557 (2000)
83. E. Rentschler, D. Gatteschi, A. Cornia, A.C. Fabretti, A-L. Barra, O.I. Shchegolikhina, A.A. Zhdanov: Inorg.Chem.   35 , 4427 (1996)
84. Y. Furukawa, A. Lascialfari, Z.H. Jang, F. Borsa: J. Appl. Phys.   87 , 6265 (2000)
85. Y. Furukawa, M. Luban, F. Borsa, D.C. Johnston, A.V. Mahajan, L.L. Miller, D. Mentrup, J. Schnack, A. Bino: Phys. Rev. B   61 , 8635 (2000)
86. M. Luban, F, Borsa, S. Budko, P. Canfield, S. Jun, J.K. Jung, P. Kogerler, D. Mentrup, A. Muller, R. Modler, D. Procissi, B.J. Suh, M. Torikachvilly: Phys. Rev. B   66 , 054407 (2002)
87. D. Procissi, B.J. Suh, J.K. Jung, P. Kogerler, R. Vincent, F. Borsa: J. Appl. Phys. 93 , 7810 (2003)
88. J.K. Jung, D. Procissi, Z.H. Jang, B.J. Suh, F. Borsa, M. Luban, P. Kogerler, A. Muller: J. Appl. Phys.   91 , 7391 (2002)
89. B.J. Suh, D. Procissi, K.J. Jung, S. Budko, W.S. Jeon, Y.J. Kim, D.Y. Jung: J. Appl. Phys.   93 , 7098 (2003)
90. M. Fardis, G. Diamantopoulos, M. Karayianni, G. Papavassiliou, V. Tangoulis, A. Konsta: Phys. Rev. B   65 , 014412 (2001)

Correlated Spin Dynamics and Phase Transitions in Pure and in Disordered 2D S = 1 / 2 Antiferromagnets: Insights from NMR-NQR

A. Rigamonti, P. Carretta and N. Papinutto

Department of Physics "A.Volta" and Unit` a INFM, University of Pavia Via Bassi n 6, I-27100, Pavia (Italy)
rigamonti@fisicavolta.unipv.it

Abstract. A recall of the phase diagram for two-dimensional quantum Heisenberg antiferromagnets (2DQHAF) and of the main issues involving phase transitions and spin dynamics in these systems is first given. After a pedagogical description of the basic aspects for the NMR-NQR relaxation rates in terms of amplitudes and decay rates of spin fluctuations, the problem of the temperature dependence of the correlation length in prototype, pure 2DQHAF (CFTD and La$_2$CuO$_4$) is addressed. Then spin- and charge-doped systems (Zn/Mg for Cu and Sr for La substitutions in La$_2$CuO$_4$) are considered and the effects on the spin stiffness, on the correlation length and on the staggered Cu^{2+} magnetic moment are reported, in particular near the percolation threshold. A critical outline of the properties of the cluster spin glass phase in Sr- doped La$_2$CuO$_4$ is given. Finally the results of a ^{63}Cu NQR-NMR relaxation study around the quantum critical point (in CeCu$_{6-x}$Au$_x$) are presented. It is pointed out how the 2D response function with anomalous exponent and energy/temperature scaling indicated by neutron scattering is basically confirmed, while the ^{63}Cu relaxation measurements reveal novel effects involving low-energy spin excitations and the role of an external magnetic field.

1 Introduction and Contents

Since the discovery that La$_2$CuO$_4$, the parent of high temperature supercon-ductors, was also the experimental realization of the model two-dimensional (2D), quantum (S = 1 / 2), Heisenberg antiferromagnet (AF), a great deal of interest was triggered towards low-dimensional quantum magnetism. The sys-tem we are going to deal with is basically a planar array of S = 1 / 2 Cu^{2+} magnetic ions onto a square lattice, as sketched in Fig. 1, in antiferromagnetic interaction, namely described by the magnetic Hamiltonian

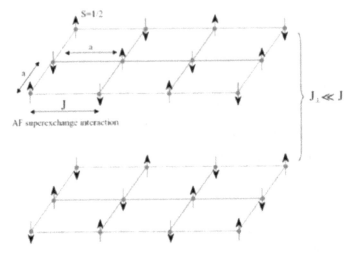

Fig. 1. Pictorial sketch of planar arrays of   S = 1 / 2 magnetic moments, in antiferromagnetic (AF) super-exchange interaction onto a square lattice, a model system
rather well describing La $_2$CuO$_4$ and Copper formiate tetrahydrate (CFTD). The
spin dilution is achieved, in La $_2$CuO$_4$, by S = 0 Zn^{2+} or Mg^{2+} for S = 1 / 2 Cu^{2+}
substitutions, while charge doping, namely injection of holes in the plane, is obtained
by substituting the La $^{3+}$ ions (out of plane) by Sr $^{2+}$

$$H = \frac{J}{2} \sum_{i,j} S_i \cdot S_j \qquad J > 0 , \qquad (1)$$

while the summation is often limited to the first nearest neighbours spin operators.

We shall devote our attention to a variety of aspects involving static and
dynamical properties of that 2D array: the temperature dependence of the in-
plane magnetic correlation length $_{2D}$ entering in the equal-time correlation
function $S_i(0) \cdot S_j(0)$; the critical spin dynamics driving the system towards
the long-range ordered state (at   T = 0 in pure 2DQHAF in the absence of
interplanar interaction   J); the validity of the dynamical scaling, where $_{2D}$
controls the relaxation rate   of the order parameter according to a law of
the form $_{2D}^{-z}$, with a given critical exponent   z. We shall also discuss
the modifications induced by   spin dilution (or spin doping) namely when
part of the S = 1 / 2 magnetic ions are substituted by non-magnetic   S = 0
ions, as well as the e ects related to   charge doping , namely the injection (for
instance by heterovalent substitutions) of   S = 1 / 2 holes creating local singlets
which can itinerate onto the plane, locally destroying the magnetic order and
inducing novel spin excitations. These aspects are of particular interest in
the vicinity of the percolation thresholds, where the AF order is about to
be hampered at any finite temperature. This can be considered a situation
similar to a quantum critical point , where no more the temperature but rather
Hamiltonian parameters can drive the transition.

In principle the problems involving the aforementioned aspects could be studied by spectroscopic techniques relaying on the electron's response. However electron paramagnetic resonance is often prevented by very broad lines, possibly due to the strong electron-electron correlation. Neutron scattering is a powerful technique but it could su er of resolution limitation for low-energy excitations or for short correlation length. Thus the nuclei have represented in recent years one of the best tools to investigate the properties of 2DQHAF, by resorting to the hyperfine interactions and to their time-dependence, as explored by NMR-NQR spectra and relaxation.

The chapter is organized as follows.

In Sect. 2 we briefly discuss the rich phase diagram of 2DQHAF, as it results from a variety of experimental studies and theoretical e orts (that will not be described in detail). Then (Sect. 3) the basic principles of the experimental approach using the nuclei as local probes are recalled, describing the sources of the hyperfine fields and giving the expressions for the relaxation rates W in terms of the amplitudes $|S_q|^2$ and decay rates $_q$ of the spin fluctuations, with an illustrative example for a 2D Ising-like system approaching the ordering transition (Sect. 4). It is shown, in particular, how the temperature dependence of the critical decay rate is extracted.

In Sect. 5 the results obtained in two 2DQHAF prototypes, namely La_2CuO_4 and CFTD, are described, showing how quantitative estimate of $_{2D}$ is derived and concluding that in a wide temperature range the predictions of the renormalized classical (RC) regime are rather well followed. The role of the "filtering factor" A_q which weights in di erent way di erent regions of the Brillouin zone, is illustrated by comparing the results in La $_2CuO_4$ and CFTD. Then (Sect. 6) the case of charge and spin doped 2DQHAF is treated and the insights derived from ^{63}Cu and ^{139}La NQR in Sr-doped and in Zn (or Mg) -doped La $_2CuO_4$ are illustrated. It is shown, in particular for spin doping, how the "dilution model" namely the simple correction in Hamiltonian (1) by the probability of presence $p_{i,j}$ of the magnetic ion, explains rather well the behaviour of $_{2D}(x, T)$, of the spin sti ness $_s(x)$ and of the ordering temperature $T_N(x)$, when the doping amount x is far from the percolation threshold. The modification occurring when the threshold is approached are then discussed in Sect. 7, by comparing the NQR results with the ones obtained by neutron scattering. The e ects on the staggered magnetic moment are also addressed.

The main properties of the cluster spin glass phase, occurring when the charge doping is above the percolation threshold in Sr-doped La $_2CuO_4$, are then briefly reviewed (Sect. 8). Finally, in Sect. 9 we present and discuss the results of a recent NQR-NMR study carried out in a system (CeCu $_{6-x}$Au$_x$) around the quantum critical point (QCP) (x = 0.1) separating Fermi liquid and AF phases, with a field-dependent magnetic generalized susceptibility of 2D character, with anomalous exponent and energy/temperature scaling. While the main results obtained by other authors by means of inelastic neutron scattering are confirmed also by our measurements based on low-energy susceptibility over all the Brillouin zone, novel aspects are pointed out in regards of the role of the magnetic field. Summarizing remarks are collected in Sect. 10.

2 The Phase Diagram of 2DQHAF

In Fig. 2 the phase diagram for a planar array of $S = 1/2$ spin in square lattice and AF interaction is shown. The diagram results from a variety of experimental studies in synergistic interplay with theoretical descriptions, that will not be recalled in detail (for a nice introduction and an exhaustive review of the studies of the magnetic properties of 2DQHAF, see [1]). In the Figure g is a dimensionless parameter measuring the strength of quantum fluctuations and it can be related to spin wave velocity c_{sw} and to the spin stiffness ρ_s:

$$g = \frac{c_{sw}}{k_B \rho_s} \frac{\hbar^2}{a} , \tag{2}$$

(a lattice parameter).

The spin stiffness ρ_s has been introduced in order to describe the properties of disordered AF's and it measures the increase in the ground state energy for a rotation by an angle θ of the sublattice magnetization ($\Delta E = \rho_s k_B \theta^2/2$). It can be written $\rho_s = \chi_\perp (c_{sw}^2/\hbar)$, $\chi_\perp$ being the transverse spin susceptibility. For classical Heisenberg spins, from usual spin wave theories one has

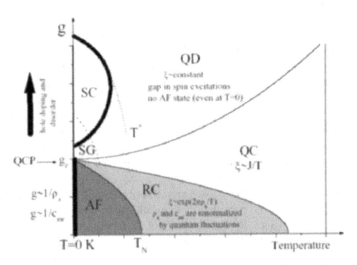

Fig. 2.   Phase diagram argued to describe the occurrence in 2DQHAF of various regimes, as a function of temperature and of the parameter g measuring the strength of quantum fluctuations (see discussion in the text)

$$\chi_s = Z_\chi (S) J S^2, \tag{3}$$

$$c_{sw} = Z_{sw} (S)(8)^{1/2} k_B J S a/h,$$

with $Z_\chi (S) \simeq 1 - 0.235/2S$ and $Z_{sw}(S) \simeq 1 + 0.158/2S$. The parameter g is expected to increase upon doping and disorder.

Starting from the seminal paper by Chakravarty et al. [2] (who mapped the Hamiltonian (1) onto the so-called non-linear σ model, that for $T \rightarrow 0$ is the simplest continuum model with the same symmetry and the same spectrum of excitations) the diagram in Fig. 2 has emerged. Below a given value g_c the ground state, at $T = 0$, is the Neel AF state, which is somewhat extended to a finite temperature T_N because of the interplane interaction $J_\perp \ll J$. The percolation threshold for the AF state at $T = 0$ as a function of g, is at g_c. For $g < g_c$, upon increasing temperature above T_N one enters in the renormalized classical (RC) regime. Here the effect of the quantum fluctuations is to renormalize χ_s and c_{sw} with respect to the mean field values for the "classical" 2D Heisenberg paramagnet. Thus the in-plane correlation length goes as $\xi_{2D} \simeq \exp[2\pi\rho_s/T]$. The prefactor (and a correction term) have been more recently evaluated [3] so that

$$\xi_{2D} = 0.493 \, e^{1.15 J/T} \left[1 - 0.43 \frac{T}{J} + O\left(\frac{T}{J}\right)^2 \right], \tag{4}$$

(hereafter $\xi_{2D} = \xi$ is expressed in lattice units a) and $1.15J = 2\pi\rho_s$, while $c_{sw} = 1.18 \, 2Jk_B a/\hbar$. Weakly damped spin waves exist for wave vectors $q \geq \xi^{-1}$ while for longer wave lengths only diffusive spin excitations of hydrodynamic character are present. For $T \gtrsim J/2\pi \simeq 2\rho_s$ instead of entering into the classical limit (that would be reached for $T \gg J$) the planar QHAF should cross, according to the above mentioned descriptions [1–3] to the quantum critical (QC) regime. In this phase, typical of 2D and 1D quantum magnetic systems, the only energy scale is set by temperature and $\hbar\omega = J/T$.

On increasing g, according to proposals by Anderson [4] and to quantum phase transitions theories [5] the increase in quantum fluctuations can inhibit an ordered state even at $T = 0$. The system is then in the quantum disordered (QD) regime, the correlation length being short and temperature independent. In the spectrum of spin excitations a gap of the order of c_{sw}/ξ opens up.

The somewhat speculative phase diagram illustrated in Fig. 2 is still under debate. In particular the validity of the non-linear σ model is questioned at large T and/or for large g regions. For instance, as we shall see later on, some experimental studies do not provide evidence of crossovers from RC to QC or QD regimes. Also the occurrence of micro-segregation and/or a phase separation in hole-depleted and hole-rich stripes, as well as the real nature of the low-energy excitations are still open questions.

We conclude the discussion of the diagram in Fig. 2 just mentioning the following. The cluster-spin-glass (SG) phase is the one for $g > g_c$ in which the experiments (primarily NQR spectra and relaxation and magnetic susceptibility) indicate the presence of mesoscopic "islands" of AF character separated by domains walls, with e ective magnetic moments undergoing collective spin freezing without long range order even at temperature close to zero. Above a given amount of charge doping (e.g. hole injection as in La $_{2-y}$Sr$_y$CuO$_4$) the systems become superconductors (SC phase), with the so-called underdoped and overdoped regimes characterized by a transition temperatures $T_c < T_c^{max}$ (the one pertaining to the optimal doping), as discovered by M" uller and Bednorz. In SG and SC underdoped phases a gap in the spin excitations at the AF wave vector $q_{AF} = (/a, /a)$ has been experimentally observed to arise at a given temperature T. The spin gap (and charge pseudo-gap) region has possibly to be related to superconducting fluctuations of "anomalous" (i.e. non-Ginzburg-Landau) character or to AF fluctuations locally creating a "tendency" towards a mesoscopic Mott insulator. Exotic excitations of various nature have been considered to occur in the regions of high g's. In this Chapter we shall not go into detail involving these aspects, which are still under debate and less settled than the ones for low g, namely for the doped non-superconducting 2DQHAF.

Finally we shall present and discuss the results around the quantum critical point in CeCu$_{6-x}$Au$_x$ that could be placed at the critical value g_c separating the AF phase from the Fermi liquid region.

3 Basic Aspects of the Experimental Approach

In the 2DQHAF considered here the electron paramagnetic resonance of Cu $^{2+}$ ions can hardly be carried out. Therefore the nuclei are used as microscopic probes to detect the Cu $^{2+}$ S $= 1/2$ spin dynamics through the time-dependent hyperfine interaction driving the NQR-NMR relaxation processes. In NQR experiments instead of the external magnetic field as in NMR is the V_{zz} component of the electric field gradient at the Cu site (perpendicular to the CuO$_2$ planar lattice) that acts as quantization axis. The recovery towards the thermal equilibrium, after the usual sequences of RF pulses, is driven by the time dependence of the fictitious hyperfine field $h(t)$ due to Cu $^{2+}$ S $= 1/2$ electrons and the relaxation rate is

$$2W = \frac{2}{2} \quad h_+(0) \, h_-(t) \, e^{-i_m t} \, dt , \qquad (5)$$

where $_m$ ($_Q$ or $_L$) is the measuring frequency in the MHz range and the correlation function ..... involves the transverse components of the field

$$h(t) = \sum_i e \quad \frac{S_i}{r_i^3} - \frac{3(r_i \cdot S_i) r_i}{r_i^5} - \frac{l_i}{r_i^3} - \frac{8}{3} S_i \, (r_i) . \qquad (6)$$

The spin dynamics is reflected in the spin operators $S_i(t)$, with collective spin components

$$S_q = \frac{1}{\sqrt{N}} \sum_i e^{-iqr_i} S_i(t) , \qquad (7)$$

yielding the effective field in the form

$$h(t) = \frac{1}{\sqrt{N}} \sum_q A_q S_q(t) , \qquad (8)$$

where A_q is the Fourier transform of the lattice functions specifying the "positions" of the magnetic ions. For the systems we are going to discuss, $h(t)$ originates from the on-site contribution (namely from the 3d electrons of the magnetic ion) and from transferred hyperfine scalar and dipolar contributions. From (5) and (8) one has

$$2W = \frac{\gamma^2}{2N} \sum_q A_q^2 \int S_q(0) S_{-q}(t) e^{-i\omega_m t} dt , \qquad (9)$$

where $\{...\}$ indicates the dynamical structure factor components $S_{\perp}(q, \omega)$ at ω_m involved in the field components perpendicular to the quantization axis. From the fluctuation-dissipation theorem

$$S_{\perp}(q, \omega) = k_B T \chi_{\perp}(q, \omega)/\omega , \qquad (10)$$

(for details see [6] and references therein).

4 NMR-NQR Relaxation Rates: Amplitude and Decay Rates of Spin Fluctuations and Critical Behaviour in 2D Systems

In the cases we shall discuss later on, the decay rates of the spin fluctuations will remain greater than ω_m and (9) can be written

$$2W = \frac{\gamma^2}{2N} \sum_q A_q^2 \frac{|S_q|^2}{\Gamma_q} , \qquad (11)$$

where $|S_q|^2$ are the amplitudes of the collective fluctuations, namely $|S_q|^2 \equiv \int S_{\perp}(q, \omega)d\omega$ and Γ_q the correspondent decay rates. It is noted that in correspondence to the critical wave vector q_c in practice the amplitude and the decay rate of the order parameter are involved. In (11) A_q^2 has the relevant role of weighting particular regions of the Brillouin zone (BZ), or, in other words, a particular symmetry of the spin fluctuations. In case of dominant on-site contribution to the effective field $h(t)$, in practice only the auto-correlation function $\langle S_i(0) S_i(t) \rangle$ is involved in (5) and thus A_q^2 is q-independent:

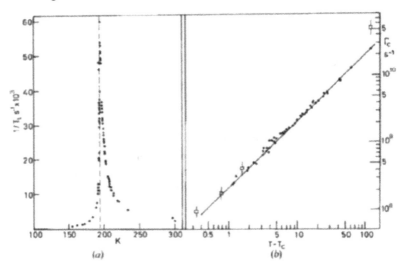

Fig. 3.   Spin lattice relaxation rate in p-terphenil on approaching the transition temperature $T_c = 194.5\,K$ (a). No dependence from the measuring frequency   m was detected. In part b) the temperature behaviour of the critical frequency (see text) is reported and compared with the few results from high resolution neutron back-scattering in the deuterated crystal (   ) (see [7])

$$2W = \frac{2}{\text{int}}\,\frac{1}{2N}\sum_{q}\frac{|S_q|^2}{q}\,, \tag{12}$$

int $=$ A being an e ective strength of the nucleus-electrons interaction. Direct access to the spin dynamics is thus obtained.

A long-lasting 2D illustrative example, for an Ising-like pseudo-spin system, is o ered by the relaxation measurement in the p-terphenil crystal [7]. The relaxation data on approaching the order-disorder transition (Fig. 3) can be analysed as follows. The pseudo-spin variable $S_z(t) = (+,-)1$ describes the time dependence of the dipole-dipole nuclear Hamiltonian and the correspondent dynamical structure factor is written

$$S(q,\ ) = |S_q|^2\ q^{-1}\,F(\frac{\ }{q})\,,$$

where F is a sharp function with width of the order of unit and unit area, that can be taken $F = 1$ when no frequency dependence of the relaxation rates is detected (see Fig. 3). The amplitudes and the decay rates of the spin fluctuations scale with the correlation length as follows:

$$|S_q|^2 = {}^{2-}\ f(q\ )\,,$$
$$_q = {}_{qc}\,g(q\ )\,,$$

where f and g are homogeneous functions of   q , which correspond to scale the generalized susceptibility involved in (10) in the form

$$\chi(q, \omega) = \chi_0 \xi^{\gamma z} f(q\xi, \omega / \Gamma \xi^z), \tag{13}$$

with $\chi_0 = S(S + 1) / 3 k_B T$.

From the first order expansion of f and g one has

$$\frac{\langle |S_q|^2 \rangle}{\xi^q} = \frac{1}{\xi^{D-2}} \int dq\, q^{(D-1)}\, fg^{-1}\xi^{2-\eta}\Big|_{q_c} \tag{14}$$

and

$$2W \propto \int q^{(D-1)} \frac{dq}{(1 + q^2 \xi^2)^2}\Big|_{q_c} \Gamma^{-1} \xi^{(D-z)-\eta}, \tag{15}$$

with $q_c \propto \xi^{-z}$, $\epsilon = (T - T_c)/T_c$, ν critical exponent for ξ, $\gamma = (2 - \eta)\nu$ critical exponent for $\langle |S_{qc}|^2 \rangle$ and z dynamical critical exponent (while in 2D $T_c \to 0$). Equation (12) is reduced to

$$2W = \xi_{int}^2 \omega_{qc}^{-1},$$

justifying the temperature behaviour of the relaxation rates and yielding the critical frequency ω_{qc} when ξ_{int} can be obtained from the spectra or from numerical estimate (see Fig. 3 b)). It should be remarked that the 2D scaling is represented by $\omega_{qc} \propto \xi^{-1}$, namely by the dynamical critical exponent $z = 1$.

5 Pure 2DQHAF: Temperature Dependence of the Correlation Length (in La$_2$CuO$_4$ and in CFTD, within Scaling Arguments)

Real systems that, to a good approximation, obey to the Hamiltonian in (1) (neglecting for the moment the interplanar coupling) are La$_2$CuO$_4$ (the parent of high-temperature superconductors) and CFTD (copper formiate tetrahydrate deuterated). The former is characterized by the in-plane super-exchange constant $J \simeq 1500$ K. Due to the weak ($J' \simeq 10^{-5} J$) interplanar coupling La$_2$CuO$_4$ orders 3D at $T_N = 315$ K. Still a wide temperature range exists where the in-plane correlation length is much larger than the lattice constant, while no 3D long range AF order sets in. However the main reason to deal with La$_2$CuO$_4$ is that charge and spin doping can easily be obtained by Sr^{2+} for La^{3+} substitutions (thus injecting itinerant holes in the CuO$_2$ plane) or by Zn^{2+} (or Mg^{2+}) $S = 0$ for Cu^{2+} substitutions, which induce spin dilution.

Instead CFTD is a good prototype of 2DQHAF (with $J'/J \simeq 10^{-4}$), with rather small $J \simeq 80$ K. Thus a large (T/J) range can experimentally be explored, searching for possible crossover to QC regime, for instance (see Fig. 2).

In the light of (11) one realizes that a quantitative estimate of the correlation length by means of NMR-NQR relaxation requires a reliable form of the

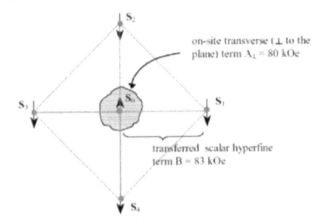

Fig. 4. Sketch of the simplified magnetic array of the Cu $^{2+}$ ions in La $_2$CuO$_4$ assumed in order to relate the ^{63}Cu NQR relaxation rates to the correlation length

A_q^2 factor. For La $_2$CuO$_4$ the fictitious field at the Cu nucleus can be written (see Fig. 4)

$$h = A S_0 + \sum_{i=1}^{4} B S_i \qquad (16)$$

and the values $A = 80$ kOe and $B = 83$ kOe have been estimated ([6] and references therein). Then for external field H_0 parallel to the c-axis or in NQR, one has

$$A_q^2 = [A - 2B(\cos q_x a + \cos q_y a)]^2 , \qquad (17)$$

with q starting from $q_{AF} = (\pi/a, \pi/a)$.

At the ^{139}La site more complicate expression holds: A_q^2 has maxima at the center and at the borders of the Brillouin zone and an average value is approximately $A_q^2 \approx (1\,\text{kOe})^2$.

In CFTD, at the hydrogen atom lying in the Cu plane one has

$$A_q^2 = A^2 + B^2(\cos q_x a + \cos q_y a) \qquad (18)$$

and from the rotation pattern of the paramagnetic shift [8] one derives $A = 2.31\,\text{kOe}$ and $B = 1.46\,\text{kOe}$.

The form factor for CFTD given in (18) is depicted in the 2D BZ in Fig. 5. It is noted that the form factor is peaked at the center of the BZ. Thus the "critical" AF fluctuation is filtered out and a kind of anti-divergence of 2W as a function of temperature can qualitatively be expected in correspondence to the enhancement and slowing down of the order parameter fluctuations. The experimental results for ^{63}Cu 2W in La$_2$CuO$_4$ and for ^{1}H 2$W \equiv T_1^{-1}$ in CFTD are reported in Fig. 6.

Now we outline how the temperature behaviour of ξ is extracted from the experimental data. Since $\xi \gg 1$, scaling arguments for $|S_q|^2$ and τ_q should hold. Then, in the light of (10)–(13) and (14) we write

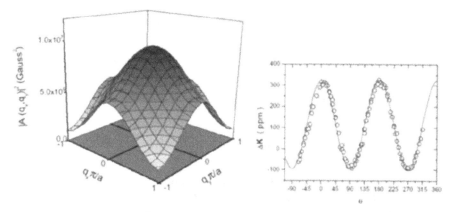

Fig. 5.   Contour plot of A_q^2 for CFTD, as derived from the rotation pattern for the paramagnetic shift   K   of the H NMR signal (from [8])

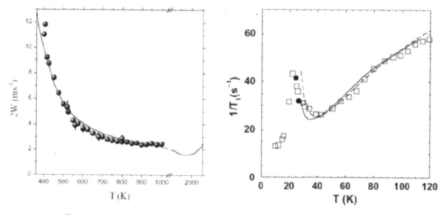

Fig. 6.   (a) ^{63}Cu NQR relaxation rate in La $_2$CuO$_4$ as a function of temperature. The solid line is the theoretical behaviour correspondent to (21), for correlation length in the RC regime (from [9]; (b) proton-spin lattice relaxation rate as a function of temperature in CFTD. The dashed line is the theoretical mode-coupling calculation while the full line is the Monte Carlo numerical estimate with the correlation length derived in the framework of the pure-quantum self-consistent harmonic approximation, in substantial agreement with (4) (From [10])

$$|S_q|^2 = |S_{q_{AF}}|^2 f(q) = p \frac{S(S + 1)}{3} \,^2 f(q) \tag{19}$$

and

$$_q = _{q_{AF}} g(q) = \frac{2 _E}{^2} \,^{-z} g(q) , \tag{20}$$

where p < 1 accounts for the zero-temperature reduction due to quantum e ects, $_E$ is the Heisenberg exchange frequency describing the uncorrelated spin fluctuations for   T   . A normalizing factor preserving the sum rule

$$\sum_q |S_q|^2 = p N \frac{S(S+1)}{3}$$

has to be introduced in $f = g^{-1} = (1 + q^2 \xi^2)^{-1}$ (see (15)).
   Thus one writes

$$2W = \xi^2 \frac{S(S+1)}{3} p^{z+2} \frac{\gamma^2 \hbar^2}{E} \cdot$$

$$\frac{a^2}{4\pi^2} \int_{BZ} dq \frac{[A - 2B(\cos q_x a + \cos q_y b)]^2}{(1 + q^2 \xi^2)^2} . \qquad (21)$$

An approximate expression is obtained by averaging A_q^2 (as given by (17)).
For ^{63}Cu in La$_2$CuO$_4$ the approximate form reads

$$2W \simeq [(4.2 \times 10^3)/(\ln q_m \xi)^2] \xi^z , \qquad (22)$$

with $q_m = 2\pi$.
   However for the quantitative evaluation of the correlation length the complete q-dependence, and therefore (21), has to be taken into account. As shown
in Fig. 7 the temperature behaviours of ξ deduced from the experimental results in La$_2$CuO$_4$ and in CFTD are consistent with (4), namely with the RC
regime and dynamical scaling with $z = 1$.

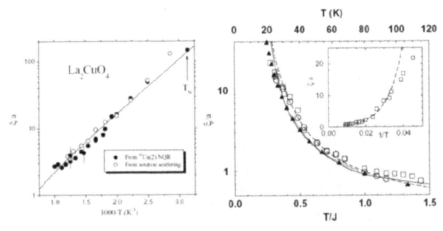

Fig. 7.   (a) In plane magnetic correlation length as a function of temperature in the
paramagnetic phase of La$_2$CuO$_4$ as extracted from the inversion of (21) in the text
(for   the value $\eta = 0.3$ was used, according to theoretical estimates and neutron
scattering data). The solid line tracks (4). For comparison some representative results from neutron scattering are also shown (for complete references see [9]). (b) In
plane magnetic correlation length for CFTD as a function of   T/J   and comparison
with the theoretical estimates. In the inset the lack of a possible crossover to the
QC regime, in the high temperature range, is evidenced (the dotted line is the theoretical behaviour within the mode-coupling theory and Kubo relaxation function;
for details see [10])

The conclusions that can be drawn from the studies in pure 2DQHAF are the following. The NMR-NQR relaxation rates allow one to derive reliable estimates of the absolute value and of the temperature dependence of the magnetic correlation length, particularly when the proper form of the factor A_q^2 is taken into account in the inversion yielding   from the experimental results. In La $_2$CuO$_4$ and in CFTD the RC regime appears to hold, up to a temperature of the order of 1 .5 J and no evidence of crossover to QC or QD regimes is observed. The in-plane magnetic correlation length follows rather well the theoretical expression given by (4).

6 Spin and Charge Lightly-Doped La $_2$CuO$_4$: E ects on the Correlation Length and on the Spin Sti ness

As already mentioned, in La $_2$CuO$_4$ charge doping is obtained by heterovalent Sr^{2+} for La^{3+} substitutions, while spin dilution is obtained by   S = 0 Zn $^{2+}$ (or Mg^{2+}) for S = 1/2 Cu^{2+} substitutions.

In La$_{2-y}$Sr$_y$CuO$_4$ the Neel temperature drops very fast with the Sr content. Analogous e ect, at a lower rate, has the spin dilution in La $_2$Zn$_x$Cu$_{1-x}$O$_4$, in terms of the Zn (or Mg) content   x. (See Fig. 8). Upon spin dilution the spin sti ness is expected to change and we shall derive insights on the doping dependence of the correlation length.

At this aim we start by analysing first the experimental results for ^{63}Cu 2W, again using the general theoretical framework leading to (21), in the limit of weak doping, when the dilution model should hold. The dilution model amounts to modify the Hamiltonian (1) simply by considering the probability p$_i$ that a given site is spin-empty:

$$H = J \quad p_i \, p_j \, S_i \cdot S_j = J(0) \, [1 - x]^2 \quad S_i \cdot S_j . \tag{23}$$
$$ {}_{i,j} {}_{i,j}$$

Then the spin sti ness should depend on doping according to   (x) = (0) [1 − x]^2, the correlation length becoming (see (4))

$$(x, T) = (0, T) e^{-(2 - x) x \, 1.15 J/T} . \tag{24}$$

Equation (21) is then rewritten

$$2W = C \quad {}^{z+2} \frac{1}{E \quad BZ} \quad dq \, \frac{A_q^2}{(1 + q^2 \, {}^2)^2} , \tag{25}$$

where C is a constant including all the quantities that do not change while A$_q$ = A − 2(1− x)B [cos q$_x$ a + cos q$_y$ b] and $_E$(x) = $_E$(0) [1 − x]^2. From the experimental results for ^{63}Cu NQR relaxation rates, the correlation lengths

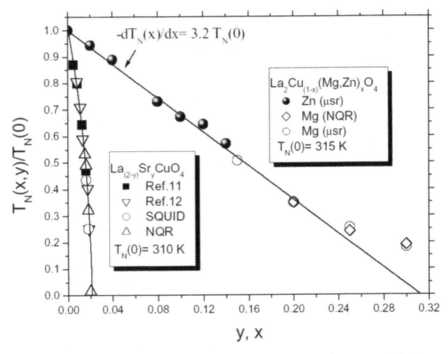

Fig. 8.   Drop of the Neel temperature T_N in La $_2$CuO $_4$ upon charge or spin doping,
as obtained from a combination of measurements (most   μSR, NQR spectra and
SQUID magnetization). The percolation threshold is around $x = 0.02$ for Sr doping
(experimental observation) while for Zn or Mg doping is expected around $y = 0.41$,
according to quantum Monte Carlo simulation and to experimental findings (see
later on). The   solid line for the Sr-doped compound is a guide for the eye, while for
Zn-doped La $_2$CuO $_4$ the line tracks the initial suppression rate (−3.2) expected from
theoretical treatments and experimental data in analogous compounds

reported in Fig. 9 have been derived. The echo signal is lost for temperature
below about 450 K, due to the shortening of the dephasing time T_2. An indication for the value of the correlation length at T_N can be obtained from the
mean field argument according to which

$$T_N(x) = {}^2(x, T_N) J (x) .$$ (26)

Then for the correlation length one has

$$(x, T_N) = (0, T_N)[1 - 4x]^{1/2}/(1 - x) .$$ (27)

The solid lines in Fig. 9 are the theoretical behaviours of (x, T) according to (24). In the inset the values estimated for the correlation lengths at
T_N from (27) are also included. As it appears from the comparison with the
experimental data it can be concluded that for doping x 0.1 and T   800 K

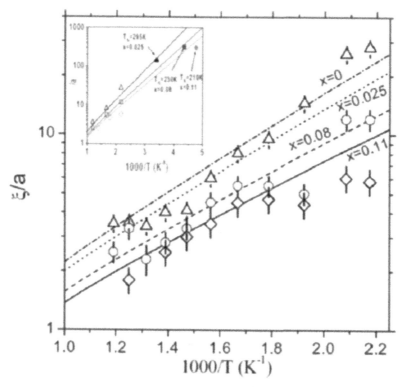

Fig. 9. Temperature dependence of the in-plane magnetic correlation length in spin doped La $_2$CuO $_4$, in the weak doping regime, in the assumption that the Hamiltonian (1) and the related quantities are modified according to the dilution model. The inset shows the extrapolation at T_N of $\xi(x, T)$ given by (24) and the comparison with the values derived according (27), with the doping dependence of T_N evidenced in Fig. 8 for $x \to 0$

the system remains in the RC regime, with reduction of the spin stiffness according to the dilution. No evidence is achieved of a crossover to QD regime, as it is also confirmed by the Gaussian contribution to the ^{63}Cu echo dephasing, which was experimentally found [9] to follow the temperature dependence T_{2G}^{-1} (ξ/T), as it is derived from the scaling conditions applied to the q-integrated static generalized susceptibility $\sum_q \chi(q, 0)$ which controls the echo dephasing rate.

Now we are going to discuss how the charge doping affects the properties of 2DQHAF by referring to Sr doped La $_2$CuO $_4$. The effect of itinerant holes is more complicate than the one due to spin vacancies, essentially because of the occurrence of charge inhomogeneities: the holes tend to segregate along stripes, leaving randomly distributed effective magnetic moments inside AF domains, which in turn are long-range interacting. Cooperative spin-freezing is expected to occur on cooling, with consequent rise of low-energy

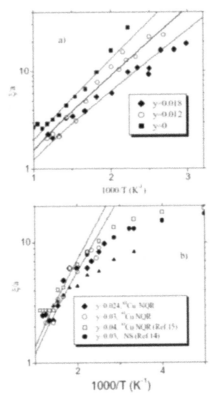

Fig. 10.   Temperature dependence of the correlation length in Sr-doped La $_2$CuO $_4$ for doping amount smaller (a) and larger (b) than the "critical" value y = 0 .02 (see Figs. 2 and 8), as deduced from [63] Cu NQR relaxation. In part b) of the Figure some representative data from neutron scattering [14] and from Imai et al. [15] are also reported. The solid lines are the best fit behaviours according to (4), with the spin sti ness as adjustable parameter

spin excitations (cluster-spin glass phase, somewhat extending into the SC phase, see Fig. 2 and discussion later on).

Let us first consider the problem of the spin dynamics in the paramagnetic phase well above T_N, where the [63]Cu dephasing rate is not so large to wipe-out the NQR signal [13]. From [63]Cu NQR relaxation measurements in La$_{2-y}$Sr$_y$CuO$_4$, with a procedure similar to the one outlined above for spin-doped compounds, the correlation length has been extracted. The experimental data for   are compared in Fig. 10 with the theoretical behaviours according the RC regime (4), having left the spin sti ness   (y) as adjustable parameter.

The conclusions that can be drawn from the Cu NQR relaxation are the following. For doping smaller than the percolation threshold the RC regime is still obeyed, although the spin sti ness decreases with the Sr content less than expected for itinerant holes (for which a dependence of the form $\xi(y)$ y^{-1} has been derived [16]). For Sr content above $x = 0.02$, while in the high temperature region one has a tendency of the correlation length to follow the RC behaviour, in the low temperature range one has $\xi(y, T \to 0)$ $1/ny$ (with n in between 1 and 2). This dependence is different from the one related to the average distance between holes, namely $1/\sqrt{y}$, and therefore is compatible with the hypothesis of charge segregation in stripes, altering the topology from 2D to 1D. It is also remarked that from the sub-lattice magnetization data different spin dynamics seems to drive the relaxation process in different temperature ranges. Well above T_N, where ξ is smaller than the average stripe distance d_s, the stripes are mobile. On the contrary, in the low temperature range where $\xi > d_s$ the stripes appear quasi-static, their hopping rate being less that the NQR line width, i.e. about 100 kHz.

We mention here that in the ordered state, in spin-diluted and in charge-doped La$_2$CuO$_4$, spin excitations different from the paramagnon spin dynamics considered until now arise, related to effective magnetic moments and their cooperative freezing. As regards instead the doping and temperature dependences of the ordered parameter, i.e. the sub-lattice magnetization, namely of the expectation value of the Cu^{2+} magnetic ion, we postpone the discussion, in order to include the effects over all the doping range.

7 Spin and Charge Doped La$_2$CuO$_4$ Near the AF Percolation Thresholds: Spin Stiffness, Correlation Length at the Transition and Staggered Magnetic Moment

Now we extend the analysis of the NQR relaxation rates to the spin doping region where the dilution model, and therefore (23), is evidently a too crude assumption. In practice only ^{139}La NQR relaxation have been measured in the temperature range of interest. By a procedure similar the one outlined in the previous Section, the correlation length has been extracted. In Fig. 11 the absolute values and the temperature dependences derived in the strong dilution condition are compared with the data for light doping and with the theoretical behaviours described by (4), by leaving the spin stiffness as adjustable parameter. Again one is led to the conclusion that the RC regime does hold also for strong dilution. It is noted that also the values for $\xi(x, T_N)$ deduced from (26), still follow the temperature trend of the NQR relaxation data.

The unexpected reliability of the RC description in spin diluted AF has been independently confirmed by Vajk et al. [18] by means of neutron scattering, up to $x = 0.35$. Only close to the percolation threshold ($x = 0.41$) for the nearest-neighbour square AF lattice and in the high temperature range, $\xi(x, T)$ seems to cross over from exponential behaviour to a power law (see Figure 3B in [18]).

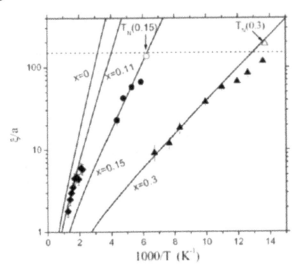

Fig. 11. In-plane magnetic correlation length in spin-doped La $_2$CuO$_4$ (x: Zn or Mg amount substituting Cu) derived from nuclear relaxation rates (close symbols) and from T$_N$ (x) (open symbols). The   solid lines   track the RC behaviour, according to (4). The dotted line is the value   (x, T   T$_N$), for x   1. Data from [17]

It should be remarked, however, that in the strong dilution regime the reduction of the spin sti ness dramatically departs from the one predicted by the dilution model, as it is evidenced in Fig. 12. Another relevant observation involves the absolute value of the in plane correlation length at the transition temperature. In spite of the drastic reduction in the spin sti ness and of the Neel temperature, still the transition to the AF state occurs when the correlation length reaches an in-plane value around 150 lattice steps, as in pure or lightly doped systems.

Analogous conclusion is obtained in regards of the charge-doped La $_2$CuO$_4$. In Fig. 13 we report the temperature behaviour of   (y, T) deduced from a combination of ^{63}Cu and ^{139}La NQR relaxation measurements in La $_{2-y}$Sr$_y$CuO$_4$, for Sr amount   y = 0.016, close to the percolation threshold (see Fig. 8). It has to be observed that the doping amount   y = 0.02, coinciding with the percolation threshold, would not provide more pertinent results. In fact, as it has been confirmed by a detailed analysis of inelastic relaxation [19] for   y close to y$_c$ a microscopic phase separation in AF and cluster-spin-glass phase is noticed. The Sr concentration   y = 0.016 was found the closer one to the percolation threshold still providing information on the charge doped La $_2$CuO$_4$ without detectable "contamination" from phase separation. Again the solid line in Fig. 13 is the theoretical behaviour according (4), in correspondence to a value of the spin sti ness reduced to almost one third of the one in pure 2DQHAF. Still one notes that the maximum of   is reached around

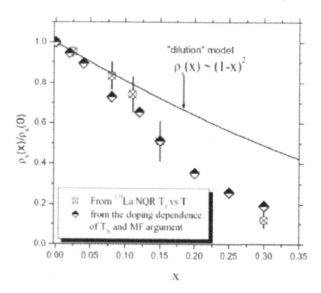

Fig. 12.   Spin sti ness   (x) in spin-diluted La $_2$CuO$_4$ and comparison with the dependence expected within the dilution model (23)

T_N = 150 K and corresponds to about 150 lattice constant, similarly to the
spin doped compound.

   Another quantity of interest for the quantum e ects in disordered 2DHQAF
is the zero-temperature staggered magnetic moment $< \mu_{Cu}$ (x, T   0) > along
the local quantization axis , namely the dependence of the sublattice magnetization on spin dilution. The staggered magnetic moment is di erent from
the classical S = 1 / 2 value because of the quantum fluctuations, that in turn
are expected to increase with spin dilution. The quantity

$$R(x, T = 0) = \frac{< \mu_{Cu} (x, 0) >}{< \mu_{Cu} (0, 0) >}$$

has been obtained to a good accuracy from the magnetic perturbation due to
the local hyperfine field on ^{139}La NQR spectra or from   µSR precessional frequencies [21, 17] and recently evaluated also close to the percolation threshold
from neutron di raction in a single crystal of Zn-Mg doped La $_2$CuO$_4$ [18]. In
Fig. 14 we report the   x-dependence of R as it results from a combination of
NQR [21, 17] and neutron di raction data [18].

   While the classical doping dependence [24] (for   S   ) as well as the
one predicted by the quantum non-linear   model [22] are not supported by
the experimental findings, the data in 14 indicate a doping dependence of the
form R = (x_c - x) , with critical exponent   = 0 .45, close to the behaviour
deduced from spin wave theory and T-matrix approach [23]. The non-classical
critical exponent   is in substantial agreement with finite -size scaling analysis
[25].

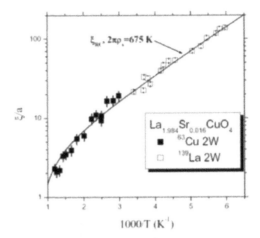

Fig. 13.   In plane magnetic correlation length in Sr doped La $_2$CuO$_4$, for Sr amount y= 0 .016, close to the percolation threshold, as a function of the inverse temperature. The values are obtained along the procedure outlined in the text, from ^{63} Cu NQR and from ^{139} La NQR relaxation rates (data from [20]). The   solid line   tracks the behaviour according to (4) and yields a spin sti ness   = 107 K. Such a reduction implies a small growth, on cooling, of the correlation length and therefore a reduction in T$_N$, the transition to the ordered state still occurring when   is about the same as in pure system

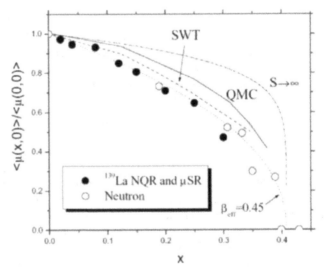

Fig. 14.   Zero-temperature normalized staggered magnetic moment   R in La$_2$CuO$_4$ as a function of spin dilution, from neutron di raction data [18] and from ^{139} La Zeeman perturbed NQR spectra or from   μSR precessional frequencies [21,17] and comparison with the behaviours expected from some theoretical approaches [22-24] (see also [18] and references therein)

As regards the temperature dependence of   R, it appears that both in the light doping regime (where it has been derived from the e ect of the AF field on the [139]La NQR spectra [21]) as well as for strong dilution (from elastic neutron di raction [18]) an universal law of the form

$$R(x = \text{const}, T) \quad [T_N(x) - T]^n$$

holds, with a small critical exponent   n that appears to be around 0.2 for light doping while on approaching the percolation   n increases to 0.3 (Fig. 15).

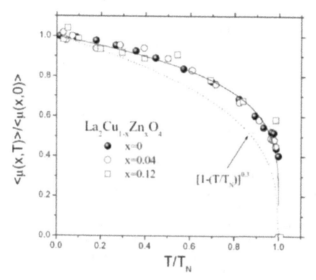

Fig. 15.   Temperature dependence of the staggered magnetic moment in spin diluted La_2CuO_4. Up to a dilution amount   x   0.12 a second order transition is detected and the critical exponent turns out about   = 0.2 (solid line, data from [21]). These data can hardly be fitted by an exponent   = 0.3 (dotted line) that instead appears to justify the results for strong dilution, according to the findings from neutron scattering reported in [18]

We do not discuss in detail the complicate and still open issue of the magnetic excitations in the ordered state, namely for $T < T_N(x, y)$, in spin and/or charge disordered La $_2CuO_4$ and only mention the following. Around the $S = 0$ impurity e ective magnetic moments μ_e have been envisaged, long-range interacting through the AF matrix and yielding anomalous low-energy excitations superimposed to almost unaltered (at least for light doping) magnons. The cooperative freezing, on cooling, of the μ_e is evidenced by a peak in the [139]La NQR relaxation rates, occurring at the temperature T_f where the fluctuation frequency of the site-dependent magnetic field at the La nucleus becomes of the order of the quadrupole coupling constant. The "freezing" temperature T_f increases linearly with the Zn content   x. For strong doping the situation in unclear.

Even more complicate picture holds for the magnetic excitations in the ordered state of Sr-doped La $_2$CuO $_4$. Below T_N (y) the holes itinerate along "rivers" (the stripes) separating small domains of almost unaltered AF. The stripes should "evaporate" when the holes tend to localize and their low-frequency "diffusion" along the stripe can be expected to imply a strong relaxation mechanism. Below the localization temperature a region of magnetic perturbation around the localized hole is induced, with a spin-texture in the AF layer, equivalent to the extra-magnetic moments. Their fluctuations cause maxima in the relaxation rates, with recovery laws characterized by a distribution of relaxation times, and a continuous freezing transition similar to the one occurring when the spin vacancies are induced by Zn doping. Again, the freezing temperature T_f increases linearly with the Sr amount. For some more details, see [6] and references therein.

8 The Cluster Spin-Glass Phase

Above the Sr concentration $y_c = 0.02$ in La$_{2-y}$Sr$_y$CuO $_4$ the AF ordered state is no longer attained at any temperature. The typical experimental observation regarding the ^{139}La NQR relaxation is reported in Fig. 16, for $y = 0.03$. One should remark that in this range of charge doping the recovery law does no longer keep the simple exponential form but it is rather given by a stretched exponential, with characteristic time $_e$. In discussing the experimental data in Fig. 16, first of all one has to realize that in this temperature and doping regime, according to Sect. 6 and Fig. 10 b), the in-plane correlation length is drastically reduced (around 10–15 lattice steps) with respect to the theoretical RC value and practically temperature independent. Thus the marked peaks in the relaxation rates observed on cooling have to be attributed to a relaxation mechanism different from the critical paramagnons imbedded in the behaviour of a divergent correlation length. It turns out that a good fit of the data (particularly above about 6 K, see inset in Fig. 16) is obtained according to a law of the form

$$_e^{-1} \quad / (1 + \quad^2 \quad_m^2) , \tag{28}$$

with $\exp(E/T)$.

The peak in $_e^{-1}$ occurs at a temperature T_g where the characteristic correlation time   reaches $_m^{-1}$, the inverse of the measuring frequency ($_m = 2$ $_Q = 2$ 12.4 MHz, with $_Q$ ^{139}La quadrupole coupling frequency, for the data in Fig. 16). Remarkably, T_g was found to decrease with   y, approximately in the form T_g y^{-1}. Since a long ago Cho et al. [27] interpreted T_g as the freezing temperature into a cluster spin glass phase. It was believed that the localization of the holes along the "stripes" generates 2D finite size

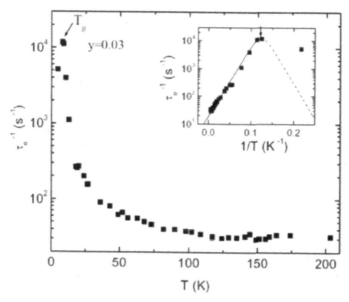

Fig. 16.   Temperature dependence of the characteristic   [139] La NQR relaxation rate τ_e^{-1} in La$_{1.97}$Sr$_{0.03}$CuO$_4$. In the inset the activated temperature behaviour is emphasized, the departure below T_g being the consequence of the distribution of relaxation rates involved in the stretched exponential recovery (data from [26])

AF domains, including an odd number of Cu^{2+} and therefore an e ective magnetic moment with enhanced fluctuations. The walls of the domains are mobile, somewhat corresponding to charge density waves. The doping dependence of T_g is compatible with a mean field argument whereby the magnetic coupling among clusters is proportional to the cluster size $L^2 \propto 1/y$. An alternative interpretation of the NQR relaxation data typically indicated by Fig. 16 could be given in terms of "fluctuating stripes" . In this case the inverse of the correlation time in (28) represents the average frequency ν_{str} for the motion of the stripes. For $T < T_g$ the stripes are "frozen" in the time scale of the NQR experiment. Although the stripes are defined as "static" in neutron scattering experiment when ν_{str} decreases below about 10^{11} rad s^{-1}, they are actually not-static down to 350 mK [28].

A fine confirmation of the spin-glass features of the low temperature phase in Sr doped La$_2$CuO$_4$ above $y_c = 0.02$ has been accomplished by Wakimoto et al [29], by means of high resolution magnetic susceptibility measurements in single crystals. Di erence between FC and ZFC magnetization were found below $T_g = 6$ K, in the crystal at $y = 0.03$. Furthermore a canonical spin-glass order parameter was detected, displaying temperature and magnetic field dependence in accordance to universal scaling relationships expected for spin glasses. The Curie constants evaluated from the in-plane susceptibility [29] resulted independent of the Sr concentration, a feature that might reflect an inhomogeneous distribution of the holes in the CuO $_2$ plane, thus compatible with the stripe structure.

Finally we should mention that in the light of μSR [30] and NMR-NQR experiments [31], the cluster spin glass phase was suggested to extend into the underdoped phase of superconducting LSCO, for hole content larger than about y = 0 .05. On cooling from about the superconducting transition temperature, divergent behaviours of the relaxation rates were noticed, reminiscent of the one reported in Fig. 16 for the spin-glass phase, both in LSCO and in Ca-doped YBCO. Thus one is led to conclude that coexistence of superconductivity and of the spin freezing process typical of the cluster spin glass phase occurs in the underdoped phases of cuprate superconductors. In reality, it seems that in those phases a variety of low-energy magnetic excitations can actually be present, for instance sliding motions of orbital currents implying fluctuating magnetic fields at the La or at the Y sites [32]. The μSR and NQR experimental observations, in particular the strengths, the distribution and the temperature dependences of the fluctuating fields, are as well compatible with the freezing processes of sliding orbital vortex-anti vortex currents [32, 33].

The highly inhomogeneous charge distribution in the CuO $_2$ plane in underdoped superconducting cuprates and in the cluster spin glass phase is proved by the phenomenon called "wipe out e ect" , namely the fact that on cooling some Cu nuclei no longer contribute to the NQR signal because their echo dephasing times become too short to allow one to detect them (see Julien et al. [31], and references therein).

9 The Quantum Critical Point in an Itinerant 2DAF – E ect of Magnetic Field

Going back to the schematic phase diagram depicted in Fig. 2 a very interesting issue involves the point at T = 0 and g = g_c. This should be considered a quantum critical point (QCP). In fact, when for T 0 the parameter g can be tuned, for instance by applying pressure, external magnetic field or chemical composition, a quantum phase transition [34] can be induced. In strongly correlated electron systems, transition of quantum character are rather ubiquitous and of prominent interest is the problem of the spin dynamics accompanying the transition and of the related low-energy excitations [5].

In charge-doped typical 2DQHAF, as in Sr-doped La $_2$CuO $_4$, it appears di cult to locate QCP's, primarily because of the simultaneous tendency to phase separation, so that in general a multiphase system occurs, with AF and cluster-spin-glass or superconducting phases. Thus the experimental investigation of the spin-dynamics driving the phase transition could hardly be carried out in the 2DQHAF leading to high temperature superconductors. In recent times groups of related strongly correlated metals have been discovered where tuning between the AF and the paramagnetic phases can easily be obtained by controlling the atomic composition in intermetallic alloys.

Au-doped $CeCu_6$ can be considered the prototype of these systems, the magnetic response being of 2D character and the electronic properties of the Cu^{2+} ions being involved, as in the more conventional 2DQHAF.

The low-temperature phase diagram for $CeCu_{6-z}Au_z$ around the Au content $z_c = 0.1$ is shown in Fig. 17. $CuCe_6$, namely in correspondence to $z = 0$, is a heavy fermion system, that below the Kondo temperature T_K 6 K, where the 4 f Ce electrons are delocalized into the Fermi sea, is a good example of Fermi liquid (FL), with itinerant pseudo-particles of fermionic character, specific heat going linearly with T and resistivity quadratic in temperature. For $z > z_c = 0.1$ one has an itinerant AF metal with transition to the 3D AF state at $T_N = 2.3$ K for $z = 1$. The Neel temperature T_N decreases on decreasing z towards z_c. On cooling along the line at $z = z_c$ one approaches the QCP, which is believed to result from a competition between Kondo mechanism, which tends to screen the Ce magnetic moments, and the long-range RKKY interaction favouring an ordered magnetic state. Also an external magnetic field can be used to tune the system around the QCP, by suppressing the AF state for $z > 0.1$.

Around the quantum criticality, inelastic neutron scattering revealed a generalized susceptibility of the form [35, 35]

$$(q,) = 1 / [f(q) + A \quad], \tag{29}$$

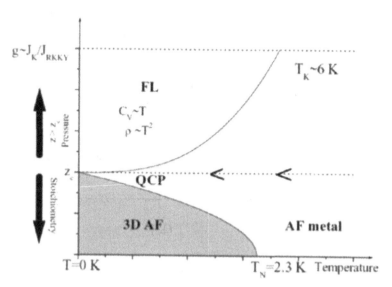

Fig. 17.   Schematic phase diagram (z, T) for $CeCu_{6-z}Au_z$, as derived from a combination of experimental data ([37]) and speculative theoretical considerations

with $f(q)$ going to zero along lines of the Brillouin zone, thus corresponding to 2D magnetic fluctuations in real space. Anomalous critical exponent   and (/T) scaling have been also pointed out.

By means of ^{63}Cu NMR-NQR relaxation the low frequency generalized spin susceptibility over all the Brillouin zone, hardly detectable by neutron scattering, can be accessed. Here we discuss the spin-lattice relaxation measurements in CeCu$_{6-z}$Au$_z$ for $z = 0$, $z = 0.1$ and $z = 0.8$, for external magnetic field ranging from zero (NQR) up to 110   kOe [38].

The ^{63}Cu NQR spin-lattice relaxation rates are reported in Fig. 18, in selected temperature ranges. As it appears from the Figure, the temperature behaviour of 2 W for $z = 0.8$ is apparently similar to the one typical of itinerant metal, with carriers in AF interaction [39]. However, for   T T_N one observes a weak temperature dependence of 2   W which contrasts with the critical behaviour that one would derive from (9)–(11) for a generalized susceptibility of a nearly AF metal [39], that would imply (2   W/T) $(T - T_N)^{-0.5}$. On the contrary the data for 2 W for T   30 K are rather accounted for by the weak logarithmic divergence expected for dipolar interactions, with anisotropic response function. Below about 20   K the almost constant value of 2   W is likely to reflect the screening of the magnetic moments for   T T_K.

For $z = 0$ (pure CeCu$_6$) the FL behaviour 2   W   T is obeyed, as expected for T T_K.

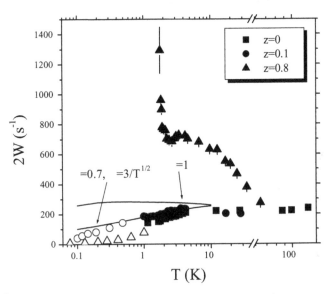

Fig. 18. ^{63}Cu NQR relaxation rates in CeCu$_{6-z}$Au$_z$ (for the resonance line at frequency around 11 .3 MHz). The low-temperature data (T   1 K) for $z = 0$ and $z = 0.1$ are from other authors (see [38] and references therein). The   solid lines correspond to (31) in the text, for   = 1 and   = 0.7

From inelastic neutron scattering [40, 41], in combination to heuristic arguments [36], the magnetic response function, of 2D character, can be envisaged in the form

$$\chi_{2D}^{-1}(k,\omega,T) = k_B \frac{(T - i\omega/a_\omega)^\gamma}{c} + f_\chi(k,T) \ ,\tag{30}$$

with anomalous exponent $\gamma = 1$, (ω/T) scaling and renormalized Curie Weiss constant c.

Let us first discuss the NQR relaxation data, namely the case of zero external magnetic field. From (30) one can derive the form of the generalized susceptibility by expanding $f_\chi(k,T)$ in even powers of q starting from the critical AF wave vector under the condition that the excitation frequencies $\omega_q \simeq \omega_{AF}$ remain larger than the measuring frequency (See (11)–(15)). Then

$$2W = \frac{\gamma^2}{2N} k_B T A^2 \int_{2D}^{BZ} \frac{c}{a_\omega} \frac{1 + (q\xi)^2}{k_B T} \omega^{\gamma+1} {}^{-2} \ ,\tag{31}$$

where $(a_\omega T)^\gamma = \omega_{AF}$ is the critical frequency and $\xi(T, H = 0)$ is the correlation length (again in lattice units). In the case that the critical exponent γ would be 1, from the 2D integration in (31) one has

$$2W = \frac{\gamma^2}{8\pi^2} A^2 \frac{S(S + 1)}{aT} \frac{q_D^2}{1 + \xi^2 q_D^2} \ ,\tag{32}$$

with the correlation length given by [42, 5] $\xi(T, H = 0) = (1/\sqrt{\pi}\bar{T}) \ln(T_K/T)$ and q_D Debye-like wave vector. In Fig. 18 the solid lines correspond to (32) for $a = 10^{10}$ rad s^{-1} K^{-1} and $A = 3.8$ kOe (according to [43]).

It is evident that the experimental results for $T \to 0$ indicate $\gamma = 1$, as already noticed from neutron scattering. Numerical integration is required to derive $2W$ from (31) for $\gamma = 1$. The line in Figure 18 has been obtained in correspondence to $\gamma = 0.7$ and $\omega_{AF} = (1.5 \times 10^{10} T)$ rad s^{-1} and for $\xi = 3$ at $T = 1$ K. This value of the correlation length is close to the one derived from neutron scattering, where it was found [40] $\xi^2 = (\rho_0/k_B T q_0^2)$ with $(\rho_0/q_0^2) = 10$ meV Å^2, yielding $\xi = 10$ Å at $T = 1$ K. One could remark that below about 1 K the critical frequency ω_{AF} slows down to less than 10^{11} rad s^{-1} and therefore hardly detectable by neutron scattering because of resolution limits.

Now we discuss the effect of the magnetic field. Since close to the quantum criticality, for both energy and field dependence, the only scale in the response function is temperature, one can reformulate the derivation leading to (31) and (32) by simply substituting T with an effective temperature

$$T_{eff}(H) = [T^2 + (g\mu_B H/k_B)^2]^{1/2} \equiv [T^2 + (T_{mag})^2]^{1/2} \ .\tag{33}$$

By considering first, for simplicity of discussion, the case of $\gamma = 1$, in the presence of the field (32) is modified in

$$(2W/T) \qquad (T^2 + T_{mag}^2)^2 (T,H)^{-1}, \qquad (34)$$

showing that for T ≈ T_{mag} no field dependence should be expected. This is in agreement with the trend of the data in Fig. 19. For strong field so that T_{mag} is dominating, (34) predicts (W/T) temperature independent and going as H^{-1}, as approximately indicated by the experimental results in Fig. 19, in the temperature range $0.5 - 2$ K.

The solid lines reported in (Fig. 19 b) are the theoretical behaviours obtained by means of numerical integration, having used in (31) for T, including in the correlation length, the e ective temperature given by (33). As it appears from the Figure, above a given temperature $T (H)$, the experimental results for representative strengths of the field justify rather well the theoretical behaviours. Below $T (H)$ a drastic departure of the data from the theoretical trends is noticed, with a sudden decrease of the scaled relaxation rates on cooling. For $T < T (H)$, W takes a temperature dependence approximately of the form W ≈ exp[− (H)/T], typical of a system with a gap in the spin excitations. To give an order of magnitude, one finds (H = 6.7 T) ≈ 0.6 K.

Recent SQUID magnetization and ^{63}Cu NMR measurements in high fields [44] indicate that the gap in the magnetic excitations results from the saturation of the magnetization, implying the quenching of the spin fluctuations.

Summarizing, one can state that ^{63}Cu NQR-NMR spin lattice relaxation measurements in CeCu$_{5.9}$Au$_{0.1}$ provide interesting new insights around QCP. On one side the k-integrated response function at low energy confirm the 2D character of the magnetic fluctuations, the anomalous critical exponent and the energy/temperature scaling. Furthermore, in the low temperature range where the critical frequency is below the resolution limit in neutron scattering new aspects involving the role of an external magnetic field and unconventional scaling [44] are pointed out.

10 Summarizing Remarks

In this Chapter it has been shown how NMR-NQR relaxation can be a valuable tool in order to study the correlated spin dynamics and the phase transitions in systems that in the last decade have called strong interest as models for quantum magnetism and as parents of high temperature superconductors, the square planar arrays of $S = 1/2$ magnetic moments in antiferromagnetic interaction (2DQHAF). Their rich phase diagram as a function of temperature and of spin dilution or hole injection, can be explored by means of NQR-NMR spin-lattice relaxation measurements. Quantitative estimate of the quantum-fluctuations-a ected correlation length, spin sti ness and order parameter can be derived by resorting to the integration of the generalized susceptibility in the Brillouin zone, once that the wave vector dependence of the electron-nuclei hyperfine interaction is properly taken into account.

First it has been shown that in pure, non-disordered, 2DQHAF the in-plane correlation length $(0, T)$ can actually be obtained from the relaxation

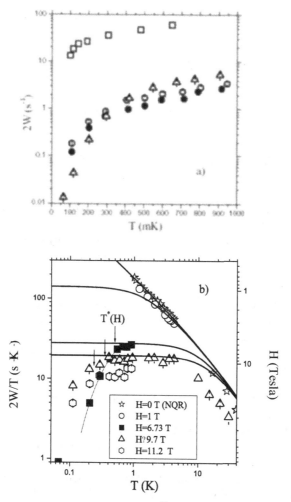

Fig. 19. a) Temperature dependence of the ^{63}Cu relaxation rates in CeCu$_{5.9}$Au$_{0.1}$
for H = 0 ( ), H = 6.7 Tesla ( ), H = 9.8 T ( ) and H = 11.2 T (●). For
H = 0 the relaxation rates have been extracted from the recovery laws for Zeeman perturbed NQR or for quadrupole perturbed NMR. Some modifications in the
hyperfine coupling term can be expected (see text). In part b) of the Figure the
relaxation rates have been scaled by   T and the solid lines represent (31) in the text,
with e ective temperatures according to (33). The arrows indicate the temperature
T (H) at which the data depart from the theoretical forms. The dashed-dotted line
tracks the behaviour of   W in a magnetic field   H = 6.7 Tesla expected in the presence
of a gap in the excitations. The departure from the behaviour predicted on the basis
of (33) could be due to the breakdown for strong field of that scaling law (that is
feasible to hold only up to T_{eff} T_K) (see [44])

rates. By working in the prototype CFTD, where the exchange constant J is rather small, it has been possible to prove that the classical regime for , with renormalization of the spin sti ness and of the spin wave velocity due to the quantum fluctuations, holds in a wide (T/J) range. No evidence of the crossover to quantum critical regime, expected on the basis of an extension of the non-linear   model, is observed. Similar conclusion holds also for La_2CuO_4, although only a more limited (T/J) range can be explored in this compounds, where $J = 1500\,K$. On the other hand La $_2CuO_4$ allows one to perform spin dilution by Zn $^{2+}$ S $= 0$ (or Mg) for Cu $^{2+}$ S $= 1/2$ substitutions or charge doping by Sr $^{2+}$ for La^{3+} substitutions, which corresponds to injecting itinerant holes in the CuO $_2$ plane. The disordered 2DQHAF is thus created and a variety of interesting e ects is observed. For moderate spin doping the dilution model is found to lead to a reliable description: the spin sti ness and therefore the correlation length are still the ones pertaining to the renormalized classical regime, once that the probability of a spin vacancy is taken into account in the AF Hamiltonian. For strong dilution the model is evidently inadequate, the spin sti ness decreasing with increasing the Zn or Mg content x with a x-dependence much stronger when the percolation threshold is approached. Still the spin doped La $_2CuO_4$ was found to remain in the RC regime, (x,T) displaying a temperature dependence similar to the one for $(0,T)$, once that the spin sti ness is renormalized to the corrected value. The transition to the 3D ordered state occurs at the temperature where (x,T) reaches about the same value as in the pure 2DQHAF, namely about 150 lattice steps.

Similar results have been found also in charge doped La $_2CuO_4$, for a Sr content $y = 0.016$, not far from the percolation value $y = 0.02$. Again the temperature behaviour of (y,T) appears almost the same as in the pure compound, although with a strongly reduced spin sti ness, and $(y = 0.016)$ at T_N turns out around 150 lattice steps.

The reduction of the expectation value of the Cu $^{2+}$ magnetic ion as a function of the spin dilution has also been derived. In accordance to neutron di raction data it has been found that the   x-dependence turns out of the form $(x - x_c)$ with $= 0.45$, close to the one expected from spin wave theories and T matrix description and in agreement with finite-size scaling. The temperature dependence of the staggered magnetic moment seems to follow a rather universal law in terms of the   x-dependent Neel temperature, with an abrupt but continuous phase transition and a critical exponent close to 0.2 for small x, possibly increasing to 0.3 for large doping amounts.

A system where a nice study of the spin dynamical properties at the disorder conditions corresponding to a quantum critical point has been possible, is $CeCu_{5.9}Au_{0.1}$. The NQR-NMR ^{63}Cu relaxation rates have provided enlightening insights on the magnetic response function, particularly when the critical frequency slows down below the resolution limit of neutron scattering and in regards of the role of an external magnetic field. On one side a 2D response, with a critical exponent di erent from 1 and the energy/temperature scaling have been confirmed from the relaxation measurements involving the k-integrated dynamical susceptibility at low energy. On the other hand it has been shown that below a certain field-dependent temperature the system crosses over to a phase of gapped spin excitations, with the quenching for T   0 of the magnetization fluctuations and unconventional magnetic field scaling.

On the whole it has been shown that the nuclei can be used as useful tools in the attempt to unravel the many static and dynamical phenomena occurring in 2DQHAF upon charge and spin doping. The powerfulness of the NQR-NMR measurements, when accompanied by a suitable analysis, has been illustrated vis a vis to inelastic neutron scattering. While in some cases fine confirmations of the data found by this technique have been obtained, in other cases the information obtained from NQR-NMR relaxation in regards of low-energy spin excitations have turned out even more subtle and novel aspects have been pointed out, in turn stimulating new scientific work.

Acknowledgments

In this Chapter some results obtained in works carried out in cooperation with the authors indicated in the References have been used. In particular F. Borsa, R. Cantelli, F. Cordero, M. Corti, M. Eremin, M.J. Graf, A. Lascialfari, M. Julien, J. Spalek, V. Tognetti and A. Varlamov are gratefully thanked for their contributions and for useful discussions.

References

1. D.C. Johnston, in   Handbook of Magnetic Materials   Vol. 10, Ed. K.H.J. Buschow (Elsevier 1997) Chapter 1
2. S. Chakravarty, B.I. Halperin, D.R. Nelson: Phys.Rev. B   39, 2344 (1989)
3. P. Hasenfratz, F. Niedermayer: Phys. Lett. B   268, 231 (1991); Z. Phys. B   92, 91 (1993)
4. P.W. Anderson: Frontiers and Borderlines in Many-particle Physics   Eds. R.A. Broglia and J.R. Schri er (North Holland 1988) and references therein.
5. S. Sachdev: Quantum Phase Transitions   (University Press, Cambridge 1999)
6. A. Rigamonti, F. Borsa, P. Carretta: Rep.Prog. Phys.   61, 1367 (1998)
7. M. Conradi, T. Guillion, A. Rigamonti: Phys. Rev.   31, 4388 (1985)
8. P. Carretta, T. Ciabattoni, A. Cuccoli, A. Rigamonti, V. Tognetti, P. Verrucchi: J. Appl. Magn. Resonance   19, 391 (2000)
9. P. Carretta, A. Rigamonti, R. Sala: Phys. Rev. B   55, 3734 (1997)
10. P. Carretta, T. Ciabattoni, A. Cuccoli, E.R. Mognaschi, A. Rigamonti, V. Tognetti, P. Verrucchi: Phys. Rev. Lett.   84, 366 (2000)
11. F.C. Chou et al: Phys. Rev. Lett.   70, 222 (1993)
12. H.H. Klauss et al: Phys. Rev. Lett.   85, 4590 (2000)
13. P. Carretta, F. Tedoldi, A. Rigamonti, F. Galli, F. Borsa, J.H. Cho and D.C. Johnston: Eur. Phys. J. B   10, 233 (1999)

14. B. Keimer et al: Phys. Rev. B   46, 14034 (1992)
15. T. Imai et al: Phys. Rev. Lett.   70, 1002 (1993)
16. M. Acquarone: Physica B   259-261, 509 (1999), Proceedings of the SCES Conference, Paris (1998)
17. P. Carretta, A. Rigamonti, E. Todeschini, L. Malavasi: Acta Physica Polonica B 34 (2003), Proceedings of the SCES Conference 2002 (Kracow)
18. O.P. Vajk, P.K. Mang, M. Greven, P.M. Gehring, J.W. Lynn: Science   295, 1691 (2002)
19. A. Paolone, F. Cordero, R. Cantelli, M. Ferretti: Phys. Rev. B   66, 094503 (2002)
20. A. Paolone, R. Cantelli, F. Cordero, M. Corti, A. Rigamonti, M.Ferretti: Inter. J. Modern. Phys. B   17, 512 (2003)
21. M. Corti, A. Rigamonti, F. Tabak, P. Carretta, F. Licci, L. Ra o Phys.Rev.B 52, 4226 (1995)
22. Y.C. Cheng, A.H. Castro Neto: Phys. Rev. B   61, R3782 (2000)
23. A.L. Cheryshev, Y.C. Chen, A.H.Castro Neto: Phys. Rev B   65, 104407 (2002)
24. A.W. Sandvik: Phys. Rev. B   66, 024418 (2002)
25. K. Kato et al: Phys. Rev. Lett.   84, 4204 (2000)
26. A. Campana, M. Corti, A. Rigamonti. F. Cordero R. Cantelli: Europ. J. Phys. B 18, 49 (2000)
27. J.H. Cho, F. Borsa, D.C. Johnston, D.R. Torgeson: Phys. Rev. B   46, 3179 (1992)
28. P.M. Singer, A.W. Hunt, A.F. Cederstrom, T. Imai: cond-mat /0302077 (2003)
29. S. Wakimoto, S. Ueki, Y. Endoh, K. Yamada: Phys. Rev. B   62, 3547 (2000)
30. Ch. Niedermayer et al: Phys. Rev. Lett.   80, 3843 (1998)
31. M.H. Julien et al: Phys. Rev. B   63, 144508 (2001) See Ref.6 and references therein for early data.
32. See M. Eremin and A. Rigamonti, Phys. Rev. Lett.   88, 037002 (2002) and references therein.
33. A. Rigamonti, M. Eremin, A. Campana, P. Carretta, M. Corti, A. Lascialfari, P. Tedesco: Intern. J. Modern Physics B   17, 861 (2003)
34. J.A. Hertz: Phys. Rev. B   14, 1165 (1976)
35. Q. Si et al: Nature   413, 804 (2001)
36. A. Schroder et al: Nature   407, 351 (2000)
37. H. v.L"ohneysen et al: Physica B   223-224, 471 (1996); H. v.L"ohneysen: J. Phys. Condens. Matter 8, 9689 (1996)
38. P. Carretta, M. Giovannini, M. Horvatic, N. Papinutto, A. Rigamonti: Phys. Rev. B 68, 220404 (2003). See also N. Papinutto, M. J. Graf, P. Carretta, M. Giovannini and A. Rigamonti: Physica B   359, 89 (2005)
39. T. Moriya: Spin Fluctuations in Itinerant Electron Magnetism   , Vol. 56 (Springer, Berlin 1985)
40. O. Stockert et al: Phys. Rev. Lett.   80, 5627 (1998)
41. A. Schroder et al: Phys. Rev. Lett.   80, 5623 (1998)
42. A.J. Millis: Phys. Rev. B   48, 7183 (1993)
43. M. Winkelmann, G. Fisher, B. Pilawa, E. Dormann: Eur. Phys. J. B   26, 199 (2002)
44. N. Papinutto, M.J. Graf, P. Carretta, A. Rigamonti, M. Giovannini, K. Sullivan: Proceedings of SCES'05 meeting (Vienna) to be published on Physica B

Two-Dimensional Exchange NMR and Relaxation Study of the Takagi Group Dynamics in Deuteron Glasses

R. Kind

Institute of Quantum Electronics, ETH-Hoenggerberg, 8093 Zurich
Switzerland
kindrd@phys.ethz.ch

Abstract. The dynamics of the deuteron glass Rb$_{1-x}$(ND$_4$)$_x$D$_2$PO$_4$ has been studied with various one dimensional and two dimensional NMR techniques. Each of these techniques provides a low dimensional homomorphous mapping of the true situation, corresponding to a small piece in a huge puzzle. To create a model which is as close as possible to the true situation it is therefore necessary to combine all available macroscopic and microscopic information on the system. In this combination the NMR-techniques play a dominant role because of the broad spectral window extending from the mHz (2D NMR) to the GHz (1D NMR) region. In this contribution we show how the combination of various NMR techniques, together with symmetry relations, geometrical constraints and model calculations is leading to a consistent model of the deuteron glass dynamics. This model is based on a random Slater lattice with a certain amount of defects i.e., Takagi pairs and unpaired Takagi groups that can propagate through the lattice analogous to OH$^-$ and OH$_3^+$ ions in hexagonal ice.

1 Introduction

In contrast to most contributions of this book, where a single investigation method is described and it's application is demonstrated on various samples, we present here the application of di erent complementary methods on a single class of mono-crystalline samples, the substitutionally disordered pseudo-spin glass system Rb$_{1-x}$(ND$_4$)$_x$D$_2$PO$_4$ (D-RADP-x). It will be shown, that for the construction of a consistent microscopic model of the glass transition, a combination of NMR methods with di erent spectral windows had to be applied and analyzed.

The dynamics of the freezing transition in spin glasses and their dielectric analogues, namely, proton and deuteron glasses, has remained one of the important problems of condensed matter physics. In the latter category, the mixed ferroelectric-antiferroelectric (FE-AFE) solid solution Rb$_{1-x}$(ND$_4$)$_x$D$_2$

R. Kind: Two-Dimensional Exchange NMR and Relaxation Study of the Takagi Group Dynamics in Deuteron Glasses, Lect. Notes Phys. **684**, 383–405 (2006)
www.springerlink.com

PO$_4$ (DRADP-x) has probably been investigated more thoroughly than any other glassy system [1, 2, 3]. A real breakthrough in the understanding was achieved when the random-bond-random-field model of Pirc et al. [4, 5] was introduced to explain the NMR line shapes of the acid deuterons in DRADP-44 [6], as well as the Tl^{2+} ESR line shapes of Tl-doped RADP-70 [7]. This analysis revealed unambiguously the glassy character of the material, and confirmed the thermally activated dynamics observed in earlier   T$_1$ measurements [8, 9]. A further proof of the glassy character was the observation of a bifurcation between the field-cooled and zero-field-cooled static dielectric susceptibility in DRADP-60 at 61 K [10].

The Edwards-Anderson order parameter   q$_{EA}$ defined in the models of [4, 5], which can be written as the second moment of the probability distribution function of the time averaged local polarization   p as

$$q_{EA} = \int_{-1}^{+1} W(p)\, p^2 dp, \quad \text{where} \quad p = \frac{1}{\ } \int_0 p(t)\, dt, \quad -1 < p < +1, \qquad (1)$$

cannot, by itself, distinguish between a random freeze-out and a locally correlated freeze-out of the protons or deuterons, respectively, in biased hydrogen bonds. The length over which the deuterons are coherently frozen in one out of the six Slater configurations (H$_2$PO$_4$) [11], has been derived from the diffuse di racted x-ray intensity in RADP. It is 2.0 nm at low temperatures [12]. This is longer than the average deuteron-deuteron distance, but considerably shorter than the typical size of a conventional domain. Similar results were obtained from a ^{87}Rb NMR line shape analysis [13]. In [14] it was shown that in the KDP framework a great variety of pure Slater lattices (lattices containing only Slater groups) can be realized, with a ratio of FE to AFE configurations ranging continuously from zero to one. Using the proper ratio the low temperature ^{87}Rb NMR line shape can be reproduced.

However, the picture presented above has two major drawbacks: First, it is assumed that the system is static at low temperatures, i.e. that the long time average of each local polarization   p$_i$ is constant and di ers from zero, and second, it is based on symmetric hydrogen bonds in the paraelectric PE phase state. The first assumption was disproved by the deuteron two-dimensional exchange NMR measurements at low temperatures in DRADA-32 of Dolinsek et al. [15], which revealed that all local polarizations vanish in the long-time (10–200 s) average, i.e., q$_{EA}$ = 0 in this time scale, whereas   q$_{EA}$ di ers from zero in the time scale of one-dimensional NMR (10^{-3} s). The second assumption is confirmed by the deuteron NMR measurements Bjorkstam [16] revealing symmetric potentials for the hydrogen bonds above   T$_c$ and asymmetric potentials below   T$_c$. This is in apparent contradiction with the Slater ice rules [11], which predict only asymmetric hydrogen bond potentials. As 1D-NMR is performing a gliding time average with an integration time of the order of T$_2$ the asymmetric potentials must be reversed stochastically or periodically with a correlation time much shorter than   T$_2$ to overcome the contradiction.

As known since the early work of Slater [11] and Takagi [17], the corresponding bias fluctuations, which include the exchange between FE and AFE Slater configurations, can take place at low temperatures exclusively via di usion of unpaired Takagi groups (HPO$_4$ and H$_3$PO$_4$). It seems that this mechanism is dominant also in the PE high temperature phase.

In the following sections we will first present model calculations showing how the motion of unpaired Takagi groups a ects the order in a KDP lattice and then some 1D and 2D NMR experiments which prove the presence of this motion and the corresponding exchange of FE and AFE Slater groups.

2 Model of the Glass Phase Dynamics in DRADP-50

Since the pioneering work of J.C. Slater 1941 on the ferroelectric FE transition in KH$_2$PO$_4$ (KDP) [11] it is known that due to the so called Pauling ice-rules only two out of the four protons linking adjacent PO$_4$ groups via O-H...O bonds are close to each PO$_4$ ion, see Fig. 1. The double well potentials of all O-H...O bonds are here asymmetric since any intra-bond proton transfer from one well to the other creates a Takagi [17] pair (HPO$_4$-H$_3$PO$_4$) which has a higher energy than the Slater pair (H$_2$PO$_4$-H$_2$PO$_4$) it is originating from. Nevertheless, for tractability reasons most order disorder model calculations for the KDP family, including the pseudo-spin model, are working with symmetric hydrogen bonds just using an FE or AFE order parameter eigenvector leading to the corresponding long range ordered Slater lattices.

However, as mentioned in the introduction, the two FE and the four AFE domains are not the only possible Slater lattices. There exists a whole variety of Slater lattices which could all serve as equivalent ground states of

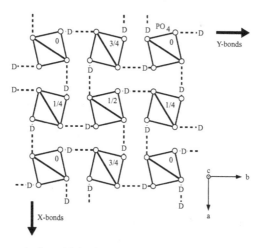

Fig. 1.   PO$_4$-tetrahedra with hydrogen bonding network of the KDP-family

an intra-bond proton or deuteron ordering. The crucial question is whether these ground states are stable or whether there exists a mechanism allowing transitions among these states. In analogy to hexagonal ice where isolated ion states HO^- and H_3O^+ can di use through the lattice without changing the internal energy, in the KDP family the di usion of unpaired Takagi groups HPO_4 (T1) and H_3PO_4 (T3) could transform an initial Slater lattice into another. An unpaired Takagi group in an otherwise perfect Slater lattice has three symmetric and one strongly asymmetric hydrogen bond. Moving the proton on one of the symmetric bonds to the opposite side corresponds to moving the T1 or T3 along this bond to the adjacent PO_4 group leaving a Slater group behind, see Fig. 2. However, moving the proton on the asymmetric bond to the opposite side would create a $T0$ (PO_4) or $T4$ (H_4PO_4) state which is highly improbable.

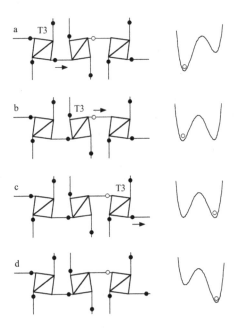

Fig. 2.   Schematic illustration of the bond bias inversion in a random Slater lattice by the passage of an unpaired Takagi defect H $_3PO_4$ (T3). The left side shows the configuration evolution in four steps (a)–(d). Corresponding double well potentials of the deuteron (indicated by the open circle) are shown on the right side

To check the e ect of the random walk of an unpaired Takagi group, we have used a rigid lattice model of 6 $\times$ 6 $\times$ 8 lattice constants in the x,y,z directions containing 1152 PO_4 groups. As initial state for all calculations a long range ordered FE Slater lattice was used. In this lattice a Takagi T1–T3 pair was created and separated at an arbitrary position. Then the unpaired

Takagi groups could start their random walk through the lattice, while the state of every PO $_4$ group was monitored. Periodic boundary conditions were used so when a T3 or T1 accidentally left the lattice on side it was reentered at the opposite side of the prism. Upon accidental recombination of a pair a new one was created again at an arbitrary position. After a large number (10^7) of individual steps a state is reached, which we call a random Slater lattice, where all six Slater configurations are about equally populated. This state, which corresponds to the situation in hexagonal ice, was used as initial state for any further calculations. The fact that such a state can be reached from an initially long range ordered state by the di usion of unpaired Takagi groups clearly demonstrates the power of the mechanism.

However, the ice-rule barrier is not the only short range interaction in the system especially when we deal with a random substitution of K or Rb ions by NH $_4$ or ND $_4$ groups as e.g., in the solid solution Rb $_{1-x}$(ND $_4$)$_x$D$_2$PO$_4$ (DRADP-x). Depending on the value of x either the FE or the AFE Slater configurations become more probable than the random walk model predicts. To take care of this fact we have introduced a biased random walk of un-paired Takagi groups. The corresponding bias energy corresponds to an aver-age Slater energy (2 $_{av}$). Starting with a random Slater lattice and depending on this average bias we can reach any ratio from 100% FE to 100% AFE Slater configurations. For symmetry reasons the two FE configurations are equally populated, as well as the four AFE configurations appear with equal proba-bilities so that the macroscopic polarization is always zero and the average tetragonal structure is established.

Locally, however, the situation is more complicated. For a given "symmet-ric" O...D-O bridge of a T3 or T1 the Slater bias energy $_d^j$ depends in first order on whether a short, or a long, or no N-D...O bridge is leading to one of the oxigens of the bridge, i.e., we have distribution of $_d^j$ that might be even quasi-continuous if also second order contributions are taken into account. The inner energy of two Slater lattices connected by a Takagi step di er thus by $_d^j$ and so sort of a fractal energy landscape is established. For none of these Slater lattices the local symmetry, is tetragonal and non-polar, but as the time evolves the local time averaged symmetry develops these properties, as the 2D-NMR experiments revealed. To reach the observed local symme-try, however, the di usion of unpaired Takagi groups is not su cient, since a static distribution of Slater energies $_d^j$ biasing the random walk of the T1 or T3 always leads to static local deviations from the average symmetry. The '
only possibility to reach also locally the observed average symmetry, is a time dependent distribution of $_d$, where each local $_d^j$ averages to zero in the long time average. In the random-bond random-field model this would correspond to a time dependent random field with constant variance. The origin of the $_d^j$ is the random Rb-ND $_4$ distribution but it is highly improbable that the symmetry is reached by a chemical exchange of Rb and ND $_4$ groups, which would, of course, do the job. A more probable mechanisms based on the fact that each ND $_4$ groups forms four N-D...O hydrogen bridges to PO $_4$ oxygens,

two of them being short, the other two long. This leads to four sites for the nitrogen deviating slightly from   A site of the lattice. An exchange between these four sites, combined with the corresponding "short-long" exchange of the N-D...O bonds would provide the requested time dependence of the random field.

At high temperatures reorientational motion of the ND $_4$-groups can be observed (ND $_4$ deuteron T_1 minimum at 160 K) [8]. For geometrical reasons these reorientations are restricted to the symmetry operations of the tetrahedron, i.e., ± 120 reorientations around the four trigonal axes, and 180   flips around the three twofold axes. The latter are less probable, since all four N-D...O bonds have to be opened and reformed, whereas for the 120   reorientations only three bonds are a ected. Anyhow, during the reformation of the bonds, the ND $_4$-groups have the occasion to adjust their positions to the state of lowest potential energy, so that all four nitrogen positions are reached during successive reorientations. At low temperatures the reorientations are supposed to be frozen-in. Nevertheless, when the Takagi motion results in local polarizations that correspond to a high energy state for the actual ND $_4$ position, there is a non-vanishing probability that the ND $_4$ group reorients to reach a site with lower energy, thus changing the $_d$ for the four O-D...O bonds involved. The next Takagi group visiting one or some of these bonds will find thus changed biases.

3 ^{87}Rb 2D Exchange-Di erence NMR Reveals a Correlated Motion

The aim of these experiments was to observe polarization fluctuations during the glass ordering process to understand the way the order is established. The gradual freeze-out predicted by the pseudo-spin random-bond random-field model should lead to a smooth static (but temperature dependent) probability distribution w(p) of local polarizations p_i where all values between -1 and $+1$ are allowed [4, 5]. This can only be the case for the time averaged value of p_i, as the instantaneous values are either -1 or $+1$, i.e., there must be a smooth static probability distribution of bond bias energies. This picture is confirmed by the ^{87}Rb 1D NMR measurements indicating a smooth distribution of local polarizations for intra-bond hopping frequencies well above the homogeneous NMR line width. On the other hand this picture is in conflict with the ice-rules as long as we do not allow a second mechanism at the same time namely the fast inversion of the ice-rule bias. Otherwise there would be a gap in   w(p) around $p = 0$. Thus the ice-rule dilemma a ects not only the PE phase of the KDP family but the whole "freezing" process of the glass phase e.g., in DRADP-50.

Assuming that the ice-bias fluctuations are much less frequent than the individual intra-bond hopping of the deuterons in the asymmetric potential wells, we have performed ^{87}Rb 2D-exchange NMR measurements at low temperatures for B_0 c. This orientation of the external magnetic field is in so far special as the signal intensity above the Larmor frequency $_L$ comes from Rb spins in a predominantly FE surrounding, whereas the signal intensity below $_L$ from spins in a predominantly AFE environment [14]. Thus polarization fluctuations should become visible in 2D-experiment provided that the ice-bias fluctuations become slower than the homogeneous line width.

The 2D-exchange difference NMR data were recorded using the two pulse sequences one having a long mixing time, the other a short one (shorter than the expected exchange time) are shown in Fig. 3. The idea is to take the difference of the two resulting 2D exchange spectra (one with exchange, the other without exchange) to resolve also off-diagonal intensity close to the diagonal. To compensate for T_1 effects both pulse sequences have the same length by introducing a T_1 weighting time which makes up for the difference in the two mixing times [18].

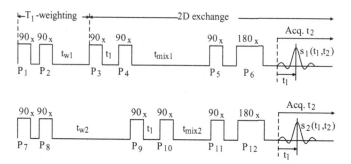

Fig. 3.   Pulse sequences used for the 87Rb 2D-exchange-difference NMR measurements

The spectra show indeed some pronounced off-diagonal intensity which consists, however, of two rather narrow rims close to the diagonal, exhibiting frequency differences   between initial and final state of less than 5 kHz, see Fig 4a. At this point the suspicion arose that the observed off-diagonal intensity could be the result of spectral spin diffusion and not of chemical exchange. To check this a separate study [19] was performed in the FE phase state of a RbH$_2$PO$_4$ single crystal well below the phase transition temperature where the system is completely frozen-out. Any off-diagonal intensity can here be only the result of spectral spin diffusion. The time constant determined for this process was about 600 ms for   = 2 kHz, so that for the case of D-RADP-0.5 a spin-diffusion time of 2.4 s can be expected in view of the 50% reduction of Rb atoms due to the substitution with ND$_4$. This time increases rapidly with   so that e.g., for   = 6 kHz we get 8 s. Thus for the mixing times $_m$ of less than 1 s used in our measurements the spin diffusion contribution can safely be neglected.

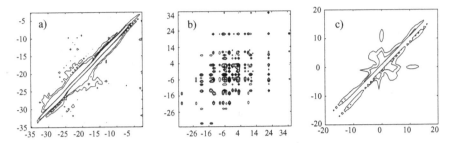

Fig. 4. Comparison of measured ^{87}Rb 2D exchange-di erence NMR spectrum at 45 K (a) with model calculations: Uncorrelated bias inversions (b) and motion of unpaired Takagi groups in a Slater lattice (c)

This leaves us with the task to understand the narrow o -diagonal exchange rims. For this we have used the rigid lattice model described in the previous section but this time by taking only the eight closest deuterons to a given ^{87}Rb into account. The EFG-tensor and the corresponding NMR frequencies were calculated for all 2^8 possible deuteron configurations like in [14]. For symmetry reasons only 35 di erent distinct frequencies were obtained, the distribution of which corresponding roughly to the measured 1D signal. This knowledge opens the path for a computer simulation of the 2D-exchange. Any of the 256 states can serve as initial state with frequency $_i$ or after a mixing time (represented by a certain amount of deuteron intra-bond jumps selected by a random generator) as final state with frequency $_f$, respectively. The resulting coordinates ($_i$, $_f$) are then accumulated in a 2D histogram for about 10^5 repetitions. Figure 4b shows the resulting 2D frequency pattern for eight steps between initial and final state. It clearly displays frequency di erences of up to 40 kHz and is entirely di erent from the measured rim structure. This clearly indicates, that independent ice-bias inversions do not explain the observed pattern [20].

On looking for a correlated mechanism with the required properties the idea arose to try out the di usion of unpaired Takagi groups. As shown in Fig. 2 they have the property of inverting the ice bias on every hydrogen bond they pass. To check this mechanism we have used the rigid lattice model described in Sect. 2, but instead of monitoring the states of the PO$_4$ groups, the positions of the sixteen closest deuterons to a given ^{87}Rb were observed and the corresponding NMR frequencies calculated. Starting every set with a freshly calculated random Slater lattice and the related initial frequency $_i$ a certain number of random Takagi steps (16 to 4064) were calculated to reach the final state with $_f$. Again the resulting coordinates ($_i$, $_f$) were accumulated in a 2D histogram for about 10^7 sets. The result is shown in Fig. 4c as 2D contour plot. At least for a small number of Takagi steps the pattern looks similar to the measurements. A better impression can be gained by looking at the cross sections through the model patterns, see Fig. 5. Here we clearly

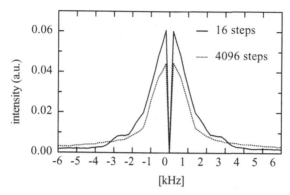

Fig. 5. Cross-Section through the calculated ^{87}Rb 2D NMR exchange spectrum for two diﬀerent numbers of Takagi steps between initial and ﬁnal state. The intensity versus frequency shift curve clearly shows that most of the intensity is in the two rims close to the diagonal of Fig. 4c

find most of the oﬀ-diagonal intensity close to the diagonal as in the experiment. This strongly supports the Takagi diﬀusion as responsible mechanism. However, for a ﬁnal proof it must be shown experimentally that the polarization ﬂuctuations of the PO$_4$ groups (between the six Slater conﬁgurations) associated with the Takagi diﬀusion are really taking place.

4 Distinction of the Six Slater Conﬁgurations by the Anisotropic ^{31}P Chemical Shift Tensor

The first ^{31}P NMR measurements in the KDP family were performed to investigate the PE-FE phase transition in KH$_2$PO$_4$ [21]. The authors have determined the anisotropic part σ of the ^{31}P chemical shift (CS) tensor in both the PE and FE phase states. Since the phosphorus atoms are located in the center of the PO$_4$ tetrahedra the symmetry of the ^{31}P CS-tensor reflects the symmetry of the B-site in the KDP lattice. Because of the symmetry element $\overline{4_z}$ of this site in the PE phase the CS-tensor is diagonal with $\sigma_{xx} = \sigma_{yy} = -\frac{1}{2}\sigma_{zz}$:

$$\sigma_{PE}\ [ppm] = \begin{array}{ccc} -9 & 0 & 0 \\ 0 & -9 & 0 \\ 0 & 0 & 18 \end{array}\ ,\tag{2}$$

whereas the remaining symmetry element 2_z in the FE phase requires the elements σ_{xz} and σ_{yz} to vanish:

$$\sigma_{FE_{1,4}}\ [ppm] = \begin{array}{ccc} 30 & 52 & 0 \\ 52 & -15 & 0 \\ 0 & 0 & -15 \end{array}\ ,\quad \sigma_{FE_{2,3}}\ [ppm] = \begin{array}{ccc} -15 & \pm 52 & 0 \\ \pm 52 & 30 & 0 \\ 0 & 0 & -15 \end{array}\ .$$

$$\tag{3}$$

The two sets are related by the diagonal glide planes of the FE space group $Fdd2$, i.e., $x \to y$, $y \to -x$ and thus exchanging σ_{xx} and σ_{yy} while σ_{xy} changes sign. Within the set the tensors are related by the lost symmetry elements 2_x and 2_y relating the two FE domains with electric polarizations up or down, respectively.

The CS tensor elements were obtained ^{31}P NMR from rotation patterns where the single crystal is rotated in the external magnetic field B_0 around its a or c-axis, respectively. This corresponds to a rotation of the CS tensor and the anisotropic part of the CS is given by the transformed element σ_{zz}.

$$\sigma = R \, \sigma \, R^{-1}, \qquad \sigma = (\sigma_{iso} + \sigma_{zz}) \, L . \tag{4}$$

This yields the following angular dependencies for the a and c-rotations:

a-rotation $\quad \sigma_{zz} = -\tfrac{1}{2}\sigma_{xx} + \tfrac{1}{2}(\sigma_{zz} - \sigma_{yy})\cos(2\varphi) + \sigma_{yz}\sin(2\varphi),$

c-rotation $\quad \sigma_{zz} = -\tfrac{1}{2}\sigma_{zz} + \tfrac{1}{2}(\sigma_{xx} - \sigma_{yy})\cos(2\varphi) + \sigma_{xy}\sin(2\varphi).$ $\qquad (5)$

For the FE CS tensor the $\sin(2\varphi)$ vanishes for the a-rotation and the base line is determined by $\sigma_{iso} - \tfrac{1}{2}\sigma_{xx}$. From (5) it is evident that that both rotation patterns (a and c) are needed to determine the whole FE CS tensor.

To determine the AFE CS tensors we have measured ^{31}P rotation patterns in a single crystal of DRADP-95 which undergoes an AFE ordering. The AFE phase state has the space group $P2_12_12_1$, i.e. there are no elements left transforming the PO_4 groups into themselves. Therefore there are no symmetry restrictions for the AFE CS tensor. Nevertheless, for unknown reasons the c-rotation does not show any angular dependence, meaning that $(\sigma_{xx} - \sigma_{yy}) = \sigma_{xy} = 0$. The a-rotation is shown in Fig. 6 yielding for the four AFE CS tensors:

$$AFE_{1,2}\ [ppm] = \begin{array}{ccc} 16.5 & 0 & \pm 16.3 \\ 0 & 16.5 & 51.8 \\ \pm 16.3 & 51.8 & -33.0 \end{array} \ ,$$

$$\tag{6}$$

$$AFE_{3,4}\ [ppm] = \begin{array}{ccc} 16.5 & 0 & \pm 51.8 \\ 0 & 16.5 & 16.3 \\ \pm 51.8 & 16.3 & -33.0 \end{array} \ .$$

Though (5) allow only the determination of $(\sigma_{zz} - \sigma_{yy})$ and σ_{yz}, the remaining elements can be obtained from the same rotation pattern because of the four AFE domains present in the lattice, which are related by the lost symmetry elements of the space group $I\bar{4}2d$ of the PE high temperature phase. The four lines shown in Fig. 6 (right side) correspond to the four AFE domains, or in other words, to the four AFE Slater configurations. For unknown reasons the two inner lines have much less intensity than the two outer lines, though the number of spins is the same for all four lines. If the crystal is rotated around the a-axis, the inner and outer lines are exchanged, but again the inner lines have much less intensity. Thus it is dangerous to draw conclusions

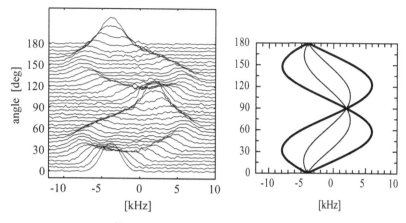

Fig. 6.   Left : Measured ^{31}P NMR rotation pattern of DRADP-95 around the   a-axis in the AFE phase state at   T = 195 K. The two inner lines are only barely visible in this scale.   Right : calculated angular dependence with fitted chemical shift tensor

alone on the basis of signal intensities, especially if known symmetries are apparently violated. We know now the CS tensors of the six Slater configurations for the long range ordered systems. It is clear, that we cannot expect that their size is the same in DRADP-50, but the symmetry relations in the glass phase are very likely the same as for the pure systems since the average tetragonal non-polar symmetry of the PE phase is maintained in the glass phase.

In DRADP-50 the chemical shift tensors of the ^{31}P give rise to an NMR spectrum with twelve separated lines for general orientation of the crystal in the external magnetic field B_0. For symmetry reasons one can find orientations where sets of FE or AFE NMR lines merge to a single line. For instance if B_0 is either perpendicular to the   c or to the a-axis of the crystal, the lines of corresponding Slater groups of the two physically non-equivalent ^{31}P sites in the primitive unit cell merge and the set of lines is reduced from twelve to six. Due to the inhomogeneous line widths of about 2 kHz the spectra are not well resolved, see Figs. 7, 8. Nevertheless, it was possible to determine the CS tensors from the low temperature rotation patterns by using the symmetry relations presented above and by fitting the orientation dependence of the corresponding set of Gaussians simultaneously to the whole rotation pattern. We can again distinguish the six CS tensors for the two di erent FE and the four AFE Slater configurations. In the   c-rotation pattern only the FE CS tensors exhibit an angular dependence (satellites), while similar to the AFE phase state above, the lines originating from AFE Slater groups are all part of the central line.

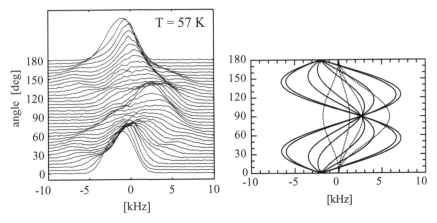

Fig. 7.   Left: Measured ^{31}P NMR rotation pattern of DRADP-50 around an axis in the a-b plane accidentally only close to the   a-axis (o set 7 .3) at 57 K. Right : Corresponding calculated angular dependence exhibiting all twelve lines as obtained from the fit. The   full lines belong to AFE Slater configurations,   the dotted to the FE ones

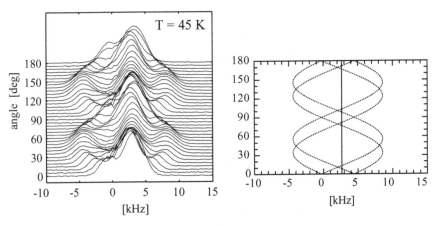

Fig. 8.   Left: Measured ^{31}P NMR rotation pattern of DRADP-50 around the   c-axis at 50 K. Right : Corresponding calculated angular dependence, where the   full line belongs to the four AFE Slater groups,   the dotted to the FE ones

$$\text{FE}_{1,4}\,[\text{ppm}] = \begin{array}{ccc} 25.0 & 49.7 & 0 \\ 49.7 & -14.1 & 0 \\ 0 & 0 & -10.9 \end{array} \quad ,$$

$$\text{FE}_{2,3}\,[\text{ppm}] = \begin{array}{ccc} -25.0 & \pm 49.7 & 0 \\ \pm 49.7 & 25 & 0 \\ 0 & 0 & -10.9 \end{array} \quad .$$

(7)

$$\text{AFE}_{1,2} \, [\text{ppm}] = \begin{matrix} 12.7 & 0 & \pm 18.2 \\ 0 & 12.7 & 51.3 \\ \pm 18.2 & 51.3 & -25.4 \end{matrix} \quad ,$$

$$\tag{8}$$

$$\text{AFE}_{3,4} \, [\text{ppm}] = \begin{matrix} 12.7 & 0 & \pm 51.3 \\ 0 & 12.7 & 18.2 \\ \pm 51.3 & 18.2 & -25.4 \end{matrix} \quad .$$

However, it was not possible to identify CS tensors belonging to Takagi configurations.

To illustrate the strong temperature dependence of the spectra during the glassy ordering we have measured the a and c-rotation patterns also at 168 K, Fig. 9. The a-rotation pattern shows the angular dependence of the PE phase, with $(\sigma_{zz} - \sigma_{yy})\cos(2\phi)$, whereas the c-rotation pattern exhibits only a modulation of the line width, indicating the onset of the glassy ordering. In Fig. 9 a set of spectra with temperature as parameter is shown for a special orientation of the external magnetic field B_0, where for symmetry reasons only three lines are observed. It corresponds to an orientation in the c-rotation pattern where the lines of the two physically inequivalent FE Slater groups (of the primitive unit cell) merge. The two satellites correspond to the two FE polarizations up or down, respectively, whereas the central line is a superposition of all AFE-lines. Though the temperature dependence of the spectra (Fig. 10) looks like a dynamic line shape transition, it is in fact the result of the pseudo-static glass ordering. It should be noted, that because of the intensity problems mentioned above, the intensity ratio $R = 1.86$ of the AFE central line and one FE satellite is very likely not equal to the corresponding ratio of AFE and FE up, or FE down Slater groups. Thus one should not draw corresponding conclusions from the ratio R.

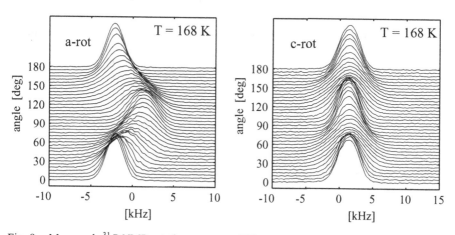

Fig. 9. Measured ^{31}P NMR rotation patterns of DRADP-50 around the a-axis (left) and the c-axis (right) at 168 K

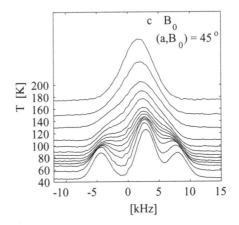

Fig. 10. Measured ^{31}P NMR spectra at various temperatures for a special orientation in the external magnetic field where for symmetry reasons only three lines are visible. The satellites correspond to the two FE polarizations up or down, respectively, whereas the central line is a superposition of the four AFE contributions

5 Slow Polarization Fluctuations of the PO$_4$ Groups Observed by ^{31}P 2D Exchange NMR

Since an unambiguous assignment of the ^{31}P NMR lines to the corresponding Slater configurations is possible a ^{31}P 2D-exchange NMR experiment can reveal transitions between the different Slater states. Thus if the random walk mechanism of unpaired T3 or T1 Takagi groups described in the previous section takes place it should be observable in the ^{31}P 2D-exchange NMR spectrum. Clearly there are no transitions possible between Slater states of different ^{31}P sites and therefore there is no loss in generality if one performs the experiment with a set of six lines only.

Following Ernst et al. [22] we start with a set of six 1D NMR lines with frequencies $\omega_1, \omega_2, \ldots, \omega_6$ and measured integral intensities (zero'th moments) denoted by M_{j0} ($j = 1, 2, \ldots, 6$) among which chemical exchange is supposed to take place. The 2D intensities recorded after a mixing time τ_m are then given by

$$I_{ij}(\tau_{mix}) = a_{ij}(\tau_{mix}) M_{j0} \qquad (9)$$

and

$$a_{ij}(\tau_{mix}) = [\exp\{L\tau_{mix}\}]_{ij} . \qquad (10)$$

Neglecting relaxation, the transition matrix L relating the six Slater States has for symmetry reasons the form

$$L = \begin{pmatrix} -2(K1+K2) & K1 & 0 & K1 & K2 & K2 \\ K1 & -2(K1+K2) & K1 & 0 & K2 & K2 \\ 0 & K1 & -2(K1+K2) & K1 & K2 & K2 \\ K1 & 0 & K1 & -2(K1+K2) & K2 & K2 \\ K3 & K3 & K3 & K3 & -4K3 & 0 \\ K3 & K3 & K3 & K3 & 0 & -4K3 \end{pmatrix}, \qquad (11)$$

where the indices of L denote $1 = AFE1$, $2 = AFE2$, $3 = AFE3$, $4 = AFE4$, $5 = FE1$, $2 = FE2$. The L_{ij} are transitions rates from state j to state i and the zeros indicate the transitions that are not directly possible for geometrical reasons, i.e., there are two successive visits of a $T3$ or $T1$ needed to perform such a transition. Provided that there is any chemical exchange taking place in the system, the intensities of the signals in the 1D spectrum are the continuous or equilibrium result of that mechanism. Therefore the transition rates L_{ij} (except for the forbidden transitions) are related to M_{j0} in the following way: $L_{ij}/L_{ji} = M_{i0}/M_{j0}$. The observed average tetragonal symmetry of the system indicates that the four AFE Slater configurations are equally populated and thus there is only a single transition rate $K1$ relating these states. Since the intensity of the two equivalent FE signals di ers from the one of the four AFE signals ($M_{50} = M_{60} = M_{10} = M_{20} = M_{30} = M_{40}$) the rate $K2$ of the FE-AFE transitions di ers from the rate $K3$ of the AFE-FE transitions. The six eigenvalues L_{jj} of L are the time constants of the system and the back-transform of $\exp(L_{mix})$ yields the time dependence of all signal intensities in the 2D spectrum.

For a crystal orientation with B_0 perpendicular to the c-axis the four AFE lines merge to a single line and the 1D spectrum consists now of an AFE central line and two FE satellites corresponding to the two anti-parallel FE polarizations. This allows the use of a reduced transition matrix, as well as the analytical calculation of the 2D intensities. Provided that the 1D spectrum is given by $M_0 = (1, R, 1)/(2 + R)$ the transition matrix becomes

$$L = \begin{matrix} -K & K & 0 \\ RK & -2RK & RK \\ 0 & K & -K \end{matrix} \quad , \quad (12)$$

where the indices of L denote $1 = FE$, $2 = AFE$, $3 = FE$. From the measured 1D spectrum we have determined $R = 1.86$. The eigenvalues of L are $[0, -(2+R)K, -RK]$ and for the normalized intensities one obtains

$$I_{11} = I_{33} = \frac{1 + 0.5[(2+R)\,e^{(-KR\ _m)} + R\,e^{(-(2+R)K\ _m)}]}{(2+R)^2}$$

$$I_{22} = \frac{R^2 + 2R\,e^{(-(2+R)K\ _m)}}{(2+R)^2}$$

$$I_{12} = I_{21} = I_{32} = I_{23} = \frac{R - R\,e^{(-(2+R)K\ _m)}}{(2+R)^2} \qquad (13)$$

$$I_{13} = I_{31} = \frac{1 - 0.5[(2+R)\,e^{(-KR\ _m)} - R\,e^{(-(2+R)K\ _m)}]}{(2+R)^2},$$

which for $_m = 0$ (no exchange) and for very long $_m$ (exchange saturation) become

$$I = \begin{matrix} 1 & 0 & 0 \\ 0 & R & 0 \\ 0 & 0 & 1 \end{matrix} \quad (2+R)^{-1} \text{ and } I = \begin{matrix} 1 & R & 1 \\ R & R^2 & R \\ 1 & R & 1 \end{matrix} \quad (2+R)^{-2} \text{ respectively . } (14)$$

If the above 2D exchange NMR saturation pattern is observed in the ex-periment, then we have the proof that all spins contributing to the 1D NMR

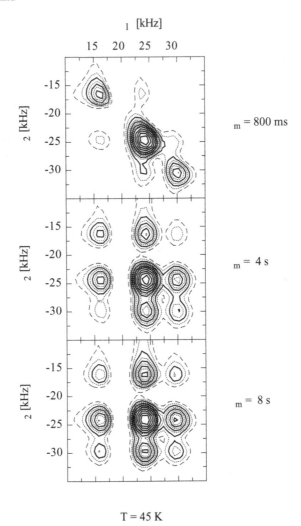

Fig. 11.   Contour plots of the ^{31}P 2D-exchange NMR spectra in DRADP-55 at 45 K for three diﬀerent mixing times (τ_m)

signal are involved in the chemical exchange mechanism, i.e. every PO$_4$ group takes on all six Slater conﬁgurations within a certain time span, so that the local time average is also reﬂecting the average non-polar tetragonal symme-try. Figure 11 shows the contour plots of the 2D exchange spectra for three diﬀerent mixing times illustrating the evolution of the cross-peaks. In Fig. 12 the evolution of the measured intensity ratios I_{12}/I_{11} and I_{13}/I_{11} is shown as a function of the mixing time τ_m for T = 45 K. The measurements clearly show that the 2D saturation pattern is reached after about 15 s. The ﬁtted value for the transition rate is K = 0.26 s^{-1}. This clearly shows the existence of slow

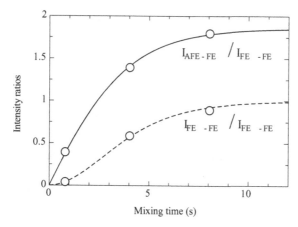

Fig. 12. Evolution of the intensity ratios between cross-peaks of Fig. 11 and one diagonal FE peak as a function of the mixing time (τ_m)

PO_4 polarization fluctuations in D-RADP-50. Together with the results of the previous section we have the unambiguous proof that these polarization fluctuations are resulting from the biased random walk of unpaired Takagi groups. These results were presented in [23, 24].

6 Interpretation of the ^{87}Rb T_1 Measurements

According to second order time dependent perturbation theory the spectral density responsible for the relaxation rate is the Fourier transform of the auto-correlation function of the Rb-EFG-tensor fluctuations. A rigid lattice point-charge model revealed that the intra-bond motions of the protons or deuterons on the O..D-O bonds yield the su ciently high fluctuation amplitudes of the EFG tensor elements to explain the observed relaxation rate at the T_1 minimum. The dynamics of this motion is coming from two sources: The motion of unpaired Takagi groups and the intra-bond motion of the protons or deuterons in the asymmetric potential wells. The autocorrelation function of the latter mechanism is well known. It decays exponentially with an auto-correlation time $\tau_c = \tau^+ \tau^- / (\tau^+ + \tau^-)$, where τ^+ and τ^- are the mean dwell times of the deuterons in the two potential wells. The amplitude is reduced by the so called depopulation factor $1 - p^2$, with

$$p = \frac{\tau^+ - \tau^-}{\tau^+ + \tau^-} = \tanh(E_b/k_B T) . \tag{15}$$

where $2E_b$ is the energy needed to create a Takagi pair. For symmetric bonds $p = 0$ and $\tau_c = \tau^+/2 = \tau_0$, where τ_0 is given by the Arrhenius law for symmetric bonds $\tau_0 = \tau_\infty \exp(E_a/k_b T)$.

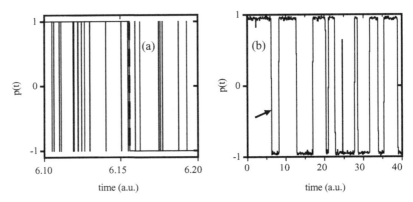

Fig. 13. Calculated time dependence of the local polarization $p(t)$ for $p = 0.95$ using an exponential distribution for the dwell times $^+$ and $^-$. Superposed is a slower mechanism inverting the bias to $p = -0.95$, corresponding to the passage of an unpaired Takagi group (a). Gliding time average $\bar{p}(t)$ of $p(t)$ exhibiting the bias inversions by successive passages of unpaired Takagi groups (b). The arrow indicates the bias inversion shown in Fig. 13a

In Fig. 13a an example for the function $p(t)$ for a given hydrogen bond is shown for $p = \pm 0.95$, where p is the gliding short time average of $p(t)$. The sign of p changes when an unpaired Takagi group is passing the bond. Such a change is indicated by the vertical dashed line. In Fig. 13b the filtered function $\bar{p}(t)$ is shown, i.e., the fast asymmetric motion is filtered out and only the effect of the Takagi group motion remains. It is calculated to reach a long-time average of zero for $\bar{p}(t)$, thus symmetrizing the bond. The arrow indicates the Takagi passing shown in Fig. 13a. The autocorrelation function of $\bar{p}(t)$ is exponentially decaying with a pre-factor p^2 and a time constant given by the average number of Takagi steps between two bias inversions n_{BI} times the autocorrelation time τ_0. Since there are twice as many bonds than PO_4-groups, n_{BI} is related to the number of unpaired Takagi groups N_{Tu} by:

$$n_{BI} = \frac{1}{-\ln(1 - 2N_{Tu})} \approx \frac{1}{2N_{Tu}}. \tag{16}$$

where $1/N_{Tu} \approx \frac{5}{2}(1 + \exp(2 E_b/k_B T))$.

The well known Fourier transform of the two autocorrelation functions yields for the normalized spectral density

$$J(\omega_L) = (1 - p^2)\frac{2\tau_0 \overline{1 - p^2}}{1 + (1 - p^2)(\omega\tau_0)^2} + p^2\frac{2n_{BI}\tau_0}{1 + (n_{BI}\omega\tau_0)^2}. \tag{17}$$

In contrast to the usual result of the BPP theory, τ_0 is not the only temperature dependent contribution in both terms of (17), so that the apparent activation energy E_{app} that is usually obtained from the slope of T_1 differs

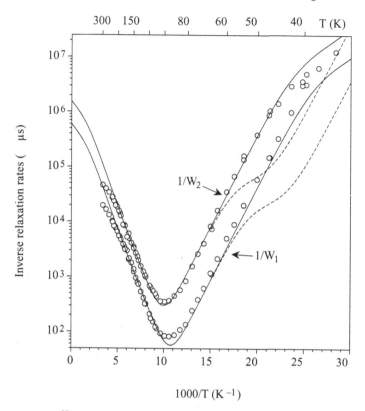

Fig. 14.   Inverse ^{87}Rb spin-lattice relaxation rates $1/W_1$ and $1/W_2$ versus inverse temperature. The solid lines are a fit of (17) and (18) to the data. The dashed line is calculated for a lower ice-rule barrier, see text

considerably from the activation energy E_a of $_0$, e.g., for the product n_{BI} $_0$ we have E_{app} $2E_b + E_a$. For $E_b > 15\,meV$ the second term is dominant down to 65 K and the observed E_{app} of 80 meV consists mainly of this sum. The first ^{87}Rb spin-lattice relaxation measurements on D-RADP-50, where the non-exponential magnetization recovery curves were systematically analyzed (to discriminate not only the mean relaxation rates at $_L$ and 2 $_L$, W_1 and W_2, respectively, but also their probability distribution functions) were performed by N. Korner [25] and first published in [26]. These measurements are shown in Fig. 14. In order to account for the reduced slope in the slow motion regime and for the distribution of the auto-correlation time, N. Korner used a Havrilak-Negami spectral density [27] for fitting the theory to the experiment.

The Havrilak-Negami spectral density is given by:

$$J_{HN}(\omega, \omega_c, \alpha, \beta) = \frac{2}{\omega} \sin\beta \; \arctan \frac{(\omega/\omega_c)^\alpha \sin(\alpha\pi/2)}{1+(\omega/\omega_c)^\alpha \cos(\alpha\pi/2)}$$
$$\times \left[1+2(\omega/\omega_c)^\alpha \cos(\alpha\pi/2)+(\omega/\omega_c)^{2\alpha}\right]^{-\beta/2}. \tag{18}$$

Four our calculations we have used this spectral density for the second term in (17) with $\omega_c = \omega_0 N_{BI}$ and $\omega = \omega_L$ or $2\omega_L$, respectively. For the calculation of the solid lines the following parameters were used: $E_{app} = 80\,meV$, $2E_b = 37.6\,meV$, $E_a = 42.4\,meV$, $\tau = 0.7 \times 10^{-13}\,s$, $\nu_1 = 98.163\,MHz$, $\alpha = 1.0$ and $\beta = 0.85$. For the fit we need also the mean square fluctuation amplitudes $A_1 = 0.60 \times 10^{13}\,s^{-2}$ and $A_2 = 0.23 \times 10^{13}\,s^{-2}$ for the transitions $\Delta m = \pm 1$ and $\Delta m = \pm 2$, respectively. The fit is excellent except for room temperature, i.e., where due to their high density the Takagi motion suffers from mutual hindering and where probably also the existence of $T0$ and $T4$ Takagi groups cannot be neglected anymore.

To show the effect of the first term of (17) we have calculated the relaxation rates for $2E_b = 30\,meV$ (dashed lines). Such a deviation clearly is not observed and whether the bending of the inverse relaxation rates at 40 K is due to this term was questionable without further low temperature data. However, such measurements exist in the literature [28] and there this effect should be visible if it exists at all. This is in fact the case: In Fig. 2 of this reference where the T_1 of the acid deuterons in DRADA-32 is displayed versus the inverse temperature, a systematic deviation from the calculated T_1 is observed that has exactly the shape we are looking for. The same feature is observed (though less pronounced because of the scale and the number of data points) in Fig. 1 for the ^{87}Rb T_1 in DRADP-50. These facts support our theory and allow for the first time to determine the ice barrier in DRADP-50: $E_b = 37.6\,meV$. Since the apparent activation energy $E_{app} = 80\,meV$, the activation energy for the symmetric bonds is $E_a = 42.4\,meV$. This clearly shows that also for the asymmetric bonds (subdued to the ice-bias) in DRADP-50 there are two localized states for the deuterons.

7 Discussion

The ^{87}Rb T_1 results show that the time scale of the Takagi visits is much faster than expected from the ^{31}P 2D-exchange NMR. The Takagi visits are in fact the mechanism that establishes the glass order. It is therefore not astonishing that the off-diagonal rims in the ^{87}Rb 2D-exchange difference spectra become visible already for mixing times as short as 10 ms. These rims can even be observed at room temperature for short mixing times [29]. The exchange time determined from ^{31}P 2D-exchange NMR is thus not the time between two Takagi visits to a given PO_4 group as erroneously stated in [25] but reflects the presence of a much slower motion, namely the reorientation of

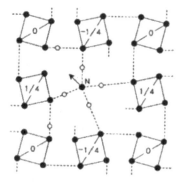

Fig. 15.   ND$_4$ hydrogen bonding scheme in DRAD-P-50 with two short and two long N-D...O bonds. The average local tetragonal symmetry can only be restored by a stochastic reorientation of the bonding scheme in   ±90   steps. The 31P 2D exchange NMR measurements revealed a correlation time for this motion of the order of seconds at 45 K

the ND$_4$ groups. A frozen-in configuration of all ND$_4$ groups (Fig. 15) acts via the corresponding Slater energies $_j$ like a maze for the Takagi group motions leading to a certain set of glassy states. With the ND$_4$ reorientation this maze is slowly changing resulting in a corresponding change of the glassy states. Within a long enough time span enough possible states are taken on so that the non-polar tetragonal tetragonal symmetry is established also in the local time average.

To conclude we want to resume all known experimental facts concerning the dynamics in the deuteron glass DRADP-50. The following facts are deduced from combining the results of all experiments performed on the system so far. They include all macroscopic techniques, as well as, x-ray and neutron scattering results, and, of course, NMR and ESR measurements.

1. Down to at least 40 K the system relaxes at any temperature to a dynamic equilibrium state. Though the responsible mechanisms are slowing down, they are far from being frozen-in.
2. This dynamic equilibrium state consists of an infinite sequence of di erent Slater lattices, so that the non-polar tetragonal symmetry, as well as the specific ratio of FE and AFE Slater groups is established in the long time average for every single PO$_4$-group.
3. Furthermore, there exists an intermediate dynamic equilibrium state, that consists of a finite sequence of di erent Slater lattices. It is observed when a gliding time average of finite length is applied to all local fluctuations of the system before taking the space average. The gliding time average can be achieved by a typical 1D quadrupole perturbed NMR experiment where the integration time is of the order of T_2 and where the deviation from the average local symmetry is changing the EFG-tensor of the nucleus under investigation linearly. This intermediate dynamic equilibrium

state is clearly not the result of a slowing down, but is resulting from a pseudo static glass ordering. This means, it can be described by a random-bond random-field model with "static" random interactions and a "static" random field. Though the local gliding time-average is time dependent on a longer time scale, its spatial distribution function (and its moments) is always time independent. This is reflected in the fact that, e.g., the NMR line shape does not depend on the absolute time.

4. We are thus faced with three mechanisms with different characteristic times.

A: The uncorrelated deuteron intra-bond jumps in the ice-rule biased double well potentials. This motion creates or annihilates the ($T1$-$T3$) Takagi pairs with the loss or gain of the energy 2 E_b.

B: Once the Takagi pairs are formed there is a non-vanishing probability that they separate and move independently through the Slater lattice as unpaired Takagi groups, hindered or aided by the Slater energies according to their signs, until they recombine with the original or another partner. At low temperatures there is also a non vanishing probability of getting trapped by a lattice defect. Such Takagi groups are then out of the game until they become loose again at higher temperatures. The motion of unpaired Takagi groups is neutralizing the ice biases of the system. However, this motion alone cannot restore the high temperature symmetry.

C: The "static" distribution of Slater energies $_d$ always leads to a "static" distribution of local polarizations corresponding to the results of the RB-RF model. Note that the $_d$ are resulting not only from the random field but also from the random interactions. As mentioned in Sect. 2 this requires the motion of the nitrogens among the four possible sites (corresponding to the N-D...O hydrogen bonding scheme) induced by the reorientational motion of the ND_4-groups. This motion must be considerably slower than the averaging performed by the Takagi group motion, otherwise the continuous glass ordering could not be observed at all and the glass order parameter q_{EA} determined from 1D-NMR would never differ from zero. In a recent ^{14}N T_1 study [30] the interpretation of the results by the authors differ considerably from our model. However, a closer analysis of the data revealed that they can as well be explained with our model.

Acknowledgments

The author would like to thank his former PhD students P.M. Cereghetti, C. Jeitziner, T. Koenig and N. Korner for performing the measurements and some of the model calculations presented here. Furthermore he is very grateful to Professor Robert Blinc – to whom this contribution is dedicated – for the stimulating scientific cooperation and his repeated hospitality over almost 40 years, as well as for his friendship.

References

1. E. Courtens: J. Phys. (Paris) Lett.   43, L199 (1982)
2. E. Courtens: Ferroelectrics   72, 229 (1987)
3. U.T. H¨ochli, K. Knorr, A. Loidl: Adv. Phys.   39, 405 (1990)
4. R. Pirc, B. Tadic, R. Blinc: Phys. Rev. B   36, 405 (1990)
5. R. Pirc, B. Tadic, R. Blinc, R. Kind: Phys. Rev. B   43, 2501 (1991)
6. R. Blinc, J. Dolinsek, R. Pirc, B. Tadic, B. Zalar, R. Kind, O. Liechti: Phys. Rev. Lett. 63, 2284 (1989)
7. R. Kind, R. Blinc, J. Dolinsek, N. Korner, B. Zalar, P. Cevc, N.S. Dalal, J. Delooze: Phys. Rev. B 43, 2511 (1991)
8. R. Blinc, D.C. Ailion, B. G¨ unther, S. Zumer: Phys. Rev. Lett.   57, 2826 (1986)
9. J. Slak, R. Kind, R. Blinc, E. Courtens, S. Zumer: Phys. Rev. B   30, 85 (1984)
10. A. Levstik, C. Filipic, Z. Kutnjak, I. Levstik, R. Pirc, B. Tadic, R. Blinc: Phys. Rev. Lett. 66, 2368 (1991)
11. J.C. Slater: J. Chem. Phys.   9, 16 (1941)
12. R.A. Cowley, T.W. Ryan, E. Courtens: Z. Phys. B   65, 181 (1986)
13. N. Korner, R. Kind: Phys. Rev. B   49, 5918 (1994)
14. R. Kind, N. Korner, T. Koenig, C. Jeitziner: J. Korean Phys. Soc.   32 S799 (1998)
15. J. Dolinsek, B. Zalar, R. Blinc: Phys. Rev. B   50, 805 (1994)
16. J.L. Bjorkstam: Phys. Rev.   153, 599 (1967)
17. Y. Takagi: J. Phys. Soc. Jpn.   3, 273 (1948)
18. J. Dolinsek, G. Papavassiliu: Phys. Rev B   55, 8755 (1997)
19. P.M. Cereghetti, R. Kind: J. Magn. Resonance   138, 12 (1999)
20. C. Jeitziner: NMR studies of the low-temperature structure and dynamics of the pseudo-spin glass D-RADP-X. PhD. Thesis, ETH, Diss. Nr. 13257, Zurich (1999)
21. R. Blinc, M. Burgar, V. Rutar, J. Seliger, I. Zupancic: Phys. Rev. Lett.   38, 92, (1977)
22. R.R. Ernst, G. Bodenhausen, A. Wokaun: Principles of nuclear magnetic resonance in one and two dimensions   (Clarendon Press, Oxford 1987)
23. P.M. Cereghetti: On the dynamics of glassy phase states: An NMR investigation. PhD. Thesis, ETH, Diss. Nr. 13806, Zurich (2000)
24. R. Kind, P.M. Cerghetti, Ch.A. Jeitziner, B. Zalar, J. Dolinsek, R. Blinc: Phys. Rev. Lett. 88, 195501-1 (2002)
25. N. Korner: From long range order to glass order: Static and dynamic properties of the solid solution Rb $_{1-x}$(ND$_4$)$_x$ D$_2$PO$_4$. PhD. Thesis, ETH, Diss. No. 9952, Zurich (1993)
26. N. Korner, Ch. Pfammatter, R. Kind: Phys. Rev. Lett.   70, 1283, (1993)
27. P.A. Beckmann: Physics Reports   171, 85–128 (1988)
28. J. Dolinsek, D. Arcon, B. Zalar, R. Pirc, R. Blinc, R. Kind: Phys. Rev. B   54, R6811 (1996)
29. Th.J. Koenig: Cluster dynamics in the solid solution D-RADP-   x investigated by ^{87}Rb NMR. PhD. Thesis, ETH, Diss. No. 12027, Zurich (1997)
30. A. Gregorovic, B. Zalar, R. Blinc, D. Ailion: Phys. Rev. B   60, 76 (1999)

Characterising Porous Media

J.H. Strange [1] and J. Mitchell [2]

[1] School of Physical Sciences, University of Kent, Canterbury, Kent, UK, CT2 7NR
   j.h.strange@kent.ac.uk
[2] Department of Physics, University of Surrey, Surrey, UK, GU2 7XH
   j.mitchell@surrey.ac.uk

Abstract. The method of Nuclear Magnetic Resonance cryoporometry has gained popularity since its inception in 1993 as a non-destructive technique for measuring pore size distributions in the nano-scale range. NMR cryoporometry is a secondary method of measuring pore sizes by observation of the depressed melting point of a confined liquid. The melting point depression constant of the absorbate has to be determined empirically although this constant is only a function of the absorbed liquid and its associated solid, not the porous matrix. Cryoporometry has the major advantage of o ering, with care, directly calibrated measurements of pore volume as a function of pore diameter, of non-destructive pore measurement, structural resolution of spatially dependent pore size distributions, and behavioural information about the confined liquid. This chapter focuses on the history of NMR cryoporometry, the basic equipment required to run an experiment, and highlights some of the major results that have been achieved by various research groups around the world using this technique.

1 Introduction

Porous media are prevalent in the natural world and are widely used for industrial applications. Porous materials exist in a vast range of forms; everything from biological cells to rocks, drying agents and catalysts in chemical reactors. It is important to be able to characterise the many properties of these systems, such as porosity, pore size, permeability, and surface morphology. Only under exceptional circumstances will porous media contain pores of a single shape and size. Normally the pores come in a range of sizes. Knowing the distribution of pore sizes in a material is particularly important when using porous media in technical applications. For example, the pore geometry can influence the use of porous catalysts where the accessible surface area (related to the open pore volume) will determine the reaction rate. There is great interest in techniques that can non-destructively, easily, and inexpensively measure pore size distributions. The technique of Nuclear Magnetic Resonance (NMR)

J.H. Strange and J. Mitchell: Characterising Porous Media , Lect. Notes Phys. 684 , 407–430 (2006)
www.springerlink.com

cryoporometry meets these requirements. It is a method for determining calibrated pore size distributions using the depression in melting point of a liquid
confined in pores compared to the bulk melting point of the same liquid. It
is known that the melting point of small crystals is depressed relative to the
bulk crystal melting point. The melting temperature is directly related to the
size of the crystal. Therefore, when the crystal size is limited by a confining
geometry, the melting point depression is related to the size of the confining
space. NMR cryoporometry relies on the established Gibbs-Thomson equation
(1) [1, 2, 3, 4, 5, 6]. The equation relates the depression in melting point (ΔT)
of a confined liquid to the diameter of the confining pore (Δx) by a constant (Δk).
It must be assumed in order to determine a calibrated pore volume that the
density of the absorbed liquid remains constant throughout the experiment,
and that the absorbate remains pure.

$$\Delta T = \frac{k}{x}. \tag{1}$$

The Gibbs-Thomson equation (1), derived in terms of the absorbate's properties, was verified by Jackson and McKenna [7] in a study of the thermal
properties of organic crystals in porous media using Differential Scanning
Calorimetry (DSC). This was exploited in conjunction with variable temperature NMR measurements by Strange et al. in 1993 [8] to measure pore size
distributions, and the method named 'NMR cryoporometry'.

Earlier NMR studies demonstrated that the molecular behaviour of absorbates confined in porous media was changed considerably from that of the
bulk material over a wide temperature range [9, 10, 11]. The mechanisms
behind the altered relaxation times were discussed in terms of surface interactions. The presence of a Non-Frozen Surface Layer (NFSL) on the pore wall
was first observed around 1973 [11]. Later molecular motion studies confirmed
the presence of a NFSL in a variety of systems and explored the general modifications in molecular mobility of confined materials [12, 13, 14]. In general
it was found that crystalline materials confined in porous media form a two-
phase system below their depressed melting point. A core lattice is normally
present of similar structure and properties to the bulk solid. A more disordered surface layer also appears, possibly being a plastic crystal phase or even
a glassy state, extending for approximately two molecular layers in depth and
having considerable molecular mobility.

The first cryoporometry experiments were conducted by simply observing
the change in NMR signal intensity of a frozen sample as it melted, with very
simple temperature control. A sample was frozen, placed in the spectrometer
and the liquid signal amplitude observed as it slowly warmed in a thermally in-
sulated container. An obvious problem was the lack of control over the sample
warming rate. Without this control, the pore size could not be reliably determined using equation (1). Subsequent versions of the cryoporometry apparatus
included programmable temperature control and warming rates of the NMR
sample. Various methods of attaining temperature control were used, offering

di erent advantages. Gas flow temperature control was most frequently employed, although this does have limitations. Improvements to the temperature control will be vital to the continued advancement of cryoporometry.

At the same time as the introduction of cryoporometry by Strange et al. [8], a paper was published on a related topic by Overloop et al. [15] using a high-resolution spectrometer. This group performed freezing studies of water in porous media, noting the hysterisis occurring due to super cooling. Although additional information on the behaviour of the confined material can be gleaned from cooling measurements (it is good practise to observe both the freezing and melting sections of each cycle), pore size distributions should be derived only from data collected on a warming run. Over small temperature changes the phase transition of a material melting in a pore can involve meta-stable states due to super-cooling e ects.

NMR cryoporometry is a secondary method of measuring pore sizes because the melting point depression constant of the absorbate (k) has to be initially calibrated using samples with known pore diameters. This constant is only a function of the absorbed liquid and its associated solid, not the porous matrix. It is commonly evaluated using model samples with narrow pore size distributions using the BJH method of (nitrogen) gas adsorption to obtain the pore surface-to-volume ratio [16]. The NMR cryoporometry experiment provides directly calibrated measurements of pore volume as a function of pore diameter. Two distinct methods of obtaining NMR cryoporometry data have been devised: scanning measurements (using a continuous temperature ramp) [17] and static measurements (using discontinuous temperature steps) [18]. In both cases the NMR cryoporometry measurement records the number of ^{1}H protons in molecules in a liquid state as a function of temperature. This is possible because NMR can be used to distinguish easily between a solid and liquid due to the very di erent nuclear magnetic relaxation of the protons in the two states. This technique therefore provides a measure of the total number of molecules that have undergone a solid-to-liquid phase transition at any given temperature. This is taken to be proportional to the total liquid volume, v_L, as a function of temperature and can be used to determine the volume distribution, d v/ dx, of pores with diameter   x; see (2) [8].

$$\frac{dv}{dx} = \frac{dv_L}{dT} \frac{k}{x^2}.$$ (2)

NMR cryoporometry can be used for studying more than just pore size distributions. In this chapter we will discuss a number of applications of cryoporometry. Combined with complimentary NMR techniques, it can be used to categorise a range of pore properties other than the geometry, including surface wetting, surface a nity and absorbate interactions. NMR cryoporometry has a unique place among the methods of porosimetry in that it can be combined with Magnetic Resonance Imaging (MRI) to provide spatially resolved pore characterisation [19, 20, 21].

2 Measurement of Liquid Fraction Using NMR

For basic NMR cryoporometry measurements, we are interested in recording the total volume of liquid in a sample as a function of temperature. The transverse relaxation time (T_2) [22] of a liquid is easily distinguishable from that of a solid. Liquids have characteristically long T_2 relaxation times, ranging from milliseconds to seconds, whereas solids usually have T_2 relaxation times in the order of microseconds. This generally allows the NMR signal from the liquid fraction of the absorbate to be measured independently from the solid fraction. Water or hydrogenous organic liquids are usually used as the absorbates in cryoporometry experiments, and are easily detected since they exhibit a strong NMR proton signal. The observed proton magnetisation is a measure of the number of hydrogen nuclei in the liquid. If the liquid density remains constant, the signal will be linearly proportional to the liquid volume [8, 9, 10, 11, 12, 13, 14, 15, 16, 17]. In cryoporometry experiments we are concerned with the precise quantity of liquid in the sample. The height of an NMR spin-echo [23] can be used as a measure of the liquid volume. This is usually the most appropriate NMR measurement of relative magnetisation since minimal post-measurement data extrapolation is required to determine the signal at zero time. In an ideal system the Free Induction Decay (FID) [22] (signal following a 90 pulse) directly provides a measure of the transverse magnetisation, $I(0)$; see (3). The signal intensity is often represented as an exponentially decaying function of time as shown in (3). However, in a typical real system the FID will depend on magnetic susceptibility changes throughout the sample and inhomogeneities in the applied magnetic field and as well as nuclear magnetic dipolar interactions. These effects cause the spins to dephase more rapidly than they would in a homogeneous static field. Unless T_2 is short ($< 300\,\mu s$ for low-resolution spectrometers) the time constant taken from a real FID signal will be strongly dependent on the spread in Larmor frequencies of the spins due to B_0 inhomogeneities; see (4). The measured transverse decay time T_2 is often a property of the spectrometer magnet and not the sample. It comes about due to a loss in coherence of the spins in the measuring frame of reference. The FID could be extrapolated back to zero time to provide the full liquid signal intensity but since the shape of the decay would be determined by the unknown magnet system and sample geometry, assumptions would have to be made.

$$I( ) = I(0)\, e^{-\ /T_2} . \qquad (3)$$

$$\frac{1}{T_2} \ \ \frac{1}{T_2}_{N} + \ \frac{B_0}{2} . \qquad (4)$$

So for cryoporometry measurements, where the liquid component is of interest and has a T_2 greater than $10\,ms$, direct FID measurements are of limited value. To overcome this problem a spin echo can be used and thus observe the liquid signal, ideally without seeing any residual solid signal. By using a

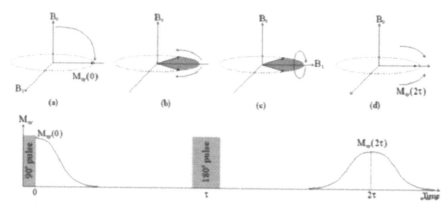

Fig. 1. Producing a spin-echo: The nuclear spins start parallel to the z-axis. (a) A 90 pulse along the x-axis shifts the spins onto the y-axis. (b) The spins dephase due to field inhomogeneities. (c) The 180 pulse rotates the spins about the y-axis and so they move back into phase (d) forming an echo

90_x − − 180_y − -echo pulse sequence, the loss in coherence of the spins due to magnetic inhomogeneities can be overcome. The 90 $_x$ pulse shifts the spins onto the y-axis where they start to recover but dephase during the time of the experiment. After a time (where > T $_2$) they have dephased in the x-y plane. Applying an 180 $_y$ pulse rotates the spins about the y-axis, causing them to rephase. A maximum signal will be measured at 2 ; see Fig. 1. This is known as a spin echo and the height of the echo is directly dependent on the volume of liquid in the sample. The echo amplitude is attenuated by transverse relaxation according to (3).

Further to this we can capture extra information by repeating the sequence to form an echo train, called the Meiboom-Gill modified Carr-Purcell (CPMG) sequence [24, 25]. A closely spaced train of echoes with short can overcome attenuation of the echo amplitude due to diffusion in magnetic field gradients. By applying the 180 pulse / 2 out of phase with the excitation pulse and taking only the even echoes we can negate signal loss due to inaccurate 180 pulses. The envelope of the echo train gives a good approximation to the FID that would be observed if all the spins were in an homogeneous magnetic field, and hence provides the true T_2; see Fig. 2. This then allows precise estimation of the signal intensity arising from liquid in the sample so quantifying the liquid content.

3 The NMR Cryoporometry Experiment

In a standard cryoporometry pore volume measurement the samples are imbibed with liquid to just overfill the pores. The extra bulk liquid between the grains provides the bulk melting point as a source of reference. Care must

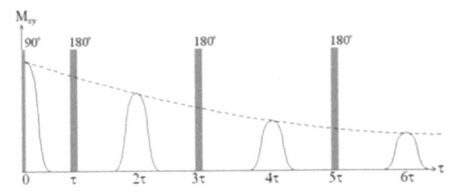

Fig. 2. To find the T_2 of a liquid, multiple spin-echoes can be plotted on the same graph. The maxima describe the transverse relaxation curve. In the CPMG measurement, this is performed in a single scan by repeating the 180 refocusing pulse

be taken to choose a suitable absorbate to fill the pores. Ideally, something should be known of the porous sample before the experiment is conducted. Water, for example, is more suited to very small pore diameters, whereas an organic liquid like cyclohexane would have to be used if the sample were hydrophobic. Also some organic absorbates are totally unsuitable because the solid sublimes easily or the liquid is very volatile. This would be revealed by a decrease of integral liquid volume during the cryoporometry experiment. Samples used in cryoporometry are normally sealed in glass tubes under vacuum. The absorbate is then completely frozen, e.g., by cooling in liquid nitrogen. For relaxation measurements the liquid should be pumped (frozen, evacuated, thawed and the process repeated several times) to remove any dissolved oxygen. Any paramagnetic oxygen dissolved in the sample will reduce the T_2 time and will therefore a ect the relaxation measurement. Under-filled samples are made in the same way, with the exception that the pore volume and density of the sample has to be predetermined to allow the correct quantity of absorbate to be added.

A cryoporometer can be used either in a scanning mode, or stepped in temperature. The scanning cryoporometer uses a continuous heating ramp. The rate of warming is simultaneously controlled and measured, providing a scale against which the pore size distribution can be calibrated. All cryoporometry experiments rely on capturing as many points as possible on the melting curve to maximise the resolution of the final pore size distribution. Improved resolution has been achieved by ramping the temperature as slowly as possible and using a logarithmic warming ramp to measure melting point depressions of less than 1 K. This increases the resolvable pore diameter since the amount of data recorded between the bulk and depressed melting points of the absorbate increases. Since the signal-to-noise ratio may be low on each point, filtering processes can be applied to allow clear pore size distributions to be determined. A greater number of data points provide a better trend for the filtering functions to resolve, producing a more accurate picture of the pore volume distribution. A recent version of the scanning cryoporometer automatically controls the warming ramp, records the height of a single spin-echo, and determines the pore size distribution without user input.

The stepped cryoporometer permits the simultaneous measurement of relaxation times as well as the liquid volume fraction as a function of temperature. The stepped cryoporometer sets the temperature but allows it to stabilise before either a single spin-echo or a CPMG echo train is recorded. Recording an echo train provides information not only on the signal amplitude but also signal decay (T_2) at each temperature. The precision of the pore size distribution determination may be improved by combining the cryoporometry results with the relaxation data. The transverse decay shown by CPMG may be fitted using an exponential decay function, such as (3). More frequently a multiple exponential function is required, (5), where the subscript n represents the relaxation time of the nth component.

$$I = \sum_n I_n(0) \, e^{\frac{-\tau}{T_{2n}}} . \tag{5}$$

The fitted curves may then be back projected to provide the liquid signal amplitude at zero time. This allows for any residual signal due to the solid or any signal attenuation effects to be overcome. Although the method does not provide as many points on the cryoporometry curve as the scanning technique, the data produced have a very high signal-to-noise ratio and require little, if any, filtering. Such stepped cryoporometry data sets are suitable for determining mean pore diameters and comparing samples, although fine structure would be indistinguishable. The temperature steps allow other NMR parameters to be measured during the heating cycle, including T_1, T_1 (spin-lattice relaxation and spin-lattice relaxation in the rotating frame) and diffusion rates.

4 Determining the Melting Point Depression Constant

The melting point depression of the absorbate is inversely related to the diameter of the confining pore, if it can be assumed that all the other terms in the full Gibbs-Thomson equation can be replaced by a constant k giving (1). To determine the k constant for a particular absorbate, a series of cryoporometry measurements are conducted on standard samples. The absorbate is added to just overfill the pores of a set of test porous silicas with well-defined pore size distributions. Each samples is then cooled to well below the depressed melting point of the absorbate to overcome any super-cooling effects. This is best performed in situ, the NMR signal being used to determine when the entire sample has frozen. The cooling should be slow to prevent the formation of metastable states that could potentially occur if the sample were quenched. The cryoporometry experiment can then be conducted as detailed above and the data used to determine the average pore melting temperature from the melting curve; see Fig. 3. The average pore melting points are plotted against the inverse of the median pore diameter of the test samples as determined by gas adsorption. The resulting data is fitted linearly to determine the   k value.

Careful analysis of the melting point depression data shows that in some cases it is not a linear relationship; rather a curve that is fitted precisely by the modified Gibbs-Thomson equation of the form:

$$T = \frac{k}{x - 2Sl} . \tag{6}$$

Introducing a surface layer thickness,   Sl, allows the curvature to be quantified. There is a factor of 2 multiplying the surface layer thickness to take into account that x is the pore diameter . This disappears if x is expressed as pore radius. The surface layer was first observed [11] and studied in detail with

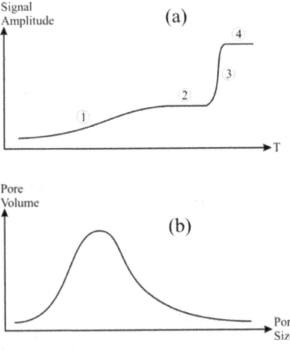

Fig. 3.   (a) Shows an ideal cryoporometry curve with four main features: (1) the pore melting step; (2) the total pore volume plateau; (3) the bulk melting step, and (4) the total liquid volume plateau. 4(b) shows the pore distribution expected from such a cryoporometry curve conducted on an artificial porous silica sample. To calibrate the pore size scale, the predetermined k value is used. Sample mass ratios are used to calibrate the pore volume (vertical) scale

water [12, 26], and also noted in organic liquids [27]. It was suggested by Stapf and Kimmich [28] that the surface layer may interfere with the method of cryoporometry. The surface layer will reduce the observed pore radius since the material at the pore wall does not undergo a sharp solid-liquid transition as defined by the Gibbs-Thomson equation. However, studies suggest the surface layer is independent of pore structure or geometry, allowing it to be included in the k constant. The fraction of non-frozen surface material can contribute to the NMR liquid signal. Normally the T_2 decay time of the surface layer is much less than the rest of the liquid due to enhanced surface relaxation. In some cases therefore, the signal component can be ignored or overcome, particularly when measuring CPMG data on a static cryoporometer.

Problems arise when the molecular size of the absorbate is large compared to the pore diameters, such that the surface layer occupies a significant volume of the pores. In this case it has been observed that organic crystals do not truly freeze, making a cryoporometry experiment impossible. Rather, they enter an amorphous state at low temperatures, similar in properties to a glassy phase. This phenomenon has been observed for cyclohexane in pores of diameter less than 40 Å [17, 29]. For naphthalene, this occurs around 50–60 Å [14, 18]. Water, on the other hand, has been successfully used to measure pore diameters below 20 Å [17, 30].

Water is perhaps the most obvious absorbate to be employed in cryoporometry experiments. It has a k value of 573 K Å [30] determined using the scanning cryoporometer without including the surface layer factor, and noted to be independent of . Cyclohexane is more widely used since it has a number of advantages: its higher k value means larger pore diameters can be resolved and its soft plastic crystal does not damage the porous matrix, allowing the sample to be reproducibly re-measured.

5 Cryoporometry Hardware

Simple NMR apparatus is required for NMR cryoporometry experiments, regardless of the technique used or modifications to the method. The arrangement consists of a magnet (either permanent or super-conducting), an NMR spectrometer (computer) and a temperature control system. Complexity arises when considering the implementation of accurate temperature control. As previously mentioned, the temperature control needs to be as precise and as stable as possible. Temperature regulated gas flow is now the most commonly used method. Assuming the gas flows at a high rate over the sample and that the sample has a suitable thermal mass, the temperature can be extremely stable. The two cryoporometers discussed in this section are given as examples of temperature-regulated NMR systems and were constructed using low-resolution (broad-line) spectrometers. The same principles of temperature control apply to high-resolution spectroscopy and imaging systems.

In early cryoporometry experiments cooling was achieved using the direct injection of liquid nitrogen into a 'splash-pot' in the probe (the Lindacot system) [31]. A heater in a Dewar forced droplets of liquid nitrogen into the probe. Thus the full latent heat of evaporation of the nitrogen was available to cool the probe and sample. The thermal mass of the splash-pot smoothed out suddenly temperature fluctuations due to the evaporation of individual liquid nitrogen droplets. A resistive heater was built directly into the probe to enable it to be warmed above room temperature. Later this probe was used e ectively with gas flow cooling [17]. Instead of forcing droplets of liquid nitrogen into the probe, a dried air supply was cooled in liquid nitrogen and passed into the splash-pot. This provided a much more uniform warming ramp than the Lindacot system, although it did consume more liquid nitrogen. An air heater was added to the gas flow just prior to entering the probe. This was used to control the temperature of the gas passing into the splash-pot. A thermocouple inside the splash-pot was used to monitor the temperature and provide feedback for the temperature control unit. Another thermocouple was soldered to an earthed copper shield in thermal contact with the sample to provide an accurate sample temperature for the cryoporometry measurement. A schematic of this probe, which is placed inside a Dewar between the poles of a permanent magnet, can be seen in Fig. 4. This probe is used for scanning cryoporometry measurements over a temperature range -100 C to 80 C, and had a repeatable temperature resolution of 1 mK.

Measuring the temperature of the sample is a fundamental problem in cryoporometry. Ideally a thermocouple should be placed in direct thermal contact with the sample. However, a thermocouple placed inside the NMR coil will conduct RF noise into the system, thus ruining any result. Electrically insulating and earthing of the thermocouple can overcome this [17], although great care is required to ensure that no RF interference reaches the NMR coil. An alternative method, not yet fully explored, is the use of an optical temperature sensor to remotely monitor the sample temperature. Current optical temperature sensors are limited in their ability to measure temperatures below ambient, but as the technology improves this may become possible. Current optical thermometers would be suitable for super-ambient cryoporometry measurements.

The stepped cryoporometer had a more modern variable temperature NMR probe (see Fig. 5) that was used in a commercial spectrometer system [18, 32]. This probe relies on the thermal insulation properties of PTFE to provide a sample chamber with a temperature range of -70 C to $+180$ C without risk to the magnet. This probe is also temperature regulated by a gas flow system, although in this case the gas flows directly over the sample and does not utilise the thermal mass of a splash-pot. As with the scanning probe (Fig. 4), dried air was passed through a Dewar of liquid nitrogen and an air process heater used to regulate the temperature. To warm the sample above room temperature, the compressed air was passed directly to the heater without the use of liquid nitrogen cooling. A thermocouple was placed at the base

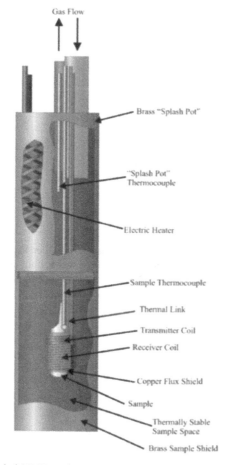

Fig. 4.   Gas-flow cooled NMR probe; custom design built by Webber et al. [17] for use in the scanning cryoporometer

of the probe where the gas enters, to simultaneously regulate the temperature and monitor the sample.

It is important in all cryoporometry experiments that the sample temperature does not fluctuate above the set temperature; Fig. 6(top). Due to super-cooling effects, the melting of the absorbate is a non-reversible process over small temperature increments. Therefore the measured liquid volume may not coincide with the measured sample temperature under such fluctuation conditions. In the cryoporometry experiment the recorded temperature should always be the maximum temperature to which the sample is exposed, see Fig. 6(bottom).

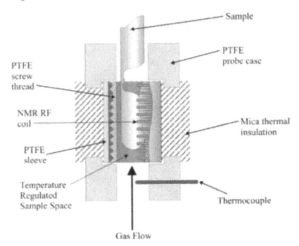

Fig. 5.   Design of the variable temperature sample chamber used in the stepped cryoporometer [18]. This probe was installed and used successfully for cryoporometry measurements in a commercial low-resolution bench-top spectrometer system

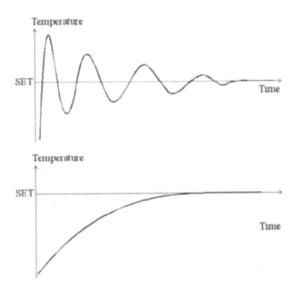

Fig. 6.   Using a PID temperature control routine usually results in a temperature change that oscillates around the SET temperature before stabilising (top). In cryoporometry experiments the maximum temperature seen by the sample must be the SET, or recorded temperature, resulting in the necessity for a smooth temperature step with no overshoot (bottom)

6 Applications of NMR Cryoporometry

The methodology of cryoporometry has been expanded to include high temperature measurements and has also been combined with other NMR techniques such as imaging [19, 20, 21], relaxometry [33], di usion [34] and NMR spectroscopy [15, 27]. By combining techniques, more information can be gained about the nature of the porous sample and the interactions at the solid/liquid interface [32]. Testing partially filled samples [29] or samples filled with more than one absorbate [35, 36] has been shown to provide information on surface interactions between the liquid and the pore wall. From these studies, the way in which liquids fill pores was deduced.

A great deal of interest in cryoporometry has come from cement manufacturer. This technique is ideal for the study of the formation of open microstructure during the hydration of cement pastes. Two examples of the use of cryoporometry in this field have been taken from papers by Jehng, Halperin et al. [26, 37]. In Fig. 7 the results from NMR cryoporometry and Mercury Intrusion Porometry (MIP) have been compared for a cement paste during drying. Significant di erences in the distributions were observed: the MIP data only contained information on the larger porous structures, whilst the NMR contained information on small pores as well. The authors explained that the mercury was unable to penetrate the small structures. Therefore the cryoporometry data yielded extra information. A similar comparison has been made (see Fig. 8) between NMR cryoporometry and NMR relaxometry. The

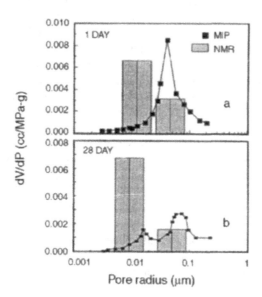

Fig. 7. Pore radius distributions of cement paste at two stages during the drying process. Both Mercury Intrusion Porometry (MIP) and NMR cryoporometry (histogram) have been used. Graphs reproduced with permission from [37]

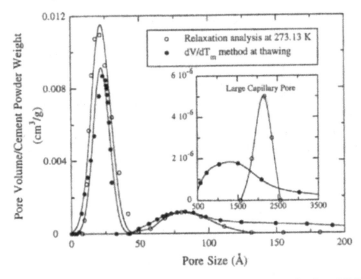

Fig. 8.   Pore distributions in cement powder. The data compare results from NMR cryoporometry (   ) and NMR relaxometry (•). Good agreement is shown at small pore sizes but not in the large capillary pores. In both techniques the absorbate used was water. Graphs reproduced with permission from [26]

two techniques diverge for the pore distributions in the large capillary pores, although cryoporometry measurements with water are limited to about 1–2 microns. For the smaller pores the two techniques agreed extremely well.

Cryoporometry has been used to study the e ect of adding a surfactant to the surface of Ultra High Pore Volume (UHPV) silicas [38]. The results from these tests were quite dramatic and can be seen in Fig. 9. The surfactant coated sample B2 had quite obviously lost its smallest pores and a significant quantity of pore volume where the space had been occupied by the surfactant. This was interpreted as a coating 25   Å deep. Sample C appeared to have lost nearly all its pore volume on coating. However, it must be stressed that the fluorinated surfactant used was potentially hydrophobic, preventing water uptake. Cryoporometry is independent of absorbate/pore wall interactions, making it a perfect choice for these types of experiments, as long as the absorbate can enter the pores. Sample C was fully categorised later by using cyclohexane as the absorbate and the surface coating thickness successfully deduced.

Experiments conducted on partially filled systems [29, 39, 40] have been used to determine the behaviour of liquids in pores. Cryoporometry experiments were conducted on porous sol-gel silicas partially filled with varying proportions of cyclohexane, Fig. 10 (a), or water, Fig. 10 (b). As the volume of cyclohexane was increased, the modal pore size in the measured distributions was seen to increase, but the minimum observed pore diameter was never less than the smallest pores in the silica. This suggested that the cyclohexane

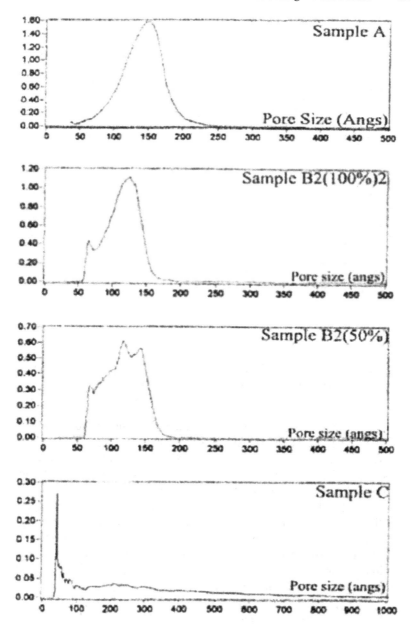

Fig. 9.   Four samples of UPHV silica: sample A – unmodified silica, nominal pore diameter 150 Å. Sample B2 – silica modified with standard surfactant. Sample C – silica modified with fluorinated surfactant. Sample B2 was tested both fully and partially (50%) filled. Water was the absorbate used in all cases. Graphs previously presented in [38]

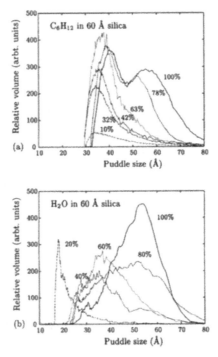

Fig. 10.   Puddle size distributions for (a) cyclohexane and (b) water in 60 Å silica at various filling fractions. Graphs reproduced with permission from ref [29]

preferentially filled the smallest pores with the highest surface-to-volume ratio. Similar experiments conducted with water showed a significant shift in the entire pore size distributions as the volume of water was increased. The smallest 'pore' diameters observed (at the lowest filling factor) were far less than the smallest pores in the sample. It was deduced, with the aid of T_2 relaxation analysis, that water forms puddles together with physisorbed layers that partially coat the silica pore walls. As the volume of water was increased, the puddles increased and joined together until the pores were eventually filled. Whilst cryoporometry on overfilled samples is independent of surface interactions, combining the technique with partial filling can provide a lot of information on absorbate behaviour and pore morphology.

The stepped cryoporometer allowed NMR cryoporometry and NMR relaxometry experiments to be combined, as demonstrated by Valckenborg, Pel and Kopinga [33]. Two silicas provided by di erent manufacturers, both having nominal pore diameters of 60 Å ± 2 Å, were compared by this technique. The silica samples were prepared by slightly over-filling with cyclohexane. A standard cryoporometry measurement was conducted as detailed above. The CPMG data sets collected at each temperature step were inverted to provide distributions of relaxation times, as in the technique of relaxometry

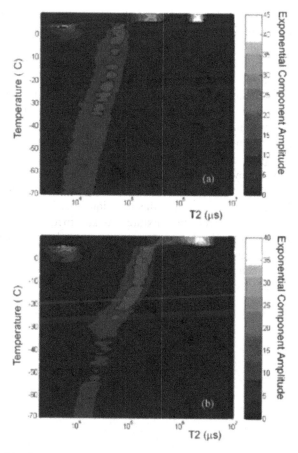

Fig. 11.   Combined cryoporometry and relaxometry of cyclohexane in two 60 Å nominal pore diameter silicas from di erent manufacturers. The surface interaction dominates over the confinement in the relaxometry measurement [18]

[41, 42, 43]. The data sets were constructed into three-dimensional graphs, plotting the temperature against T_2 relaxation time and showing the component amplitude at each point. The results from the two silica gels can be seen in Fig. 11 (a) and (b). Ideally both the signal amplitude (gray scale) and T$_2$ relaxation time (x-axis) should exhibit the same temperature dependence in both samples. It can easily be seen that the relaxation time as a function of temperature for the cyclohexane in the two silicas varied greatly.

Although the nominal pore diameter of the two silicas was quoted to be 60 Å, a slight deviation was observed in gas adsorption measurements (± 2 Å). However, such a deviation would not result in the di erent relaxation times seen in the two samples. The concentration of paramagnetic ions in the silicas was suggested as a possible cause of the variation in relaxation times. Since the cyclohexane in silica (a) exhibited shorter relaxation times, silica (a) was expected to contain more paramagnetic impurities than silica (b). Silicon NMR relaxation studies and XFD measurements were also conducted on the same silica gels. These confirmed that silica (a) contained more paramagnetic impurities than the silica (b), as expected [18]. Such impurities can be introduced at the time of manufacture and may vary with the production method or the batch, depending on the source of the trace components. By combining the NMR cryoporometry and relaxometry experiments it was clear that surface relaxation times for liquids confined in porous media can be modified by more than just the pore geometry.

Although this review has concentrated mainly on low-resolution NMR cryoporometry, high-resolution measurements can provide alternative information. Valiullin and Fur′ o have used this technique to study a dual phase absorbate [35]. A nitrobenzene/hexane mixture was absorbed into porous glass and studied using high-resolution cryoporometry to resolve the two phases. An example of the results obtained can be seen in Fig. 12. From these results it was determined that the nitrobenzene coalesces into small droplets surrounded by hexane when in the pores.

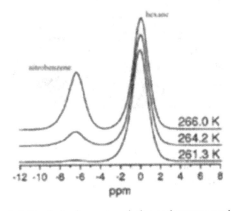

Fig. 12.   The chemical shift of nitrobenzene relative to hexane provides cryoporometry information on the two absorbates. Graph reproduced with permission from [35]

Various applications of Nuclear Magnetic Resonance (NMR) have been used to study the behaviour of oil and water in natural rocks [44, 45, 46, 47, 48]. This has been a source of considerable interest to those studying the removal of oil from oil-bearing rocks by flushing the systems with salt water [49]. Normally the natural rock samples have extremely broad pore size distributions, ranging from microns to millimetres [50], limiting the utility of NMR cryoporometry for studying the behaviour of the confined liquids [17]. The most frequently used NMR technique has been relaxometry [44]. Although this method can provide estimates of the pore size distributions in the rocks [51], it is highly sensitive to surface interactions [32] and can be di cult to interpret accurately [52]. These studies all used model immiscible liquid mixtures in real rock systems. In a recent study cryoporometry has been employed to analyse binary mixtures of immiscible liquids, water and decane, in a model system of porous sol-gel silicas [36]. The pore geometry had already been characterised by gas adsorption and standard NMR cryoporometry measurements [17], allowing the interactions between the water, decane and pore surface to be investigated.

The experiments were conducted in 100 Å nominal pore diameter silica. The samples were made by first slightly over filling the silica with decane. The decane was allowed to soak into the pores without assistance. Water was then added incrementally to the top of the sample and a cryoporometry measurement made without delay. The addition of known amounts of water, a little at a time, was conducted until a bulk water melting step was observed in the cryoporometry measurement. The cryoporometry melting curves for the binary mixtures can be seen in Fig. 13. The added water entered the pores and no bulk water signal was detectable until half of the absorbate mix (by volume) was water. As the water entered the pores, the decane shifted from the pores into the bulk. This can be seen by a shift in the ratio of the pore melting step and the bulk melting step for the decane. The total decane signal remained approximately constant. No more water was added once 50% of the absorbate mix was water. The sample was re-measured after twenty-four hours where no signal could be observed for confined decane. Instead, the

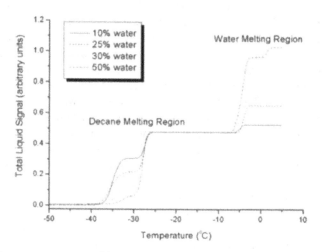

Fig. 13.   The cryoporometry melting curves of decane and water mixtures in 100   Å nominal pore diameter silica. The decane melting region is clearly separate from the water melting region. As the fraction of water increases the decane shifts from the pores into the bulk

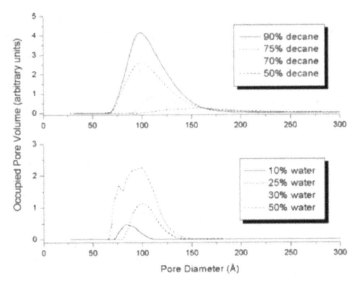

Fig. 14. The pore volume distributions for decane (top) and water (bottom) generated from the cryoporometry melting curves in Fig. 13. It can be seen that water
preferentially displaces decane from the smallest pores

water occupied all of the open pore volume, indicating that the water could
displace the decane completely over time. The pore size distributions taken
from the cryoporometry results in Fig. 13 can be seen in Fig. 14. Whereas the
cryoporometry melting curve only shows that the water displaces the decane,
the pore size distributions show where the water goes. The water preferentially displaced the decane from the smallest pores. For a mix containing 10%
water, pores with a nominal diameter of 80 Å were occupied by water. When
the quantity of water was increased to 50% of the mixture, the nominal pore
diameter of pores occupied by the water increased to 95 Å. At the same time
the width of the pore size distribution occupied by the water increased dramatically. By contrast the decane occupied pores of a nominal diameter of
97 Å when 10% of the mixture was water; this increased to 160 Å when 50% of
the mixture was water. That meant the remaining decane only resided in the
largest pores. The results of this study clearly show that water can displace
decane from porous sol-gel silicas, given suitable time for the displacement to
occur. It has been shown elsewhere that water has a far stronger a nity to the
surface in these silicas than non-polar organic liquids [32, 53, 54]. This is because the silica surface contains hydroxyl groups to which the water molecules
can form hydrogen bonds. It was therefore reasonable to expect the water to
preferentially enter the smallest pores (with the greatest surface to volume
ratio) first, displacing the decane.

   The last application of cryoporometry discussed here is the most visu-
ally impressive. By conducting a cryoporometry experiment in a spectrometer

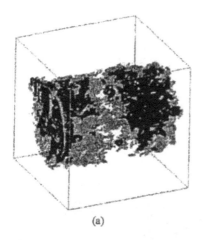 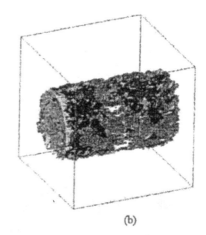

(a)   (b)

Fig. 15.   Cryoporometry combined with 3D NMR imaging as used to study reservoir rock core samples at (a) 268 K and (b) 295 K. Imagine intensity is related to porosity: black for high porosity down to   light grey for low porosity. Images previously displayed in Report to CPMT (2000) and presented in [38]

fitted with imaging gradients, spatially resolved pore size distributions were obtained. To demonstrate this method, a special silica sample was constructed, consisting of a 60 Å nominal pore diameter silica column 'wearing' a collar of 500 Å nominal pore diameter silica [21]. The intended sample shape and the actual 3D rendered volumes can be seen in Fig. 16. This technique has since been used practically to image oil reservoir rock samples [38]; Fig. 15. The results, combined with di usion measurements, illustrated the interconnectedness of the pores in the rock. This type of research o ers great potential for petrology, catalyst and construction materials industries.

7 Conclusion

In this chapter we have attempted to cover the important practical aspects involved with accurate NMR cryoporometry measurements. It has been shown that quantitative pore volume distributions can be obtained easily from a simple NMR spectrometer system. Although cryoporometry was born on a custom NMR system, it has spread to be used on a wide range of equipment and for many purposes. It has been shown to produce results equal to, and in same cases better, than those from well-established pore volume measurement methods. Cryoporometry has the major advantage of o ering direct volumetric measurements that few other techniques can match and variations of the technique permit the detailed study of fluid-surface interactions.

Useful in its own right, cryoporometry has been combined with other NMR techniques (relaxometry, di usion, imaging, and spectroscopy) to provide a

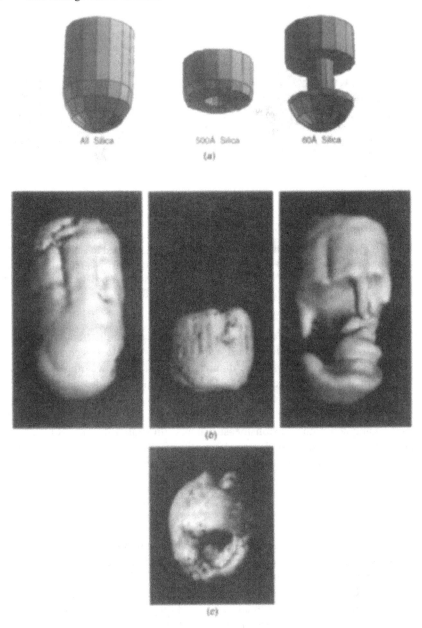

Fig. 16.   Spatially resolved silica test sample via combination of NMR cryoporometry and 3D NMR imaging. (a) The intended shapes of the silica sample. (b) Horizontal view of the rendered volumes. (c) Vertical view of the 500   Å silica volume,
clearly showing the hollow centre. Images reproduced with permission from [21]

toolbox that can fully characterise almost any porous system. The technique can be used to characterise pore surfaces using partially filled samples, and compare surface a nities of absorbates using binary liquids. Now used for the measurement of porous rocks, cements, biological cells, and in many other research areas, NMR cryoporometry is proving to be invaluable in the field of porous media exploration.

References

1. J. W. Gibbs: The Scientific Papers of J. Willard Gibbs, volume 1:Thermody-namics , new dover edition (Dover Publications Inc., Constable and Co., New York, London 1906)
2. J. W. Gibbs: Collected Works (Longmans, Green and Co., New York 1928)
3. J. Thomson: Trans. Roy. Soc. xi, (1849)
4. J. Thomson: Proc. Roy. Soc. xi, (1862)
5. J. J. Thomson: Application of Dynamics to Physics and Chemistry (Macmillan & Co., London 1888)
6. W. Thomson: Phil. Mag., Ser. 4 42 , 282 (1871)
7. C. L. Jackson, G. B. McKenna: J. Chem. Phys. 93 , 12 (1990)
8. J. H. Strange, M. Rahman, E. G. Smith: Phys. Rev. Lett. 71 , 21 (1993)
9. H. A. Resing, J. K. Thompson, J. J. Krebs: J. Phys. Chem. 68 , 7 (1964)
10. H. A. Resing: J. Chem. Phys. 43 , 2 (1965)
11. R. T. Pearson, W. Derbyshire: J. Colloid Interf. Sci. 46 , 2 (1973)
12. E. W. Hansen, M. Stocker, R. Schmidt: J. Chem. Phys. 100 , 6 (1996)
13. H. F. Booth, J. H. Strange: Molec. Phys. 93 , 2 (1997)
14. J. Mitchell, J. H. Strange: Molec. Phys. In Press , (2004)
15. K. Overloop, L. v. Gervan: J. Magn. Reson. Ser. A 101 , (1993)
16. E. P. Barrett, L. G. Joyner, P. P. Halenda: J. Am. Chem. Soc. 73 , (1951)
17. J. B. W. Webber: Characterising Porous Media. PhD Thesis Thesis, University of Kent, Canterbury (2000).
18. J. Mitchell: A Study of the Modified Behaviour of Organic Macromolecules in Confined Geometry. PhD Thesis, University of Kent, Canterbury (2003).
19. J. H. Strange, J. B. W. Webber, S. D. Schmidt: Magn. Reson. Imaging 14 , 7/8 (1996)
20. J. H. Strange, J. B. W. Webber: Appl. Magn. Reson. 12 , 2-3 (1997)
21. J. H. Strange, J. B. W. Webber: Meas. Sci. Technol. 8, 1-7 (1997)
22. E. L. Hahn: Physics Today 4, November (1953)
23. E. L. Hahn: Phys. Rev. 80 , 4 (1956)
24. H. Y. Carr, E. M. Purcell: Phys. Rev. 94 , (1954)
25. S. Meiboom, D. Gill: Rev. Sci. Instrum. 29 , (1958)
26. J. Y. Jehng, D. T. Sprague, W. P. Halperin: Magn. Reson. Imaging 14 , 7/8 (1996)
27. E. W. Hansen, R. Schmidt, M. St" ocker: J. Phys. Chem. 100 , (1996)
28. S. Stapf, R. Kimmich: J. Chem. Phys. 103 , 6 (1995)
29. S. G. Allen, P. C. L. Stephenson, J. H. Strange: J. Chem. Phys. 108 , 19 (1998)
30. J. B. W. Webber, J. H. Strange, J. C. Dore: Magn. Reson. Imaging 19 , 3-4 (2001)

31. M. O. Norris, J. H. Strange: J. Phys. E   2, 2 (1969)
32. J. H. Strange, J. Mitchell, J. B. W. Webber: Magn. Reson. Imaging   21, 3-4 (2003)
33. R. M. E. Valckenborg, L. Pel, K. Kopinga: J. Phys. D: Appl. Phys.   35, (2002)
34. A. V. Filippov, V. D. Skirda: Colloid Journal   62, 6 (2000)
35. R. Valiullin, I. Furo: J. Chem. Phys.   116, 3 (2002)
36. S. M. Alnaimi, J. Mitchell, J. H. Strange et al.: J. Chem. Phys.   120, 2075 (2004)
37. S. Bhattacharja, M. Moukwa, F. D'Orazio et al.: Advanced Cement Based Materials   1, (1993)
38. J. H. Strange, L. Betteridge, M. J. D. Mallett: Characterisation of Porous Media by NMR. In: NATO ASI series II: Mathematics, Physics and Chemistry   , vol Magnetic Resonance in Colloid and Interface Science, ed by J. Fraissard (Kluwer Academic Publishers, Dordrecht 2002)
39. S. M. Alnaimi, J. H. Strange, E. G. Smith: Magn. Reson. Imaging   12, 2 (1994)
40. S. G. Allen, P. C. L. Stephenson, J. H. Strange: J. Chem. Phys.   106, 18 (1997)
41. J. R. Zimmerman, W. E. Brittin: J. Phys. Chem.   61, (1957)
42. K. R. Brownstein, C. E. Tarr: J. Magn. Reson.   26, (1977)
43. K. R. Brownstein, C. E. Tarr: Phys. Rev. A   19, 6 (1979)
44. R. J. S. Brown, I. Fatt: Petroleum T. AIME   207, (1956)
45. J. J. Tessier, K. J. Packer, J. F. Thovert et al.: AICHE J.   43, 7 (1997)
46. S. Godfrey, J.-P. Korb, M. Fleury et al.: Magn. Reson. Imaging   19, (2001)
47. G. C. Borgia, E. Mesini, P. Fantazzini: J. Appl. Phys.   70, 12 (1991)
48. P. J. Barrie: Ann. R. NMR S.   41, (2000)
49. W. G. Anderson: J. Petrol. Technol.   38, 11 (1986)
50. F. A. L. Dullien: Porous Media. Fluid Transport and Pore Structure   (Academic Press, London 1979)
51. M. H. Cohen, K. S. Mendelson: J. Appl. Phys.   53, 2 (1982)
52. R. J. S. Brown, G. C. Borgia, P. Fantazzini et al.: Magn. Reson. Imaging   9, (1991)
53. T. Zavada, S. Stapf, U. Beginn et al.: Magn. Reson. Imaging   16, 5/6 (1998)
54. J.-P. Korb, M. Whaley-Hodges, T. Gobron et al.: Phys. Rev. E   60, 3 (1999)

Index

Lecture Notes in Physics

For information about earlier volumes
please contact your bookseller or Springer
LNP Online archive: springerlink.com